Fermented Foods of Latin America

Food Biology Series

Fermented Foods of Latin America

From Traditional Knowledge to Innovative Applications

Editors

Ana Lúcia Barretto Penna
Department of Food Engineering and Technology
UNESP - Sao Paulo State University
Sao Jose do Rio Preto, Sao Paulo
Brazil

Luis Augusto Nero
Universidade Federal de Viçosa
Veterinary Department
Viçosa, Minas Gerais
Brazil

Svetoslav Dimitrov Todorov
Universidade Federal de Viçosa
Veterinary Department
Viçosa, Minas Gerais
Brazil

CRC Press
Taylor & Francis Group
Boca Raton London New York

CRC Press is an imprint of the
Taylor & Francis Group, an **informa** business
A SCIENCE PUBLISHERS BOOK

CRC Press
Taylor & Francis Group
6000 Broken Sound Parkway NW, Suite 300
Boca Raton, FL 33487-2742

First issued in paperback 2020

© 2017 by Taylor & Francis Group, LLC
CRC Press is an imprint of Taylor & Francis Group, an Informa business

No claim to original U.S. Government works

ISBN-13: 978-1-4987-3811-8 (hbk)
ISBN-13: 978-0-367-78279-5 (pbk)

Library of Congress Cataloging-in-Publication Data

Names: Barretto Penna, Ana Lúcia, editor. | Nero, Luis Augusto, editor. | Todorov, Svetoslav Dimitrov, editor.
Title: Fermented foods of Latin America / Ana Lúcia Barretto Penna and Luis Augusto Nero and Svetoslav Dimitrov Todorov, editors.
Description: Boca Raton : Taylor & Francis, 2017. | Series: Food biology | "A CRC title." | Includes bibliographical references and index.
Identifiers: LCCN 2016006870 | ISBN 9781498738118 (hardcover : alk. paper)
Subjects: LCSH: Fermented foods– Latin America. | Food– Latin America. | Fermented beverages– Latin America. | Beverages– Latin America.
Classification: LCC TP371.44 .F4775 2017 | DDC 664/.024– dc23
LC record available at http://lccn.loc.gov/2016006870

Visit the Taylor & Francis Web site at
http://www.taylorandfrancis.com

and the CRC Press Web site at
http://www.crcpress.com

Preface to the Series

Food is the essential source of nutrients (such as carbohydrates proteins, fats, vitamins, and minerals) for all living organisms to sustain life. A large part of daily human efforts is concentrated on food production, processing, packaging and marketing, product development, preservation, storage, and ensuring food safety and quality. It is obvious therefore, our food supply chain can contain microorganisms that interact with the food, thereby interfering in the ecology of food substrates. The microbe-food interaction can be mostly beneficial (as in the case of many fermented foods such as cheese, butter, sausage, etc.) or in some cases, it is detrimental (spoilage of food, mycotoxin, etc.) The *Food Biology* series aims at bringing all these aspects of microbe-food interactions in form of topical volumes, covering food microbiology, food mycology, biochemistry, microbial ecology, food biotechnology and bio-processing, new food product developments with microbial interventions, food nutrification with nutraceuticals, food authenticity, food origin traceability, and food science and technology, Special emphasis is laid on new molecular techniques relevant to food biology research or to monitoring and assessing food safety and quality, multiple hurdle food preservation techniques, as well as new interventions in biotechnological applications in food processing and development.

The series is broadly broken up into food fermentation, food safety and hygiene, food authenticity and traceability, microbial interventions in food bio-processing and food additive development, sensory science, molecular diagnostic methods in detecting food borne pathogens and food policy, etc. Leading international authorities with background in academia, research, industry and government have been drawn into the series either as authors or as editors. The series will be a useful reference resource base in food microbiology, biochemistry, biotechnology, food science and technology for researchers, teachers, students and food science and technology practitioners.

Ramesh C. Ray
Series Editor

Preface

A wide variety of fermented food products are produced in Latin America as a result of the heritage of Pre-Colombian indigenous tribes and the technological knowledge of the immigrants colonising the New World. The specific demographic fusion in the Latin American region has resulted not only in an extremely rich human genetic diversity, but also resulted in various gastronomical habits and traditions, including preparation of different fermented food products. Regional, climatic, socio-cultural conditions, and a mix of different cultures have resulted in the development of the specific Latin American cuisine and preparation of new world traditional fermented food products. The immigrants to Latin America have looked for new economic opportunities, but at the same time brought with them their food habits and products as part of the of new Latin American tradition. However, in time, these products have acquired a new "face" and nowadays can be considered as traditional Latin American products. Fermentation processes were used as a base for conservation of milk, fruits, vegetables and meat since ancient time. Some fermented food products, besides of their nutritional characteristics, are a part of the spiritual habits of the tribes as well.

During the great geographical discoveries of 14th to 16th centuries, fermented meat products were essential part of the diet of the explorers of the New World. However, it was wrongly believed that animal proteins were an important part of the diet and had been associated with the social prosperity. Fruits and vegetables were ignored. This was the reason for the vitamin deficiency especially at the time of the expeditions. After discovering South America, new settlers focused primarily on the meat diet due the rich nutritional value and easy access to it. Even nowadays, more than 500 years later, in the regions explored, discovered and civilised during the greats geographical discovery, meat plays an important role in the social events and is still considered as symbol of prosperity. However, nowadays, besides raw and fermented meat, different fermented milk products, fruits and vegetables are part of Latin American balanced nutritional habits presenting rich fusion of European, Asian, African and Pre-Colombian heritage.

The editors hope the book will be cohesive and readable, and readers will learn about traditional fermented food products from Latin America, with special attention to the Pre-Colombian nutritional heritage and products based on the local food ingredients. Knowledge of traditional fermented food products is essential to bridge between past and the future and the development of new food products.

Ana Lucia Barretto Penna
Luis Augusto Nero
Svetoslav Dimitrov Todorov

Contents

1

Fermented Dairy Beverages in Latin America

Sabrina Neves Casarotti, Vivian Ribeiro Diamantino,
Luana Faria Silva, Cecília Loyola Afonso dos Santos, Aline
Teodoro de Paula and Ana Lúcia Barretto Penna*

Abstract

Fermented dairy beverage is mainly produced by a mixture of milk, cheese whey and other ingredients fermented by lactic acid bacteria and/or *Bifidobacterium* sp. Cheese whey is an important by-product of the dairy industry and a significant source of minerals and vitamins, mainly riboflavin. Despite its biological value, whey is not commonly used in human nutrition. In many Latin America countries, only part of total whey produced by the industry is used in food products. The use of whey, juice and vegetable extracts in the manufacture of fermented beverages shows great potential due to its nutritional properties, contributing to incorporate this by-product in human diet. These types of beverages have been well accepted due to their low acidity level, creamy consistency and high commercial value. Furthermore, the increased consumer demand for yogurt and other fermented dairy products, such as whey beverages, due to their health benefits, has resulted in an increase in the onsumption of these types of foods. Therefore, this chapter will discuss the characteristics and fundamentals of production, nutritional value, biochemistry of fermentation, and innovations as well as trends of fermented dairy beverages. Some reports about the development of novel non-dairy beverages have also been included.

Fermented Dairy Beverages General Characteristics

The terms *milk beverages, dairy beverages, dairy-type beverages* or *whey beverages* are used worldwide to cover a wide range of products made with milk and other ingredients, which may be fermented or not (Thamer and Penna 2006). Consequently, for purposes of this chapter, "fermented dairy beverages" will be discussed as fermented, drinkable, and non-alcoholic beverages-based primarily on whey or containing a significant amount of other ingredients (whey, juice or vegetable

Department of Food Engineering and Technology, São Paulo State University, Rua Cristóvão Colombo, 2265. 15054-000, São José do Rio Preto (SP), Brazil.
* Corresponding author: analucia@ibilce.unesp.br

extract) mixed with milk. Despite of many options available for production, cheese whey and milk are the most used ingredients to formulate dairy beverages (Penna 2010). In order to show some trends of the beverage market, some reports about the development of novel non-dairy beverages are also included in this chapter.

Fermented dairy beverage can be defined as a product obtained from the mixture of milk and other dairy or non-dairy products (cheese whey, juices, vegetal extracts, etc.), fermented by lactic acid bacteria (LAB) and/or *Bifidobacterium* sp. (Castro et al. 2009). In Brazil, fermented dairy beverages are produced from milk or its by-products, with or without the addition of other ingredients, but the dairy base needs to represent at least 51% (vol/vol) of its formulation (Brazil 2005).

Cheese whey is the liquid by-product that remains from casein precipitation during cheese manufacturing (Lievore et al. 2015). In the past, it was considered just a by-product of dairy industries and, generally, was disposed in the environment, causing serious pollution problems. The whey contains a high amount of lactic and other organic acids, besides ammonium nitrogen, which present a high biological (40,000 mg/L) and chemical (60,000 mg/L) oxygen demand. In the last decade, with the increase in environmental awareness and economic competitiveness, this concept changed. Cheese whey has been extensively studied as a low cost carbon source fermentation substrate for producing a variety of added value products, such as lactose, single cell protein, whey beverages, whey protein concentrate, whey powders, alcohols, organic acids, biopolymers, human food and food additives, animal feed, fertilizers, and deicers as well as anti-icers (Mahmoud and Ghaly 2004).

Nowadays, the intrinsic human-health benefits of cheese whey are well known and so the dairy beverages are accepted positively by consumers. In this context, the production of fermented dairy beverages has been increasing worldwide, not only because of their simple production technologies and their interesting nutritional value, but also because they are an alternative for reusing whey. During cheese manufacture, the whey produced can be added in different types of beverages, which, in general, have good acceptance and wide consumption (Castro et al. 2013a,b; Gomes et al. 2013; Silveira et al. 2015).

Products that include cheese whey in their formulation increase their added value, since cheese whey has the ability to improve food color, texture and flavor characteri-stics. Moreover, the whey contains a large quantity of nutritionally rich components and can be used as a functional ingredient. Thus, cheese whey plays an important role in the formulation of food products (Kabašinskienė et al. 2015). Cheese whey composition can vary according to several factors, including the type of cheese being processed, the method of casein precipitation, milk thermal treatment and its storage after milking, among others (Lucas et al. 2006).

Basically, there are two types of cheese whey: acid and sweet. The first one presents pH under 5.0 and is generated from the manufacture of cheeses produced by the acid coagulation of milk, a combination of acid and rennet, or acid and heating. The second one is obtained by the manufacture of cheeses produced by enzymatic coagulation and has higher pH value (Lievore et al. 2015).

The acid whey has higher acidity, calcium and ash contents, and lower levels of protein when compared to sweet whey. So, its addition as an ingredient in food formulations is limited, due to its acid and salty taste (Lievore et al. 2015). In general, cheese whey contains about 55% of the nutrients present in the original milk (Abdolmaleki et al. 2015). This product is composed of water (93%), lactose (5%), protein (0.85%), a minimum quantity of fat (0.36%) and minerals (0.53%), including sodium chloride, potassium chloride, calcium salts (especially phosphate) and others (Pescuma et al. 2010). Other components are also present in cheese whey, such as lactic (0.5 g/L) and citric acids, non-protein nitrogen compounds (urea and uric acid) and B vitamins (Dragone et al. 2009).

Whey proteins present almost all essential amino acids in concentrations higher than the recommendations of dietary reference intake, except for the aromatic amino acids (phenylalanine and tyrosine) which do not appear in excess; however, they attend the recommendations for all ages. They have high concentrations of tryptophan, cysteine, leucine, isoleucine and lysine amino acids (Sgarbieri 2004).

Many studies show that cheese whey has bioactive compounds which can offer specific health benefits. Individual whey protein components and their peptide fragments, for example, stimulate the immune system and have anticarcinogenic activity and other metabolic features. Furthermore, the whey protein contains high amounts of branched-chain amino acids, such as leucine, isoleucine and valine, which are used as a muscle-building supplement for athletes during intensive exercise (Kabašinskienė et al. 2015).

Many components of cheese whey have also antioxidant activity (Vavrusova et al. 2015), which increase the benefits provided by the addition of whey in fermented dairy beverages. Consequently, cheese whey, mainly because of its whey proteins, provides excellent nutritional value in foods formulated for children, adults and elderly people, or as supplements for body health maintenance (Kabašinskienė et al. 2015).

A variety of factors can affect the physical characteristics of dairy products, such as the composition and heat treatment of milk, the use of additives such as stabilizers, the culture used for the fermentation process, the breaking and stirring of the gel, and the storage conditions until the end of shelf-life. In the production of dairy beverages, the texture and mouth feel are parameters that need some attention, since they tend to be poor and watery when compared with that of fermented milk. From the technological point of view, the main difference between yogurt and fermented dairy beverages is the addition of whey to the latter, which results in lower viscosity, due to the low percentage of total solids of the liquid whey (ca. 6 g/100 g) (Castro et al. 2013b; Buriti et al. 2014).

The addition of fruits and fiber ingredients appears to be an alternative that improves the sensory acceptability of the beverages, besides enriching with nutrients that are not present in milk, such as dietary fiber (Buriti et al. 2014).

The use of enzymes has been studied as another alternative to improve the physical properties of yogurt and dairy beverages. The use of transglutaminase enzyme in their production can reduce the syneresis rate, increase the consistency index, and compensate for the possible changes in rheological and texture parameters caused by the addition of whey in dairy beverages (Gauche et al. 2009).

LAB that produce exopolysaccharides (EPS) can also be incorporated in dairy products to improve its rheology, texture and mouth feel properties (Li et al. 2014). The EPS performs other functions such as thickeners, stabilizers, emulsifiers, bodying and gelling agents or fat replacers in several food products (Degeest et al. 2001).

Supplementation with probiotic bacteria and prebiotic ingredients provide additionals value to dairy beverages, and provide health benefits, such as the decrease in blood pressure. In this case, the cheese whey has an important role on the growth and viability of the microorganisms, once its addition increases the levels of lactose and other nutrients, creating better conditions for microbial survival (Castro et al. 2013 a,b).

Consumers are increasingly concerned about their health, and they are looking for healthy foods that, preferably, are able to reduce the risk of diseases (Siqueira et al. 2013). Fermented dairy beverages can match the demands of today's market, providing not only benefits to human health but also to the environment by avoiding the disposal of whey that is often improperly done.

Biochemistry of Fermentation

The conversion of milk base and whey to fermented dairy beverage is achieved by metabolic activity of the fermenting microorganisms, mainly LAB and bifidobacteria. The term LAB describes a group of functionally and genetically related bacteria which belong to the phylum *Firmicutes*. They are generally defined as a cluster of lactic acid-producing, low percentage of guanine and cytosine (G-C), non-spore-forming, Gram-positive rods and cocci, aerotolerant, facultative anaerobe, catalase negative and non-motile microorganisms. According to their optimum growth requirements they are divided into mesophilic and thermophilic starter cultures which can grow at 20–30 °C or 37–45 °C, respectively. Bifidobacteria were classified in the genus *Lactobacillus* until the 1970s. However, it is now obvious that they are phylogenetically distinct from the LAB, and are included in the *Actinobacteria* phylum. Bifidobacteria are Gram-positive, non-motile, non-spore-forming rods with a high G-C content (55 to 67%). Cells often occur in pairs with a V or Y-like appearance. They are strictly anaerobic and catalase negative, with optimum growth temperature between 37 °C and 41 °C (Hutkins 2006).

The fermentation of the milk base and whey is characterized by the conversion of lactose to lactic acid, and other metabolites, such as aroma compounds, which contribute to the flavor and texture, and increase carbohydrate solubility and sweetness of the product. Since whey contains lactose as the main solute, the fermentation of this matrix by LAB is an interesting way to increase its use and also to develop novel food. Nevertheless, pure whey has proved to be a nutritionally incomplete medium for some LAB, leading to a slower growth. For this reason, some additional nutrients should be added prior to fermentation of this type of fermented beverage (Jelen 2009). Additionally, some LAB have proteinases, peptidases and a specific peptide and amino acid transport system, which can increase the digestibility of amino acids and hydrolyze allergenic peptides present in whey (Pescuma et al. 2008).

Carbohydrate metabolism

Considering the sugar metabolic pathway, LAB can be classified as homofermentative or hereterofermentative, including the bifidus pathway, based on the main products formed from glucose. The homofermentative bacteria use the Embden-Meyerhoff-Parnas pathway (or glycolytic pathway), whereas the heterofermentative bacteria use the phosphoketolase pathway (Fig. 1.1). The essential feature of LAB metabolism is fermentation of carbohydrates coped with substrate phosphorylation. The ATP formed is then used for biosynthetic purposes. LAB have an enormous capacity to degrade different carbohydrates and related compounds. The final product is usually lactic acid, which accounts for over 50% of total production (Axelsson 2004).

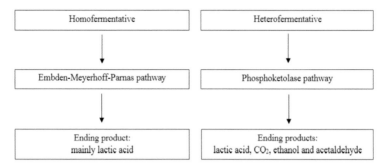

FIGURE 1.1   General differences between homo- and heterofermentative bacteria.

The Embden-Meyerhoff-Parnas pathway is characterized by the conversion of glucose into fructose-1,6-diphosphate (FDP), an intermediate key of this route, which is divided by an aldolase into dihydroxyacetone phosphate (DHAP) and glyceraldehyde 3-phosphate (GAP). Afterwards, these compounds are converted to pyruvate in a sequence that includes phosphorylation of the substrate at two sites. Under normal conditions, i.e., excess sugar and low oxygen content, pyruvate is reduced to lactic acid by a NAD-dependent lactate dehydrogenase, oxidizing the NADH formed during the early stages of the glycolytic pathway. Thus, the redox balance is obtained and the lactic acid is, substantially, the only final product (Fig. 1.2). For this reason, this type of metabolism is called homolactic fermentation. This fermentation results in a net gain of two ATP molecules and two molecules of lactic acid per fermented glucose molecule. *Enterococcus, Streptococcus, Lactococcus* and some species of *Lactobacillus* are examples of the homofermentative bacteria group (Axelsson 2004; Madigan et al. 2008).

Pyruvate catabolism can be redirected to produce acetate, acetaldehyde and ethanol *via* the common intermediate acetyl-CoA, using other enzymes. This happens when the bacteria are exposed to certain environmental conditions, such as aerobic condition, energy starvation and/or growth on sugars that are slowly hydrolyzed, like maltose or galactose. The products that will be formed from the acetyl-CoA depend mainly on the intracellular redox state. The formation of both acetaldehyde and ethanol is coupled with the regeneration of NAD^+ under anaerobic conditions. When the redox balance is maintained *via* other reactions such as NADH oxidase, which is active under aerobic conditions, acetyl-CoA is degraded *via* acetate kinase, producing acetate in a reaction coupled with ATP formation (Tamime et al. 2006).

The phosphoketolase pathway is characterized by the initial dehydrogenation with the formation of 6-phosphogluconate, followed by decarboxylation, forming a pentose-5-phosphate. This compound is divided by phosphoketolase in GAP and acetyl phosphate. GAP is metabolized in the same way as it occurs in the glycolytic pathway, resulting in the formation of lactic acid. When there is no available additional receiver electron, acetyl phosphate is reduced to ethanol via acetyl-CoA and acetaldehyde (Fig. 1.3). Such fermentation is known as heterolactic because it allows the formation of significant amounts of other end products of fermentation, such as CO_2 and ethanol, in addition to the lactic acid. This process also results in a net gain of one molecule of ATP and one molecule of each of the final products (lactic acid, ethanol and CO_2) for each molecule of fermented glucose. *Leuconostoc, Weisella* and some *Lactobacillus* are examples of heterofermentative bacteria (Gaspar et al. 2013; Prückler et al. 2015). Among the LAB, lactose metabolism in *Lactococcus* is probably the best understood in terms of its biochemical and genetic make-up and also in terms of its regulation (Shiby and Mishra 2013).

Bifidobacteria possess a metabolic pathway called *via* bifidus pathway or fructose 6-phosphate pathway, which allows the production of acetic and lactic acids in a molar ratio of 3:2. The key enzyme of this pathway is fructose 6-phosphate phosphoketolase (F6PPK), and therefore, it can be used as taxonomic marker in genera identification. The presence of this enzyme and the absence of aldolase and glucose 6-phosphate dehydrogenase enzymes differentiate the *Bifidobacterium* and *Lactobacillus* genera (Axelsson 2004; Oliveira et al. 2012); however, it does not allow differentiation between Bifidobacteria species.

Taking into account that different LAB species have the ability to form distinguishing compounds through lactose fermentation, the sensory profile of fermented dairy beverage will be directly influenced by the bacteria used for its production (Serra et al. 2009).

The compounds which influence the sensory characteristics of fermented milk and fermented dairy beverages can be divided into four main categories (Tamime and Robinson 1999):

• non-volatile acids (lactic, pyruvic, oxalic and succinic);
• volatile acids (acetic, propionic and butyric);
• carbonyl compounds (acetaldehyde, acetone, acetoin and diacetyl) and
• other compounds (certain amino acids and / or components formed by thermal degradation of protein, fat and lactose).

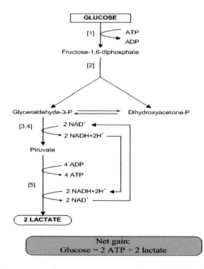

FIGURE 1.2   Schematic steps of glucose fermentation in the Embden-Meyerhoff-Parnas pathway of lactic acid bacteria. [1] Glucokinase; [2] fructose-1,6-diphosphate aldolase; [3] glyceradehyde-3-phosphate dehydrogenase; [4] pyruvate kinase; and [5] lactate dehydrogenase. Adapted from Axelsson (2004).

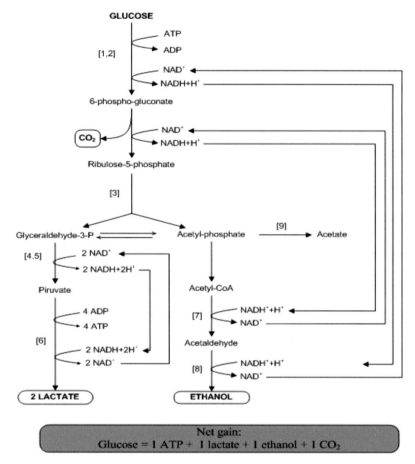

FIGURE 1.3 Schematic steps of glucose fermentation in the phosphoketolase pathway of lactic acid bacteria. [1] Glucokinase; [2] glucose-6-phosphate dehydrogenase; [3] phosphoketolase; [4] glyceradehyde-3-phosphate dehydrogenase; [5] pyruvate kinase; [6] lactate dehydrogenase; [7] acetaldehyde dehydrogenase; [8] alcohol dehydrogenase. Adapted from Axelsson (2004).

Organic acids are the main metabolites present in fermented dairy products; they are found in greater quantities and are responsible for decreasing pH values (Leite et al. 2013). Besides the effect of organic acids on sensory characteristics, they are also important for biopreservation, resulting in the contribution to an extended shelf life and enhancing the safety of the product. The antimicrobial (bactericidal or bacteriostatic) effect of organic acids in fermented milks lies in the reduction of pH, as well as the nature of the undissociated form of the organic acid, which inhibits the growth of spoilage and pathogen microorganisms (Chandan and Kilara 2013; Ghosh et al. 2015; Penna et al. 2015). In addition, it is also essential to highlight that low pH values are important for casein precipitation during the production of fermented milk.

The most commonly found organic acids in fermented milk during fermentation and storage are orotic, citric, pyruvic, acetic, propionic, uric and lactic acids (Østlie et al. 2003); however, lactic and acetic acids are the main organic acids produced from glucose while the others are produced in small quantities (Chandan and Kilara 2013).

Lactic acid is a non-volatile organic acid that presents a smooth flavor (Carvalho et al. 2005) and has been traditionally used in dairy products as preservative and flavor additive (Gaspar et al. 2013). The acetic acid is responsible for the vinegar flavor and has fungicide and bactericide action (Dang et al. 2009; Baek et al. 2012; Crowley et al. 2013).

A recent study showed that the amount of whey added did not influence the acetic acid level in probiotic fermented dairy beverage during storage. According to the authors, this finding can be an advantage from the industrial standpoint because high levels of whey reduce production costs and do not contribute to an increase in the synthesis of acetic acid, which would cause decreased sensory acceptance by the consumers. Nevertheless, a difference in the lactic acid content was observed; when 80% of cheese whey was added the levels of this compound were lower compared with 0, 20, and 35% of cheese whey, suggesting that even if the peptides contained in the whey had a greater availability, there is a limit in which it could be metabolized by the cultures. With regard to the compounds responsible for the aroma of the final product, an overall increase in the concentrations of diacetyl and acetaldehyde was observed, proportionally to the amount of whey in the beverage formula (Castro et al. 2013a).

The propionic acid is used in food as a fungistatic agent, and it is also effective in controlling the growth and reducing the viability of both Gram-positive and Gram-negative bacteria (Suomalainen and Mayra-Makinen 1999). Orotic acid, also called vitamin B_{13}, is normally associated with fermented milks and the growth of LAB and starters (Fernandez-Garcia and Mcgregor 1994). This acid is an intermediate agent in the synthesis of nucleic acids and is claimed to have a hypocholesterolemic effect that is being found in probiotic fermented milks (Alhaj et al. 2007). Citric acid has a pleasant taste, easily assimilated by the human body and shows small toxicity. Due to these characteristics, it is widely used in dairy products as an acidulant agent to intensify flavor and stability (Izco et al. 2002). The antibacterial effect of citric acid is linked not only to its acidic action, but also to its chelating activity of Ca^{2+} ions (Graham and Lund 1986).

Citrate present in milk can be used by positive-citrate (CIT$^+$) LAB to produce important aromatic and antimicrobial compounds in fermented milk. The ability to metabolize citric acid is dependent on the permease citrate enzyme (CitP) and can be used by LAB as the sole energy source or be co-metabolized. For heterofermentative LAB, citrate is converted into pyruvate, which is further reduced to lactate. In this case, when citrate is present, more energy is generated during sugar degradation (Schmitt et al. 1992). Particularly for *Lactobacillus casei* (facultative heterofermentative *Lactobacillus*), some strains are able to metabolize citrate when the presence of glucose is limited. Differently from homofermentative LAB, pyruvate is the common intermediate formed during sugar and citrate metabolism. The subsequent dissipation of pyruvate leads to the production of four carbon aroma compounds (C4 compounds - acetaldehyde, acetoin, 2,3-butanediol, diacetyl) (Smid and Kleerebezem 2014). Additionally, metabolism of lactose and citrate by LAB can also produce significant amounts of ethanol, acetic acid, and other specific flavor related molecules (Milesi et al. 2010).

Protein metabolism

The breakdown of milk proteins (caseins) by LAB occurs by extracellular and intracellular proteolytic enzymes during the fermentation process. For example, the caseins are hydrolyzed by a lactococcal, extracellular, cell envelope–bound proteinase (PrtP). This enzyme has a broad specificity and can produce more than 100 different oligopeptides from caseins (Steele et al. 2013), which subsequently release free amino acids in the cytoplasm (Kunji et al. 1996). These amino acids first feed the protein synthesis machinery, thereby supporting LAB growth and maintenance. On the other hand, the intracellular amino acid catabolism leads to the formation of a broad array of volatile aroma compounds and depends on the presence and activity of various enzymes and metabolic pathways such as transaminase and lyase pathways, and also non enzymatic conversions (Smit et al. 2005).

LAB have a very limited capacity for synthesizing amino acids using inorganic sources of nitrogen. They are therefore dependent on pre-formed amino acids in the growth medium as a nitrogen source. It should be noted, however, that the amino acid requirement differs between species and between strains of the same species. Some strains of *Lactococcus lactis* subsp. *lactis* are able to use most of the amino acids, while strains of *Lactococcus lactis* subsp. *cremoris* and *Lactobacillus helveticus* require from 13 to 15 specific amino acids (Chopin 1993).

When using whey as a substrate for fermentation it is important to take into account the allergenicity of its proteins. The main whey proteins are β-lactoglobulin (58%) and α-lactalbumin (13%). The β-lactoglobulin is the most abundant whey protein and is highly resistant to gastric digestion. This protein is the major cause of milk intolerance and/or allergenicity in humans since it is present in smaller quantities in human milk. The industrial processes use some treatments to enhance whey proteins digestibility, such as sterilization, heating, or hydrostatic high pressure. Additionally, in these processes whey proteins are hydrolyzed in order to reduce their allergenic content as well as to improve their digestibility, since the released peptides could be better absorbed in the intestinal tract than intact proteins. This degradation is carried out using enzymes, acids and alkalis, but the process tends to be difficult to control. Moreover, it generates products with reduced nutritional qualities besides forming toxic substrates such as lyso-alanine (Sinha et al. 2007; Pescuma et al. 2008).

As an alternative to these industrial processes, LAB can be used as agents to increase whey digestibility and hydrolyze allergenic peptides, as reported by Pescuma et al. (2008). The concentration of free amino acids in milk and whey are very limited, thus the sustained growth of LAB depends on the production of proteinases, peptidases, and specific peptide and amino acid transport systems. The authors evaluated the proteolysis, the release of free amino acids, and the hydrolysis of whey protein by three strains of LAB. They reported that *Lactobacillus delbrueckii* subsp. *bulgaricus* CRL 454 was the most proteolytic strain (91 µg Leu/mL) and released the branched chain amino acids leucine and valine. In contrast, *Lactobacillus acidophilus* CRL 636 and *Streptococcus thermophilus* CRL 804 consumed most of the amino acids present in whey. The strains were able to degrade the major whey proteins, α-lactalbumin being degraded in a greater extent (2.2–3.4-fold) than β-lactoglobulin.

In another study, a fermented whey beverage was produced using LAB and WPC35 (WPC containing 35% of proteins). During fermentation, single cultures produced 25.1–95.0 mmol/L of lactic acid as consequence of lactose consumption

(14.0–41.8 mmol/L) after 12 h of fermentation. *Lactobacillus delbrueckii* subsp. *bulgaricus* CRL 656 was the most proteolytic strain (626 μg Leu/mL) and released the branched-chain essential amino acids leucine (16 μg/mL), isoleucine (27 μg/mL), and valine (43 μg/mL). All strains were able to degrade β-lactoglobulin in a range of 41–85% after 12 h of incubation (Pescuma et al. 2010).

Fat metabolism

Lipolysis is an important biochemical event that occurs in some fermented dairy beverage, such as fermented milk and yogurt. The lipases produced by bacteria are relatively diverse in their properties and substrate specificities and they have been extensively used in the dairy industry for the hydrolysis of milk fat (Hasan et al. 2006; Ramyasree and Dutta 2013). Apart from the action of lipases, the temperature and agitation held in some technological processes can also cause lipolysis of milk fat (Chandan 2006). In general, lipolysis activity results in the hydrolysis of milk fat during fermentation and in the production of glycerol and free fatty acids (Cogan and Beresford 2002).

Lipolytic enzymes can produce free fatty acids of different carbon atom lengths (C4–C20) (Cheng 2010; Beermann and Hartung 2012). Moreover, the free fat acid concentration, released from the lipolytic activity, is strictly related to sensory characteristics, flavor, and aroma formation in fermented milk (Masuda et al. 2005). In contrast, some chemical substances which are detrimental to flavors (rancid, off flavors, and odors) of fermented milk can be enzymatically released from the acylglycerides of milk lipids (Chandan 2006).

In yogurt, the oxidation of fatty acids is one major pathway for the production of flavor compounds. The short-chain fatty acids that are strong flavor contributors to cultured products are produced from saturated fatty acids. Unsaturated fatty acids are oxidized in the presence of free radicals to form hydroperoxides, which rapidly decompose to form hexanal or unsaturated aldehydes (McGorrin 2001). Unsaturated fatty acids also lead to the formation of 4- or 5-hydroxyacids, which readily cyclizes to γ- or δ-lactones, and odd-carbon methyl ketones by decarboxylation of β-ketoacids (Cheng 2010). Nonetheless, the production of such compounds is not very well described in fermented milk (Hasan et al. 2006). Furthermore, the effect of lipolytic enzymes on flavor and texture properties is more important in cheese ripening, butter fat, and cream than in fermented beverages.

Production of Fermented Dairy Beverages

In the last decades, there has been a worldwide growth in the production of dairy beverages. In addition, all over the world, different dairy-type fermented beverages are produced either by adding liquid whey, whey components, soy extract or juice at previous cultured milks (i.e. fermented milks, yogurt, sour milk, butter milk, kefir, and other similar products), or by the fermentation of liquid whey, independently or in combination with milk (Jelen 2009; Penna 2010). In Brazil, fermented dairy beverages represent a significant part of the market for yogurt and other dairy-based milk beverages.

Important production aspects from fermented dairy beverages

The production of fermented dairy beverages includes several technological operations, which are basically divided in two steps: a) the pre-processing of liquid cheese whey, which generally includes centrifugation or filtration and thermal treatment; and b) the subsequent production of the fermented beverage, which generally includes: formulation, mixing of the ingredients, fermentation, cooling, and packaging.

Pre-processing of cheese whey

For the pre-processing of cheese whey, some basic processes such as filtration or centrifugation to standardize milk fat and to remove curd particles, besides a thermal treatment to ensure its microbiological safety may be required. Additionally, specialized operations, such as membrane processes to separate and concentrate the whey protein, and electrodialysis, ion exchange, or nanofiltration to a partial or full demineralization of the whey can be used (Jelen 2009).

Separation of fat and casein fines from the whey. The whey resulting from the cheese-making process, with adequate characteristics (i.e. maximum acidity of 13 °D, free from inhibitory substances, without washing water), should be filtered or clarified to eliminate casein fines and fat traces, which are undesirable in fermented dairy beverages (SBRT 2007; Brazil 2005). Casein fines present in whey affects the fat separation and needs to be removed. Different separation systems may be used for the casein separation, such as cyclones, centrifugal separators or rotating filters. Then, the fat is removed and recovered in centrifugal separators. The fines may be used in processed cheese production and the whey cream can be used to standardize cheese milk (Huffman and Ferreira 2011; Bylund 1995).

For the production of beverages, the whey may be only pasteurized, if used in a liquid form, or may be subsequently concentrated, and/or dried, or isolated to produce whey-based products (i.e. whey powders, whey protein concentrates, whey protein isolates), which are also frequently added after or before the fermentation step in beverages production.

Whey concentration. For the production of whey concentrates or whey powders, some processing steps are required; they are basically concentration and drying to produce a powder. Whey concentration commonly occurs under vacuum in a falling-film evaporator with up to seven stages. After evaporation (until 45–65% of total solids), the concentrate is rapidly cooled in a plate heat exchanger and transferred to a tank with constant stirring (for 6–8 hours in order to maintain the temperature between 15–20 °C) to obtain small crystals and consequently, a non-hygroscopic product when spray dried (Bylund 1995). The limited solubility of lactose is frequently a problem in the production of whey concentrates or powders, due to its tendency to crystallize. If whey concentrates containing considerable amounts of lactose are conventionally spray dried after evaporation, they will contain a large proportion of amorphous lactose which will lead to a very hygroscopic final product that may cake during the storage period (Huffman and Ferreira 2011; Jelen 2009).

Additionally, methods of separation and concentration of whey based on membrane separation are frequently used. Among them, ultrafiltration (UF) is probably the most used processing method in whey beverages production and the main whey-based ingredients. The UF process is effective for separation and concentration of soluble whey proteins in the retentate which may be used as a base material for whey beverages (Jelen 2009). The retentate can also be dried for the production of whey protein concentrates (WPC). When a whey product presents more than 25% of protein content in dry matter it is classified as a WPC, which is described in terms of its protein content, ranging commonly from 35% to 80% (Whey Protein Institute 2015; Yada 2004).

Furthermore, when the degree of concentration is greater than 90% (on a dry basis), the product is known as whey protein isolate (WPI), which is the purest form of whey protein available and is also a good protein source for individuals with lactose intolerance as it contains little or no lactose (Whey Protein Institute 2015). The composition of the main whey products, including whey powders, WPC, and WPI are described in Table 1.1. Those whey products are frequently added in a wide range of products for protein enrichment, including fermented dairy beverages.

In addition, owing to the functional properties of the whey proteins (e.g. α-lactalbumin, β-lactoglobulin, bovine serum albumin, immunoglobulins, glycomacropeptide, lactoferrin, proteose-peptone, and lactoperoxidase), there is an increasing interest in whey protein separation (Yadav et al. 2015). The UF method is the most used technique to recover whey functional proteins. The components of low molecular mass (lactose, salts, and water) permeate through the UF membranes, which retain whey proteins which are subsequently spray dried. Chromatography methods are also used with greater results of purification; however, they are expensive methods from an industrial point of view (Fox and McSweeney 1998). Despite the production of individual whey proteins, those proteins may also be transformed into bioactive peptides via enzymatic or fermentation processes. Bioactive peptides are protein fragments that show a positive influence on body functions and present a great impact on health. Whey proteins can be converted into various bioactive peptides, which include: ACE-inhibitory peptide, bioactive peptides with opioid activity, iron-binding bioactive peptides, and hydrolyzed whey protein (Yadav et al. 2015).

Demineralization and lactose hydrolysis. Whey has a considerable high salt content (about 8%–12% on dry matter), which may limit its usefulness in whey products. The presence of monovalent ions (Na^+, K^+, Cl^-) may result in negative sensory properties. In consequence, the demineralization of those whey powders and concentrates enables its use in a range of products. Partially demineralized whey (25%–30%) is frequently used in dairy and bakery products, whereas highly demineralized whey (90%–95%) is more used in infant formulas (Huffman and Ferreira 2011; Bylund 1995).

The principle of demineralization is the removing of inorganic salts and reducing the content of organic ions, e.g. lactates and citrates. For a partial demineralization of whey, nanofiltration (NF), another membrane process, may be used. However, for a more intense demineralization, specialized processes such as electrodialysis or ion exchange should be used (Jelen 2009). Electrodialysis is recommended for demineralization levels up to 50%–60%, whereas ion exchange may remove almost all the minerals present in whey (Huffman and Ferreira 2011).

In the same way, a reduction in lactose content is needed in some occasions, such as those related to nutrition (to avoid problems of lactose intolerance), sensory desirability (to increase sweetness, or to avoid Maillard reaction), or technology (to avoid sandiness due to the lactose crystallization or if using specific microorganisms such as yeasts, which do not ferment lactose). Reduction of lactose frequently involves partial lactose removal by crystallization or by a chromatographic process (Jelen 2009).

The composition of the main whey products, including demineralized and delactosed whey products are described in Table 1.1.

TABLE 1.1 Composition of the main whey products

Whey products	Moisture (%)	Protein (%)	Lactose (%)	Ash (%)	Fat (%)
Liquid sweet whey	93.0–95.0	0.7–1.2	3.8–5.0	0.5–0.8	0.04–0.4
Whey powder	3.0–8.0	10.0–15.0	63.0–75.0	7.0–12.0	1.0–1.5
Delactosed whey powder	3.0–4.0	18.0–24.0	52.0–58.0	11.0–22.0	1.0–4.0
Demineralized whey powder	3.0–4.0	11.0– 5.0	70.0–80.0	1.0–7.0	0.5–1.8
Whey protein concentrates*	3.0–4.6	35.0–80.0	4.0–55.0	2.0–8.0	1.0–7.0
Whey protein isolate	3.5–4.5	90.0–95.0	0.5–1.0	2.0–3.0	0.5–1.0

* Based on WPC concentrations of 35, 50, 65, and 80%. Adapted from: Bylund (1995), Pescuma et al. (2010), Yada (2004), Whey Protein Institute (2015), Canadian Dairy Comission (2015).

Thermal treatment of whey. For whey beverages based on liquid whey or on a mix of milk and whey, the whey should be previously heat treated to guarantee its microbiological quality. In general, pasteurization at low temperatures (below 70 °C) is used due to the problem of insolubilization of whey protein upon thermal processing (SBRT 2007). This occurs because whey proteins denature when exposed to temperatures at about 70 °C. Whey proteins produce insoluble complexes in the absence of casein, due to their tendency to self-association. This could be a problem resulting in undesirable sedimentation at higher temperatures of pasteurization (Jelen 2009). After thermal treatment, the whey is cooled to the desired inoculation temperature (typically 40–45 °C for thermophilic cultures, and 30–37 °C for mesophilic cultures) for the fermentation step.

Pre-processing of milk

For the production of fermented whey beverages, milk may be from different species of mammals; however, bovine milk is the most used. Moreover, the composition and quality of the milk will affect the overall quality of the beverage (Nilsson et al. 2006). Additionally, it is important to take into account that for fermented products, milk must be free from antibiotics, sanitizer residues, or any type of inhibitors (SBRT 2007).

The fat content of bovine milk normally ranges from 3.0 to 3.5 g/100 mL, while the fat content of most yogurts and fermented beverages ranges from 1.0 to 4.5 g/100 mL; therefore, the milk fat needs to be reduced (e.g. by using standardizing centrifuges or mixing whole milk with skimmed milk). The increase of solids-not-fat (SNF) can be achieved by adding skim milk powder or by evaporation (EV) or ultrafiltration (UF), which are processes that remove water and raise the levels of fat and SNF in the dairy base (Robinson et al. 2006).

Furthermore, to avoid microbial contamination, milk needs to be heat treated, which is commonly achieved in two different ways. The high temperature/short time (HTST) system involves passing the milk through a plate heat exchanger, and raising the milk temperature to 90–95 °C (for 5–10 min); on the other hand, the low temperature/long time (LTLT) system involves heating the milk to 80–85 °C (for 30 min). The HTST system is commonly used in industries, whereas the LTLT system is more effective in developing the desired textural properties. Higher temperatures could be employed, e.g. the temperatures used in ultra high temperature (UHT) systems (above 100 °C); however, fermented dairy beverages made from UHT-treated milk present lower viscosity (Robinson et al. 2006).

Mixing of the ingredients

The main ingredients frequently added in dairy beverages are sugars, sweeteners, stabilizers, fruit syrup, flavors, and/or colorants, and they can be added either before fermentation or after this stage. Other additives used are vitamins, calcium, inulin, and special fatty acids, such as omega-3 (Nilsson et al. 2006).

The addition of powder ingredients to dairy beverages is performed after or before the thermal treatment of the dairy base (mix of milk, whey, juice, or vegetable extract). If added after the thermal treatment, the ingredients may be thermal treated previously or an additional treatment in the dairy base (approximately 80 °C for 10 min) may be applied to guarantee microbial safety (SBRT 2007).

Sugar and sweeteners. Sugars (e.g. sucrose or glucose) or sweeteners may be added. Frequently, the recommendation of sugar addition is below 7–10 g/100 g milk (Penna et al. 1997; Nilsson et al. 2006); however, there is a growing tendency to reduce these contents due to the health benefits promoted by the reduction in the consumption of sugars. Additionally, adding higher concentrations of sugar before the inoculation (e.g. more than 10 percent) may negatively influence the growth of LAB. An alternative method is to add sugar after the fermentation stage, but only for long shelf-life fermented dairy beverages and before the final heat treatment step that guarantees microbiological safety (Nilsson et al. 2006). However, in some countries, this final heat treatment step is not allowed. For instance, in Brazil, it is established that fermented dairy beverages must not be submitted to any thermal treatment after fermentation, in order to maintain the viability of the microorganisms cells in the product (minimum of 10^6 CFU/g in the final product) (Brazil 2005). Consequently, the addition of sugar should necessarily happen before the fermentation stage.

Fruit pulps, juice and syrup, aroma and colorants. Although a considerable number of fermented dairy beverages are consumed in their natural state, in Latin America and in other parts of the world, most of these products contain fruit pulps, juices, or syrups (which may contain small fruit pieces), fruit aromas and colorants (Nilsson et al. 2006).

Fruit syrups and pulps are usually expensive and consequently they are not added excessively. The acidity level of the fermented beverage influences in the amount of fruit required (e.g. the higher the acidity of the fermented milk, the greater the amount of fruit syrup or pulp used); therefore, it is possible to reduce the amount of fruit syrup or pulp by using mild acidity starter cultures (Nilsson et al. 2006).

Several fruit pulps are used for the production of fermented dairy beverages in Latin America. Tropical fruits are frequently used, such as guava pulp, soursop pulp (Buriti et al. 2014), guava jelly (Gomes et al. 2013); pineapple juice (Masson et al. 2011), and passion fruit pulp (Uribe et al. 2008); mango pulp (Santos 2008), umbu pulp (Santos 2006), and others. However, subtropical fruits, such as strawberry pulp (Tranjan et al. 2009) and peach pulp (Tranjan et al. 2009; Pescuma et al. 2010) are commonly well accepted by the consumers, and hence their use is quite significant.

Fruit aroma has an important effect in less acidic products and is mainly added to products with long shelf-life. It is normally obtained in a concentrated form and can be added, as well as the colorants, in the buffer tank or in-line before the fermentation stage, if these components tolerate heat treatments. However, many aromas and colors present a negative effect when heat treated; in consequence, these ingredients should be aseptically added after the heat treatment and before the aseptic filling (Nilsson et al. 2006).

Stabilizers. The higher acidity of fermented dairy beverages promotes an impact on the sensory and functional characteristics of the product, since the casein becomes insoluble at a pH of 4.8 or less, and forms a coagulum. Thus, to prevent the sedimentation of casein in the final product, stabilizers are often added to these beverages (Jelen 2009).

Stabilizers are hydrophilic colloids that increase the viscosity and help prevent whey separation in a wide range of products, including fermented dairy beverages. Different types of stabilizers have been used by the food industry with this purpose, including pectin, sodium carboxymethyl cellulose, guar gum, gelatin, starch (native or modified), and blends of these stabilizers (Nilsson et al. 2006).

Products based on the fermentation of pure whey will not present the problem of coagulum formation at low pH values; however, the effect of subsequent thermal treatments in pure whey is higher than in whey combined with milk, causing sedimentation of the whey proteins (Jelen 2009). In this context, stabilizers may also help to avoid sedimentation in those products.

Other ingredients. Traditionally, whey beverages have been usually formulated with addition of vitamins and minerals and recognized as rich sources of these nutrients. Most of them are high in calcium and fortified with vitamin D (Jelen 2009).

Some functional ingredients have also been used in a wide variety of foods, including fermented dairy beverages, such as dietary fibers. Inulin and fructooligosaccharides (FOS) are natural dietary fibers, found in fruits and vegetables, e.g., in asparagus, artichoke, chicory root, garlic, onion, and banana. Their functionality is related to the prebiotic action of these ingredients, as they stimulate the activity and growth of beneficial bacteria in the gut. In addition, they are also low-caloric, which does not affect blood-sugar levels (Thamer and Penna 2006; Nilsson et al. 2006).

Homogenization

Homogenization process (using 15–20 MPa at 70 °C) is frequently performed to ensure dissolution of dry ingredients (e.g. sugars, stabilizers, and flavors). In addition, the homogenization process before the fermentation step also reduces the size of the fat globules (below 2 μm), minimizing the risk of coalescence during fermentation, which promotes a uniform distribution of the fat globules in the protein matrix during the curdling process (Robinson et al. 2006).

Additionally, the consistency of fermented dairy beverages, based on a mix of liquid ingredients and milk may be enhanced by the homogenization process, because casein and whey proteins become attached to the fat globule surfaces, increasing the viscosity and the stability of the system (Robinson et al. 2006).

Fermentation

Despite the characteristic flavor and aroma promoted by the fermentation process, the major difference observed in fermented dairy beverages, when compared to unfermented dairy beverages, is the pH of the final product. While unfermented beverages show a pH closest to neutral (pH 6.2–6.5), typical from cow's milk and whey, fermented ones have a higher acidity level (pH 4.5–4.8) due to their high content of organic acids (mainly lactic acid), produced by the conversion of lactose by the action of LAB cultures. As a consequence, this higher acidity has an impact on the sensory and functional characteristics of the product (Jelen 2009).

Considering that liquid whey and whey-like UF permeates contain lactose as their main solute, consequently the fermentation of these liquids by LAB (β-galactosidase producing strains) is absolutely feasible for the production of fermented dairy beverages based on whey or containing whey as a significant component (Jelen 2009).

Nonetheless, whey has been shown to be a nutritionally incomplete medium for LAB, their growth tends to be slower there, when compared with milk. However, despite the increased time of incubation, the fermentation process of whey generally occurs without problems. This lower growth is probably due to the buffering capacity of the whey protein and the co-concentration of calcium and phosphorus in the UF process (Jelen 2009). After the incubation time, when the required pH (normally about 4.2–4.5) has been reached, the beverage must be cooled to 15–22 °C in order to stop the increase in acidity (Nilsson et al. 2006).

For fermented drinking products, low viscosity culture strains are most frequently used. In addition, commercial blends of yogurt starter cultures are commonly used by the industry for the production of fermented dairy beverages (Nilsson et al. 2006). Additionally, when using specific strains of LAB, the fermented dairy beverage may present the benefit of the presence of bioactive peptides (Jelen 2009).

Subsequent post-fermentation steps

When the acidity level is reached, the fermented beverage is rapidly cooled and the coagulum is disrupted until a homogenous viscosity is obtained. The fermented product may be then filtered to remove precipitated cellular mass, whey protein, and minerals (for fermented beverages with very low viscosity) (SBRT 2007).

The fermented dairy beverage may be supplemented in this stage (if it is not supplemented before fermentation) with sugar (8–10%), frequently in the form of 50% clear sugar syrup, added or not with flavors, and pasteurized. Then, the beverage is chilled (4 °C) and packaged. Some products may be pasteurized either before packing or in the container. However, in some countries, this final pasteurization step is not allowed, as previously mentioned (Brazil 2005; Nilsson et al. 2006).

The basic processes of the main types of fermented dairy beverages are summarized in Fig. 1.4.

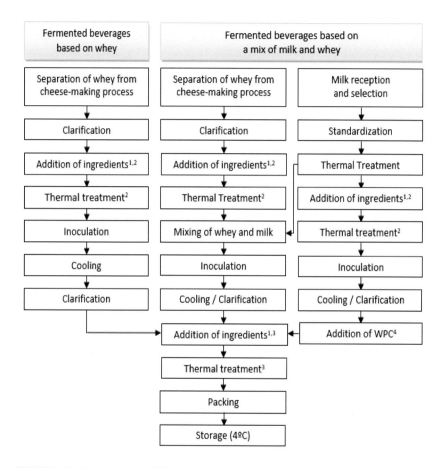

FIGURE 1.4 Basic processes of fermented dairy beverages. [1] Ingredients may include fruit, fruit syrups, juices and pulps, sugar, sweeteners, colorants, flavors, and stabilizing agents. [2] In some countries, the thermal treatment after fermentation is not allowed. In those cases, all ingredients should be added before thermal treatment (Brazil 2005). [3] Optional steps if the ingredients are not added before the inoculation. [4] If the ingredients are added before inoculation, WPC is added together with ingredients and is then thermal treated.

Innovations of Fermented Dairy Beverages in Latin America

Food industries have been investing more and more in fermented dairy beverages, thus exploring the development of innovative products and processes. New consumers are looking for desirable sensory quality, thirst-quenching effectiveness, favorable prices, and health-promoting attributes (Jelen 2009). Given the perspectives, researchers and food industries are focusing on the manufacture of new tendencies for fermented beverages (dairy or non-dairy products), which claim to be probiotic, prebiotic, and nutraceutical, with an important impact on human health and nutrition.

Probiotics are live microorganisms which confer benefits to the host when consumed in adequate amounts (FAO/WHO 2002). Many researchers have described miscellaneous effects of therapeutic properties from the intake of probiotics, indicating that the consumption of these microorganisms promotes beneficial effects on the immune system and the gut, reduces side effects associated with antibiotic use, reduces symptoms associated with the irritable bowel syndrome, helps alleviate lactose intolerance, and has antimicrobial and anticancer properties (Fontana et al. 2013). It is noteworthy that some studies focusing on health benefits have not yet been completely clarified. The literature describes the production of fermented dairy beverages using different probiotic cultures, which can be commercial or isolated from different sources. However, with the exception of Brazil, the scientific literature is scarce regarding the production of probiotic fermented dairy beverages in Latin America, making this area an interesting one to explore (Table 1.2).

Prebiotics, on the other hand, are non-digestible food ingredients that beneficially affect the host by selectively stimulating the growth and/or activity of one or a limited number of bacteria in the colon and, thus, improvingthe host's health (Gibson and Roberfroid 1995). Oligosaccharides such as lactulose, galacto-oligosaccharides, fructo-oligosaccharides, inulin, oligofructose, polydextrose and other food carbohydrates are some well-known examples. The new trend in the market is the production of fermented dairy beverages as a symbiotic product, which contains a combination of probiotics and prebiotics. This interaction can contribute for technological and sensory characteristics bringing advantages for industries and consumers (Coman et al. 2013).

Different LAB can produce antimicrobial compounds, such as organic acids, hydrogen peroxide, diacetyl antifungal compounds, bacteriocins, and bacteriocin-like inhibitory substances (Reis et al. 2012). These compounds can cause membrane injury, pore formation, release of ions, and consequently, cellular collapse (Penna et al. 2015). Bacteriocins are ribosomally synthesized antimicrobial compounds and have become the focus of many biomedical and food-based research groups (Penna et al. 2015). Nowadays, the bacteriocins produced by LAB can be used as biopreservatives in food industries, which can inhibit foodborne pathogens and spoilage microorganisms present in fermented dairy beverages. Moreover, the production of bacteriocins by LAB is relatively common, which may contribute to their colonization in the gastrointestinal system and their competitive edge over other bacteria (Garriga et al. 1993).

Over the last few decades, the consumption of products which claim to be functional has increased the production of fermented dairy beverages that are enriched with bioactive compounds, such as fatty acids (omega 3), conjugated linoleic acid (CLA), isoflavones, phytosterols, vitamins (D, E and C), minerals (calcium, magnesium, iron) and peptides (Peng et al. 2006; Oliveira et al. 2009; Le Blanc et al. 2011; De Dea Lindner et al. 2013; Radulović et al. 2014). Functional food promotes health and/or has potential to promote health through mechanisms that are not common in a conventional nutrition. This effect is restricted to promoting health, and not for curing diseases. Moreover, the consumption of functional products should be safe, without any prescription. The greatest interest and knowledge among consumers about the impact of food on health and well-being is the key point of success for the functional food market.

Functional beverages containing bioactive compounds are the new trend in food industries. The flagship of this product is improving the consumer's health conditions. Different fermented dairy beverages have been launched frequently in order to supply this new market. Ameal-S manufactured by Calpis (Japan) is a fermented milk-based product that relies upon the generation of tripeptides by *Lactobacillus* sp. strains used in the fermentation. These tripeptides have a clinically proven effect on lowering blood pressure. In the same perspective, Evolus was developed in the Scandinavian region of Europe and it is characterized as a yogurt beverage fortified with bioactive peptides and minerals, which can reduce hypertension. The development of new functional beverages have been extensively stimulated around the world, since consumers are looking for tasty, convenient, cheap, varied, and healthy products (Kelly et al. 2009). However, the development of functional beverages in Latin America is still in its infancy. In addition, there are many functional beverages that are being produced only in low-scale.

Most of the fermented dairy beverages described in this chapter are still in the early stages of commercial development or have been developed by food industries or research centers (Marsh et al. 2014). Moreover, in order to make new functional fermented dairy beverages it is important to comprise some aspects, such as: (i) the effect of probiotic cultures and prebiotics ingredients in these products, (ii) the microbiology aspects, (iii) the production of bioactive compounds, (iv) their physicochemical and rheological properties, and (v) their sensory characteristics.

Functional Dairy Beverages

Functional dairy beverages have become known for their health-promoting attributes. The literature has shown different new products that have been developed by research centers. Whey-based goat beverages fermented by *Streptococcus thermophilus* TA-40 as a starter culture, with added guava or soursop pulps, and with or without addition of partially hydrolysed galactomannan from *Caesalpinia pulcherrima* seeds (PHGM), showed to be good vehicles for probiotics cultures (Table 1.2), maintaining their viability above 7 log CFU/mL during 21 days. PHGM increased the dietary fiber content and enhanced the instrumental texture and sensory features of both guava and soursop dairy beverages, especially texture, appearance, and overall acceptability (Buriti et al. 2014).

The physicochemical parameters and sensory features of chocolate goat dairy beverages containing goat cheese whey, probiotic culture (*Bifidobacterium lactis* BLC 1) and prebiotics (inulin and oligofructose) were evaluated. The best formula of functional chocolate goat dairy beverage (45 mL/100 mL of whey and 6 g/100 mL of prebiotics) presented the highest median sensory attributes for flavor and aroma, which may be related to the larger amounts of prebiotics and whey in this formulation. Moreover, the probiotic viability maintained above 7 log CFU/mL and the viscosity was likely improved in presence of whey and prebiotic ingredients (Silveira et al. 2015). The sensorial acceptance of non-fermented beverages with addition of whey, inulin and acerola pulp was evaluated. The sensorial acceptance of samples with 0, 20, and 40% whey did not differ significantly and varied between 6 and 8 (in a 9-level scale). Developed beverages with up to 60% whey can be a good alternative for harnessing this by-product in dairy companies (Bosi et al. 2013).

Fermented whey beverages were developed using potassium sorbate (0.03%), probiotic (*Lactobacillus acidophilus*), and starter culture (*Streptoccocus thermophilus*). The viable acid-lactic bacteria count was above 7 log CFU/mL. Consumer tests resulted in an average score of 6 ("I really like it"). Moreover, the use of potassium sorbate as a preservative lengthened the shelf life of the beverage from 7 to 28 days (Miranda et al. 2014).

The development of new functional dairy beverages using different matrices has been described. Examples of such beverages include cow's and goat's milk, and whey with addition of guava jelly (Gomes et al. 2013); probiotic fermented dairy beverages with yellow mombin flavor (Ramos et al. 2013); dairy beverages enriched with oregano extract and oregano essential oil (Boroski et al. 2012); symbiotic dairy beverages containing inulin, oligofructose, *Lactobacillus paracasei* (Lpc-37), and starter cultures (Fornelli et al. 2014).

Many reports have focused on the technological aspects of the development of novel dairy beverages. The development of fermented dairy beverages using different stabilizers/thickeners (guar gum, powdered gelatin, corn starch, whey protein concentrate, carboxymethylcellulose) showed that gelatin (50%) stood out with best acceptance for appearance, aroma, color, and texture in this matrix (Costa et al. 2013a). Different concentrations of whey (0, 20, 35, 50, 65, and 80%, vol/vol) in fermented dairy beverages showed that high concentrations of whey did not result in higher counts of probiotic bacteria (Table 1.2), suggesting that there is a natural limit in the degree to which the whey constituents can be metabolized by these microorganisms (Castro et al. 2013a). In another study, the starch extracted from almond of *Tommy Atkins* mango was used as a thickener in dairy beverages. The prepared dairy beverages showed potential for commercialization by presenting good sensory and physicochemical characteristics, using 0.3% starch as a thickener (Silva et al. 2013). The literature also reported the production of fermented dairy beverages containing breast-feed serum (10, 20 and 30%) in combination with milk and stabilizers (Obsigel 8AGT, Obsigel 955B and CC 729, all to the 0.1 percent). The sensorial analysis of all treatments showed very good acceptance and samples were considered statistically equal (Arias and Zambrano 2013).

Functional Non-Dairy Beverages

Non-dairy beverages can be introduced as a new vehicle for the consumption of functional products, expanding this market to vegans, vegetarians, and lactose intolerant consumers (Prado et al. 2015). Considering these perspectives, there have recently been innovative efforts to develop non-dairy probiotic fermented beverages from a variety of substrates, including soymilk, whey, cereals and vegetable, and fruit juices (Marsh et al. 2014). A potential antimicrobial and probiotic strain (*Lactobacillus plantarum* AC-1) was isolated from natural fermentation of coconut water. In the production of the coconut water beverage, *Lactobacillus plantarum* AC-1 has fulfilled the recommended daily ingestion of potential functional microorganisms (6 log CFU/mL). The fermented beverage supplemented with sucrose and artificial coconut flavor had the most success in both sensory evaluation and acceptability among tasters in the sensorial analysis (Prado et al. 2015).

The nutraceutical-fermented beverage using herbal mate extract and probiotic strain *Lactobacillus acidophilus* ATCC 4356 was developed by Lima et al. (2012). Acceptable levels of caffeine and a high antioxidant capacity were observed for the formula when compared to other antioxidant beverages. An advantage of this product is the compliance to organic claims, while providing caffeine, other phyto-stimulants and antioxidant compounds without the addition of synthetic components or preservatives in the formula. The herbal mate beverage presented good consumer acceptance in comparison to the other two similar commercial beverages.

The production of a fermented peanut-soymilk beverage containing probiotics, yogurt starter and yeast cultures: *Lactobacillus rhamnosus* (LR 32), *Lactobacillus acidophilus* (LACA 4), *Lactobacillus delbrueckii* subsp. *bulgaricus* (LB 340), *Pediococcus acidilactici* (UFLA BFFCX 27.1), *Lactococcus lactis* (CCT 0360), and *Saccharomyces cerevisiae* (UFLA YFFBM 18.03) was also evaluated. When inoculated as a pure culture, *Lactobacillus acidophilus* (LACA 4) was the unique probiotic culture that produced significant amounts of lactic acid (3.35 g/L) and rapidly lowered the pH (4.6). In general, the strains were more efficient in the use of available carbohydrates and release of metabolites in co-culture than in single culture fermentations. Further studies are needed to evaluate the survival of these cultures over a longer period of post-acidification and the sensory acceptance of the final beverage (Santos et al. 2014).

The use of sonicated pineapple juice as a substrate for producing a probiotic beverage by *Lactobacillus casei* NRRL B442 showed that, after 42 days of storage under refrigeration (4 ℃), the microbial viability was 6.03 log CFU/mL in the non-sweetened sample, and 4.77 log CFU/mL in the sweetened sample. Moreover, the characteristic color of the juice was maintained throughout its shelf life and no browning was observed (Costa et al. 2013b).

The beneficial effects of fermented vegetal beverage containing fructooligosaccharide, *Lactobacillus casei* Lc-01, aqueous extracts of soy and quinoa were evaluated using the Simulator of the Human Intestinal Microbial Ecosystem (SHIME®), a dynamic model of the human gut. The symbiotic beverage showed the best microbiological results in the ascending colon compartment, stimulating the growth of *Lactobacillus* spp. and *Bifidobacterium* spp., and reducing *Clostridium* spp., *Bacteroides* spp., enterobacteria and *Enterococcus* spp. populations in this compartment (Bianchi et al. 2014).

The composition of probiotics (*Lactobacillus acidophilus* and *Bifidobacterium lactis*) and non-probiotics (*Streptococcus thermophilus*, *Lactobacillus delbrueckii* subsp. *bulgaricus*) cultures, and the substrate (cow milk and soymilk) have a significant influence on the production of fermented beverages, when lactic acid content, acidity, pH, and rheological behavior were evaluated. The microorganisms present different metabolic pathways according to the substrate available, which significantly affects the beverages properties (Martín and Cuenca 2009).

Functional non-dairy beverages using different matrices have been described in the literature, such as prebiotic beverage made from cashew nut kernels and passion fruit juice (Rebouças et al. 2014), and cashew juice containing prebiotic oligosaccharides (Silva et al. 2014).

It is important to emphasize that there are different non-dairy matrices to be explored in the development of non-dairy beverages. Furthermore, the feasibility and the development of adaptable technologies for these products do not occur at the same rate compared with fermented dairy beverages (Kumar et al. 2015). Hence, there is a scope for further research in this area.

TABLE 1.2 Types of dairy and non-dairy beverages using probiotic cultures

Fermented beverages	Potential probiotic cultures	Country	References
Milk and whey (cow)	*Lactobacillus acidophilus* (La–14) and *Bifidobacterium lactis* (Bl-07)	Brazil	Castro et al. 2013a, b
Milk and whey (cow)	*Lactobacillus acidophilus* LA–5, *Bifidobacterium bifidum* BB–12	Brazil	Ramos et al. 2013
Milk and whey (goat)	*Bifidobacterium animalis* subsp. *lactis* BB–12 and *Lactobacillus rhamnosus* Lr–32	Brazil	Buriti et al. 2014
Milk and whey (cow)	*Lactobacillus paracasei* (Lpc–37)	Brazil	Fornelli et al. 2014
Whey (cow)	*Lactobacillus acidophilus*	Cuba	Miranda et al. 2009
Milk and whey (goat)	*Bifidobacterium lactis* BLC 1	Brazil	Silveira et al. 2015
Soymilk	*Bifidobacterium lactis* 205 and 207; *Lactobacillus acidophilus*	Colombia	Martín and Cuenca 2009
Herbal mate extract	*Lactobacillus acidophilus* ATCC 4356	Brazil	Lima et al. 2012
Pineapple juice	*Lactobacillus casei* NRRL B442	Brazil	Costa et al. 2013b
Fermented vegetal beverage (soy and quinoa)	*Lactobacillus casei* Lc–01	Brazil	Bianchi et al. 2014
Peanut-soy	*Lactobacillus rhamnosus* (LR 32), *Lactobacillus acidophilus* (LACA 4), *Pediococcus acidilactici* (UFLA BFFCX 27.1), *Lactococcus lactis* (CCT 0360)	Brazil	Santos et al. 2014
Coconut water	*Lactobacillus plantarum* AC–1	Brazil	Prado et al. 2015

Conclusion

Over the last few decades, significant improvements have been made in fermentation technology for the production of various types of fermented dairy beverages, which resulted in a variety of products with high nutritional value.

A lot of innovations have been employed to the classical processes of fermented dairy beverage production. The use of selected starter cultures for the fermentation, aiming the production of biomolecules, such as organic acids, aromatic compounds, bioactive peptides, and antimicrobial compounds, has recently been applied by dairy companies in the manufacturing processes of fermented beverages.

It is necessary to continue to expand our knowledge for the development of innovative functional beverages for specific health purposes, while at the same time meet the demand of consumers for safety and extended shelf life of fermented beverages that retain their sensory qualities and nutritional value.

References

Abdolmaleki, F., M. Mazaheri Assadi and H. Akbarirad. (2015). Assessment of beverages made from milk, soya milk and whey using Iranian kefir starter culture. Int J Dairy Technol 68: 441–447

Alhaj, O.A., O.A. Kanekanian and A.C. Peters. (2007). Investigation on whey proteins profile of commercially available milk-based probiotics health drinks using fast protein liquid chromatography (FPLC). Brit Food J 109: 469–480.

Arias, C.G.Z. and J.R.Z. Zambrano. (2013). Bebida láctea fermentada utilizando lactosuero como sustituto parcial de leche y diferentes estabilizantes comerciales. Undergraduation Thesis, Escuela Superior Politécnica Agropecuaria de Manabí Manuel Félix López, Calceta, Ecuador.

Axelsson, L. (2004). Lactic Acid Bacteria: classification and physiology. *In:* S. Salminen, A. von Wright and A. Ouwehand (eds.), Lactic Acid Bacteria - microbiological and functional aspects. Marcel Dekker, New York, pp. 1–66.

Baek, E., H. Kim, H. Choi, S. Yoon and J. Kim. (2012). Antifungal activity of *Leuconostoc citreum* and *Weissella confusa* in rice cakes. J Microbiol 50: 842–848.

Beermann, C. and J. Hartung. (2012). Current enzymatic milk fermentation procedures. Eur Food Res Technol 235: 1–12.

Bianchi, F., E.A. Rossi, I.K. Sakamoto, M.A.T. Adorno, T. Van de Wiele and K. Sivieri. (2014). Beneficial effects of fermented vegetal beverages on human gastrointestinal microbial ecosystem in a simulator. Food Res Int 64: 43–52.

Boroski, M., H.J. Giroux, H. Sabik, H.V. Petit, J.V. Visentainer, P.T. Matumoto-Pintro and M. Britten. (2012). Use of oregano extract and oregano essential oil as antioxidants in functional dairy beverage formulations. LWT-Food Sci Technol 47: 167–174.

Bosi, M.G., B.M. Bernabé, S.M. Della Lucia and C.D. Roberto. (2013). Bebida com adição de soro de leite e fibra alimentar prebiótica. Pesquisa Agropec Bras 48: 339–341 (in Portuguese).

Brazil (2005). Ministry of Agriculture, Livestock, and Supply (MAPA) Legislation. SISLEGIS: Legislation Consultation System. Normative Instruction n. 16, 23 August 2005. (Technical rules of identity and quality of whey-based drinks.) Accessed on 29 June 2015. Available at http://sistemasweb.agricultura.gov.br/sislegis/action/detalha Ato. do? method = consultar Legislacao Federal (in Portuguese).

Buriti, F.C.A., S.C. Freitas, A.S. Egito and K.M.O. Dos Santos. (2014). Effects of tropical fruit pulps and partially hydrolysed galactomannan from *Caesalpinia pulcherrima* seeds on the dietary fibre content, probiotic viability, texture and sensory features of goat dairy beverages. LWT-Food Sci Technol 59: 196–203.

Bylund, G. (1995). Dairy Processing Handbook. Tetra Pak Processing Systems AB, Lund, Sweden.

Canadian Dairy Commission. (2015). Dairy ingredients profile - Whey Powder, Whey Protein Concentrate. Available at <http://www.milkingredients.ca/index-eng. php?id=170>, Accessed on: 30 June 2015.

Carvalho, W., D.D.V. Silva, L. Canilha and I.M. Mancilha. (2005). Aditivos alimentares produzidos por via fermentativa parte 1: ácidos orgânicos. Rev Analytica 1: 70–76 (in Portuguese).

Castro, F.P., T.M. Cunha, P.J. Ogliari, R.F. Teófilo, M.M.C. Ferreira and E.S. Prudêncio. (2009). Influence of different content of cheese whey and oligofructose on the properties of fermented lactic beverages: Study using response surface methodology. LWT-Food Sci Technol 42: 993–997.

Castro, W.F., A.G. Cruz, M.S. Bisinotto, L.M.R. Guerreiro, J.A.F. Faria, H.M.A. Bolini, R.L. Cunha and R. Deliza. (2013b). Development of probiotic dairy beverages: Rheological properties and application of mathematical models in sensory evaluation. J Dairy Sci 96: 16–25.

Castro, W.F., A.G. Cruz, D. Rodrigues, G. Ghiselli, C.A.F. Oliveira, J.A.F. Faria and H.T. Godoy. (2013a). Short communication: Effects of different whey concentrations on physicochemical characteristics and viable counts of starter bacteria in dairy beverage supplemented with probiotics. J Dairy Sci 96: 96–100.

Chandan, R.C. (2006). History and consumption trends. *In*: R.C. Chandan (ed.), Manufacturing Yogurt and Fermented Milks. Blackwell Publishing Professional, Ames, Iowa, pp. 3–14.

Chandan, R.C. and A. Killara. (2013). Antimicrobial activity and gastrointestinal infections. *In:* R.C Chandan and A. Killara (eds.), Manufacturing yogurt and fermented milk. John Wiley & Sons, Oxford, United Kingdom, pp. 460–465.

Cheng, H. (2010). Volatile flavor compounds in yogurt: A review. Crit Rev Food Sci 50: 938–950.

Chopin, A. (1993). Organization and regulation of genes for amino acid synthesis in lactic acid bacteria. FEMS Microbiol Rev 12: 12–21.

Cogan, T.M. and T.P. Beresford. (2002). Microbiology of hard cheese. *In*: R.K. Robinson (ed.) Dairy Microbiology Handbook, John Wiley & Sons, Inc., New York, pp. 515–557.

Coman, M.M., M.C. Verdenelli, C. Cecchini, S. Silvi, A. Vasile, G.E. Bahrim, C. Orpianesi and A. Cresci. (2013). Effect of buckwheat flour and oat bran on growth and cell viability of the probiotic strains *Lactobacillus rhamnosus* IMC 501®, *Lactobacillus paracasei* IMC 502® and their combination SYNBIO®, in symbiotic fermented milk. Int J Food Microbiol 167: 261–268.

Costa, A.V.S.C., E.S. Nicolau, M.C.L. Torres, P.R. Fernandes, S.I.R. Rosa and R.C. Nascimento. (2013a). Desenvolvimento e caracterização físico-química, microbiológica e sensorial de bebida láctea fermentada elaborada com diferentes estabilizantes/espessantes. Semina Cienc Agrar 34: 209–226 (in Portuguese).

Costa, M.G.M., T.V. Fonteles, A.L.T. De Jesus and S. Rodrigues. (2013b). Sonicated pineapple juice as substrate for *L. casei* cultivation for probiotic beverage development: Process optimisation and product stability. Food Chem 139: 261–266.

Crowley, S., J. Mahony and D.V. Sinderen. (2013). Current perspectives on antifungal lactic acid bacteria as natural bio-preservatives. Trends Food Sci Technol 33: 93–109.

Dang, T.D.T., A. Vermeulen, P. Ragaert and F. Devlieghere. (2009). A peculiar stimulatory effect of acetic and lactic acid on growth and fermentative metabolism of *Zygosaccharomyces bailii*. Food Microbiol 22: 320–327.

De Dea Lindner, J., A.L.B. Penna, I.M. Demiate, C.T. Yamaguishi, R.S.M. Prado and J.L. Parada. (2013). Fermented foods and human health benefits of fermented functional foods. *In*: C.R. Soccol, A. Pandey and C. Larroche (eds.), Fermentation processes engineering in the food industry. CRC Press, Boca Raton, Florida, pp. 263–297.

Degeest, B., B. Janssens and L. De Vuyst. (2001). Exopolysaccharide (EPS) biosynthesis by *Lactobacillus sakei* 0-1: Production kinetics, enzyme activities and EPS yields. J Appl Microbiol 91: 470–477.

Dragone, G., S.I. Mussatto, J.M. Oliveira and J.A. Teixeira. (2009). Characterization of volatile compounds in an alcoholic beverage produced by whey fermentation. Food Chem 112: 929–935.

FAO/WHO. Food and Agriculture Organization of the United Nations, FAO, World Health Organization, WHO (2002). Guidelines for the evaluation of probiotics in food. FAO/WHO: London, Ontario, Canada.

Fernandez-Garcia, E. and J.U. McGregor. (1994). Determination of organic acids during the fermentation and cold storage of yogurt. J Dairy Sci 77: 2934–2939.

Fontana, L., M. Bermudez-Brito, J. Plaza-Diaz, S. Muñoz-Quezada and A. Gil. (2013). Sources, isolation, characterization and evaluation of probiotics. Brit J Nutr 109: S35–50.

Fornelli, A.R., N.S. Bandiera, M.D.R. Costa, C.H.B. Souza, E.H.W. Santana, K. Sivieri and L.C. Aragon-Alegro. (2014). Effect of inulin and oligofructose on the physico-chemical, microbiological and sensory characteristics of symbiotic dairy beverages. Semina Cienc Agrar 35: 3099–3112.

Fox, P.F. and P.L.H. McSweeney. (1998). Dairy chemistry and biochemistry. Blackie Academic & Professional, London, United Kingdom.

Garriga, M., M. Hugas, T. Aymerich and J.M. Monfort. (1993). Bacteriocinogenic activity of Lactobacilli from fermented sausages. J Appl Bacteriol 75: 142–148.

Gaspar, P., A.L. Carvalho, S. Vinga, H. Santos and A.R. Neves. (2013). From physiology to systems metabolic engineering for the production of biochemicals by lactic acid bacteria. Biotechnol Adv 31: 764–788.

Gauche, C., T. Tomazi, P.L.M. Barreto, P.J. Ogliari and M.T. Bordignon-Luiz (2009). Physical properties of yoghurt manufactured with milk whey and transglutaminase. LWT-Food Sci Technol 42: 239–243.

Ghosh, K., M. Ray, A. Adak, S.K. Halder, A. Das, A. Jana, S. Parua Mondal, C. Vagvolgyi, P.K. Das Mohapatra, B.R. Pati and K.C. Mondal. (2015). Role of probiotic *Lactobacillus fermentum* KKL1 in the preparation of a rice based fermented beverage. Bioresour Technol 188: 161–168.

Gibson, G.R. and M.B. Roberfroid. (1995). Dietary modulation of the human colonic microbiota: introducing the concept of prebiotics. J Nutr 125: 1401–1412.

Gomes, J.J.L., A.M. Duarte, A.S.M. Batista, R.M.F. de Figueiredo, E.P. de Sousa, E.L. de Souza and R.C.R.E. Queiroga. (2013). Physicochemical and sensory properties of fermented dairy beverages made with goat's milk, cow's milk and a mixture of the two milks. LWT-Food Sci Technol 54: 18–24.

Graham, A.F. and B.M. Lund. (1986). The effect of citric acid on growth of proteolytic strains of *Clostridium botulinum*. J Appl Bacteriol 61: 39–49.

Hasan, F., A.A. Shah and A. Hameed. (2006). Industrial applications of microbial lipases. Enzyme Microb Tech 39: 235–251.

Huffman, L. and L. B. Ferreira. (2011). Whey-based Ingredients. *In*: R.C. Chandan and A. Kilara (eds.), Dairy Ingredients for Food Processing. Wiley-Blackwell, Ames, Iowa, pp.179–198.

Hutkins, R.W. (2006). Microbiology and Technology of Fermented Foods. Blackwell Publishing Professional, Ames, Iowa.

Izco, J.S.M., M.N. Tormo and R.J. Nez-Flores. (2002). Development of a CE method to analyze organic acids in dairy products application to study the metabolism of heat-shocked spores. J Agr Food Chem 50: 1765–1773.

Jelen, P. (2009). Whey-based functional beverages. *In:* P. Paquin (ed.), Functional and Speciality Beverage Technology. CRC Press, Boca Raton, Florida, pp. 259–280.

Kabašinskiene, A., A. Liutkevičius, D. Sekmokiene, G. Zaborskiene and J. Šlapkauskaite. (2015). Evaluation of the physicochemical parameters of functional whey beverages. Food Technol Biotech 53: 110–115.

Kelly, P., B.W. Woonton and G.W. Smithers. (2009). Improving the sensory quality, shelf-life and functionality of milk. *In:* P. Paquin (ed.), Functional and speciality beverage technology. CRC Press, Boca Raton, Florida, pp. 170–231.

Kumar, B.V., S.V.N. Vijayendra and O.V.S. Reddy. (2015). Trends in dairy and non-dairy probiotic products - a review. J Food Sci Technol (in press).

Kunji, E.R.S., Mierau, I., Hagting, A., Poolman, B. and Konings, W.N. (1996). The proteolytic systems of lactic acid bacteria. A Van Leeuw J Microb 70: 187–221.

Le Blanc, J.G., E. Laiño, M. Juarez del Valle, V. Vannini, van Sinderen, D., Taranto, M.P., Font de Valdez, G., Savoy de Giori, G. and Sesma, F. (2011). B-group vitamin production by lactic acid bacteria. Current knowledge and potential applications. J Appl Microbiol 111: 1297–1309.

Leite, A.M.O., D.C.A. Leite, E.M. Del Aguila, T.S. Alvares, R.S. Peixoto, M.A.L. Miguel, J.T. Silva and V.M.F. Paschoalin. (2013). Microbiological and chemical characteristics of Brazilian kefir during fermentation and storage processes. J Dairy Sci 96: 4149–4159.

Li, W., Ji, X. Rui, J. Yu, W. Tang, X. Chen, M. Jiang and M. Dong. (2014). Production of exopolysaccharides by *Lactobacillus helveticus* MB2-1 and its functional characteristics in vitro. LWT-Food Sci Technol 59: 732–739.

Lievore, P., D.R.S. Simões, K.M. Silva, N.L. Drunkler, A.C. Barana, A. Nogueira and I.M. Demiate. (2015). Chemical characterisation and application of acid whey in fermented milk. J Food Sci Technol 52: 2083–2092.

Lima, I.F.P., J. De Dea Lindner, V. Thomaz-Soccol, J. L. Parada and C.R. Soccol. (2012). Development of innovative nutraceutical fermented beverage from herbal mate (*Ilex paraguariensis* A. St.-Hil.) extract. Int J Mol Sci 13: 788–800.

Lucas, A., E. Rock, J.-F. Chamba, I. Verdier-Metz, P. Brachet and J.B. Coulon. (2006). Respective effects of milk composition and the cheese-making process on cheese compositional variability in components of nutritional interest. Lait 86: 21–41.

Madigan, M.T., J.M. Martinko and J. Parker. (2008). Microbiologia de Brock. Pearson Prentice Hall, São Paulo.

Mahmoud, N.S. and A.E. Ghaly. (2004). On-line sterilization of cheese whey using ultraviolet radiation. Biotechnol Progr 20: 550–560.

Marsh, A.J., C. Hill, R.P. Ross and P.D. Cotter. (2014). Fermented beverages with health-promoting potential: Past and future perspectives. Trends Food Sci Tech 38: 113–124.

Martín, M.A.B and M.C.Q. Cuenca. (2009). Valoración de diferentes indicadores de la fermentación de bebida de soya y de leche de vaca utilizando cultivos probióticos. Braz J Food Technol 1: 100–106 (in Spanish).

Masson, L.M.P., A. Rosenthal, V.M.A. Calado, R. Deliza and L. Tashima. (2011). Effect of ultra-high pressure homogenization on viscosity and shear stress of fermented dairy beverage. LWT - Food Sci Technol 44: 495–501.

Masuda, T., A. Hidaka, N. Kondo and T. Itoh. (2005). Intracellular enzyme activities and autolytic properties of *Lactobacillus acidophilus* and *Lactobacillus gasseri*. Food Sci Technol Res 11: 328–331.

McGorrin, R.J. (2001). Advances in dairy flavor chemistry. *In*: A.M. Spanier, F. Shahidi, T.H. Parliment, C.T. Ho (eds), Food Flavors and Chemistry: Advances of the New Millennium. Cambridge, United Kingdom, pp. 67–84.

Milesi, M.M., I.V. Wolf, C.V. Bergamini and E.R. Hynes. (2010). Two strains of nonstarter lactobacilli increased the production of flavor compounds in soft cheeses. J Dairy Sci 93: 5020–5031.

Miranda, O.M., P.L. Fonseca, I. Ponce, C. Cedeño, L.C. Rivero and L.M. Vázquez. (2014). Elaboración de una bebida fermentada a partir del suero de leche que incorpora *Lactobacillus acidophilus* y *Streptococcus thermophilus*. Rev Cubana Aliment Nutr 24: 7–16 (in Spanish).

Nilsson, L.E., S. Lyck and A.Y. Tamime. (2006). Miscellaneous fermented milk. *In:* A.Y. Tamime (ed.), Fermented milks products. Blackwell Science Ltd, Oxford, United Kingdom, pp. 217–236.

Oliveira, M.N. (2009). Características funcionais de leites fermentados e outros produtos lácteos. *In:* M.N. Oliveira (ed.), Tecnologia de produtos lácteos funcionais. Atheneu Editora, São Paulo, Brazil, pp. 277–320 (in Portuguese).

Oliveira, R.P.S, P. Perego, M.N. Oliveira and A. Converti (2012). Growth, organic acids profile and sugar metabolism of *Bifidobacterium lactis* in co-culture with *Streptococcus thermophilus*: The inulin effect. Food Res Int 48: 21–27.

Østlie, H.M., M.H. Helland and J.A. Narvhus. (2003). Growth and metabolism of selected strains of probiotic bacteria in milk. Int J Food Microbiol 87: 17–27.

Peng, Y., G.E. West, and C. Wang. (2006). Consumer attitudes and acceptance of CLA-enriched dairy products. Can J Agr Econ 54: 663–684.

Penna, A.L.B. (2010). Bebidas lácteas. *In:* Venturini Filho, W. (ed.), Bebidas não alcoólicas, v.2. Blucher, São Paulo, Brazil, pp. 89–119 (in Portuguese).

Penna, A.L.B., M.N. Oliveira and R. Baruffaldi. (1997). Analysis of consistency of yogurt: correlation between sensorial and instrumental evaluation. Food Sci Technol 17: 98–101.

Penna, A.L.B., A.T. Paula, S.N. Casarotti, L.F. Faria, V.R. Diamantino and S.T. Todorov. (2015). Overview of the functional lactic acid bacteria in fermented milk products. *In:* V.R. Ravishankar and A.B. Jamuna (eds.), Beneficial microbes in fermented and functional foods. CRC Press, Boca Raton, pp. 113–148.

Pescuma, M., E.M. Hébert, F. Mozzi and G.F. Valdez. (2010). Functional fermented whey-based beverage using lactic acid bacteria. Int J Food Microbiol 141: 73–81.

Pescuma, M., E.M. Hébert, F. Mozzia and G.F. Valdez (2008). Whey fermentation by thermophilic lactic acid bacteria: Evolution of carbohydrates and protein content. Food Microbiol 25: 442–451.

Prado, F.C., J. De Dea Lindner, J. Inaba, V. Thomaz-Soccol, S.K. Brar and C.R. Soccol. (2015). Development and evaluation of a fermented coconut water beverage with potential health benefits. J Funct Foods 12: 489–497.

Prückler, M., C. Lorenz, A. Endo, M. Kraler, K. Dürrschmid, K. Hendriks, F.S. Silva, E. Auterith, W. Kneifel and H. Michlmayr. (2015). Comparison of homo- and heterofermentative lactic acid bacteria for implementation of fermented wheat bran in bread. Food Microbiol 49: 211–219.

Radulović, Z., D. Paunović, M. Petrušić, N. Mirković, J. Miočinović, D. Kekuš and D. Obradović. (2014). The application of autochthonous potential of probiotic *Lactobacillus plantarum* 564 in fish oil fortified yoghurt production. Arch Biol Sci 66: 15–22.

Ramos, A.C.S.R., T.L.M. Stamford, E.C.L. Machado, F.R.B. Lima, E.F. Garcia, S.A.C. Andrade and C.G.M. Da Silva. (2013). Elaboração de bebidas lácteas fermentadas: aceitabilidade e viabilidade de culturas probióticas. Semina Cienc Agrar 34: 2817–2828 (in Portuguese).

Ramyasree, S. and J.R. Dutta. (2013). The effect of process parameters in enhancement of lipase production by co-culture of lactic acid bacteria and their mutagenesis study. Biocatalysis Agric Biotechnol 2: 393–398.

Rebouças, M.C., M.P. Rodrigues and M.R.A. Afonso. (2014). Optimization of the acceptance of prebiotic beverage made from cashew nut kernels and passion fruit juice. J Food Sci 79: S1393–1398.

Reis, J.A., A.T. Paula, S.N. Casarotti and A.L.B. Penna. (2012). Lactic acid bacteria antimicrobial compounds: characteristics and applications. Food Eng Rev 4: 124–140.

Robinson, R.K., J.A. Lucey and A.Y. Tamime. (2006). *In:* A.Y. Tamime (ed.), Fermented milks products. Blackwell Science Ltd, Oxford, pp. 53–75.

Santos, C.C.A.A., B.S. Libeck and R.F. Schwan. (2014). Co-culture fermentation of peanut-soy milk for the development of a novel functional beverage. Int J Food Microbiol 186: 32–41.

Santos, C.T., A.R. Costa, G.C.R. Fontan, R.C.I. Fontan and R.C.F. Bonomo. (2008). Influência da concentração de soro na aceitação sensorial de bebida láctea fermentada com polpa de manga. Alim Nutr 19: 55–60 (in Portuguese).

Santos, C.T., G.M.R. Marques, G.C.R. Fontan, R.C.I. Fontan, R.C.F. Bonomo and P. Bonomo. (2006). Elaboração e caracterização de uma bebida láctea fermentada com polpa de umbu (*Spondias tuberosa sp.*). Rev Bras Prod Agroind 8: 111–116 (in Portuguese).

SBRT- Serviço Brasileiro de Respostas Técnicas. (2007). Dossiê Técnico Ricota e Bebida Láctea. Available at http://www.sbrt.ibict.br/dossie-tecnico/downloadsDT/MTQw, Accessed on: 30 June 2015 (in Portuguese).

Schmitt, P., C. Diviès and R. Cardona. (1992). Origin of end-products from the co-metabolism of glucose and citrate by *Leuconostoc mesenteroides* subsp. *cremoris*. Applied Microbiol Biotechnol 36: 679–683.

Serra, M., A.J. Trujillo, B. Guamis and V. Ferragut. (2009). Flavour profiles and survival of starter cultures of yoghurt produced from high-pressure homogenized milk. Int Dairy J 19: 100–106.

Sgarbieri, V.C. (2004). Propriedades fisiológicas-funcionais das proteínas do soro de leite. Rev Nutr 17: 397–409 (in Portuguese).

Shiby, V.K. and H.N. Mishra. (2013). Fermented milks and milk products as functional foods- A review. Crit Rev Food Sci Nutr 53: 482–496.

Silva, G.A.S., M.T. Cavalcanti, M.C.B.M. Almeida, A.S. Araújo, G.C.B. Chinelate and E.R. Florentino. (2013). Utilização do amido da amêndoa da manga Tommy Atkins como espessante em bebida láctea. Rev Bras Eng Agric Ambient 17: 1326–1332.

Silva, I.M., M.C. Rabelo and S. Rodrigues. (2014). Cashew juice containing prebiotic oligosaccharides. J Food Sci Technol 51: 2078–2084.

Silveira, E.O., J.H. Lopes Neto, L.A. Silva, A.E.S. Raposo, M. Magnani and H.R. Cardarelli. (2015). The effects of inulin combined with oligofructose and goat cheese whey on the physicochemical properties and sensory acceptance of a probiotic chocolate goat dairy beverage. LWT - Food Sci Technol 62: 445–451.

Sinha, R., C. Radha, J. Prakash and P. Kaul. (2007). Whey protein hydrolysate: functional properties, nutritional quality and utilization in beverage formulation. Food Chem 101: 1484–1491.

Siqueira, A.M.O, E.C.L. Machado and T.L.M. Stamford. (2013). Bebidas lácteas com soro de queijo e frutas. Cienc Rural 43: 1693–1700 (in Portuguese).

Smid, E.J. and M. Kleerebezem. (2014). Production of aroma compounds in lactic fermentations. Annu Rev Food Sci Technol 5: 313–326.

Smit, G., B.A. Smit and W.J. Engels. (2005). Flavour formation by lactic acid bacteria and biochemical flavour profiling of cheese products. FEMS Microbiol Rev 29: 591–610.

Steele, J., J. Broadbent and J. Kok. (2013). Perspectives on the contribution of lactic acid bacteria to cheese flavor development. Curr Opin Biotech 24: 135–141.

Suomalainen, T.H. and A.M. Mayra-Makinen. (1999). Propionic acid bacteria as protective cultures in fermented milks and breads. Le Lait 79: 165–174.

Tamime, A.Y. and R.K. Robinson. (1999). Yogurt: science and technology. CRC Press, Boca Raton, Florida.

Tamime, A.Y., A. Skriver and L.E. Nilsson. (2006). Starter cultures. *In:* A.Y. Tamime (ed.), Fermented milks. Blackwell Publishing, Ames, Iowa, pp. 11–52.

Thamer, K.G. and A.L.B. Penna. (2006). Characterization of functional dairy beverages fermented by probiotics and with the addition of prebiotics. Food Sci Technol 26: 589–595 (in Portuguese).

Tranjan, B.C., A.G. Cruz, E.H.M. Walter, J.A.F. Faria, H.M.A. Bolini and M.R.L. Moura. (2009). Development of goat cheese whey-flavoured beverages. Int J Dairy Technol 62: 438–443.

Uribe, M.M.L., J.U.S. Valencia, A.H. Monzón and J.E.P. Suescún. (2008). Fermented fresh cheese milk whey beverage inoculated with *Lactobacillus casei.* Rev Fac Nal Agr 61: 4409–4421.

Vavrusova, M., H. Pindstrup, L.B. Johansen, M.L. Andersen, H.J. Andersen and L.H. Skibsted. (2015). Characterisation of a whey protein hydrolysate as antioxidant. Int Dairy J 47: 86–93.

Whey Protein Institute (2015). Whey protein types. Available at http://www.wheyoflife. org/types, Accessed on: 30 June 2015.

Yada, R.Y. (2004). Protein in Food Processing. Woodhead Publishing, Cambridge, UK.

Yadav, J.S.S., S. Yan, S. Pilli, L. Kumar, R.D. Tyagi and R.Y. Surampalli. (2015). Cheese whey: A potential resource to transform into bioprotein, functional/nutritional proteins and bioactive peptides. Biotechnology Advances (in press).

2

Brazilian Artisanal Coalho Cheese: Tradition, Science and Technology

Karina Maria Olbrich dos Santos[a]*, Maria do Socorro Rocha Bastos[b] and Antônio Silvio do Egito[c]

Abstract

The artisanal Coalho cheese is a traditional food product of the Brazilian Northeastern Region, whose simple mode of production on farms has been going through many generations, since its introduction in Brazil as a legacy of Portuguese colonization. Its name derived from the Portuguese word for rennet, and are related to the use of strips or extracts of the stomach of, mainly, young goats or calves for milk coagulation in the cheese-making process.

It is a popular cheese variety in the Brazilian Northeastern, widely accepted and consumed by the local population, but also around the country. In addition to its cultural relevance, the manufacture of artisanal Coalho cheese is an important source of income for many small family farmers.

Traditionally made from raw cow milk, Coalho cheese is a semi-hard cheese commonly market after a few days of ripening. It is usually described by its slightly acid and salty taste, resembling curd, and its rubbery texture, being able to be roasted without melting. Besides its main typical characteristics, and due to the wideness of its production area, there are notable differences among Coalho cheeses from different localities related to the local environmental factors and procedures addressed by the concept of terroir, which influence milk and cheese composition, and endogenous microbiota.

The diverse and rich microbiota of the artisanal Coalho cheese plays an important role in defining its typical features. Studies aiming to characterize its lactic acid bacteria population identified Enterococcus, Lactobacillus, Lactococcus, and Streptococcus species. Among them, some *Lactobacillus rhamnosus* and *Lactobacillus plantarum* strains were distinguished as probiotic candidates, and some *Enterococcus faecium* showed to be bacteriocinogenic. Besides, proteomic

[a] Embrapa Food Technology, Av. das Américas, n. 29.501, Guaratiba 23020-470, Rio de Janeiro - RJ, Brazil.
[b] Embrapa Tropical Agroindustry, R. Dra. Sara Mesquita, n. 2270, Planalto Pici. 60511-110, Fortaleza - CE, Brazil. E-mail: socorro.bastos@embrapa.br
[c] Embrapa Goats and Sheep, Estrada Sobral-Groaíras km 4. 62010-970, Sobral - CE, Brazil. E-mail: antoniosilvio.egito@embrapa.br
* Corresponding author: karina.dos-santos@embrapa.br

studies evidenced Coalho cheeses content of bioactive peptides with properties related to health benefits for consumers.

This chapter presents an overview of the current knowledge about the history, technological process, and biochemical and microbiological characteristics of the Coalho cheese variety. A special focus is given to the artisanal Coalho cheese produced in the semi-arid region of Jaguaribe, Ceará State, where studies for a protective Geographical Indication are in progress.

Introduction

All over the world, traditional cheeses are produced with their unique organoleptic attributes related to historical, cultural and environmental aspects of the region of origin. The use of locally produced raw milk and specific cheese-making procedures confer to the cheeses a selected microbiota, mainly composed by lactic acid bacteria, responsible for the biochemical transformations that ultimately define the flavor and texture of each cheese variety.

In Brazil, cheese-making practices were introduced during the Portuguese colonization, becoming part of the cultural practices in some regions with the installation of cattle farms. A small number of Brazilian cheeses are recognized as traditional foods. Among them, the artisanal Coalho cheese is the most important traditional cheese of the northeastern Region of Brazil. Its simple method of production on farms has been passed down through generations, since the colonial period. It is a popular variety of cheese that is not only appreciated and consumed by the local population but also all around the country (Brazil). In addition to its cultural relevance, the production of artisanal Coalho cheese is an important source of income for many small family farms.

Traditionally made from raw cow milk, Coalho cheese is a semi-hard cheese, with a medium to high moisture content, and is commonly on the market after a few days of ripening. It is usually described by its slightly acidic and salty taste, and rubbery texture. It is also known for its ability to be fried until it becomes crusty on top without melting, a feature that is greatly appreciated by the consumers.

Even though the main characteristics of Coalho cheese are typical, some important variations arise due to the large span of its production area. These variations are related to the local environmental factors and procedures addressed by the concept of *terroir*, which influences milk and cheese composition and endogenous microbiota. Nowadays, Coalho cheese is produced on an artisanal as well as industrial scale. Some characteristics such as taste, texture and appearance differ between the processes.

Several studies has investigated the microbiological, biochemical, and organoleptic aspects of artisanal Coalho cheeses from different localities, unraveling the complexity of their microbiota and chemical composition. The cheeses produced in specific regions, such as the Jaguaribe valley in the state of Ceará, and the Agreste Pernambucano in the state of Pernambuco, Brazil are distinguished for its sensory attributes, and enjoy a solid reputation among consumers. On the other side, the occurrence of food-borne pathogens and other microbial contaminants in Coalho cheeses from several localities is also reported, indicating the need to improve the sanitary conditions of milk and cheese production.

Over the time, innovative versions of the Coalho cheese has been developed based on the incorporation of spices and aromatic oils, probiotic bacteria, or using milk from others species, such as goats and buffaloes.

The objective of this chapter is to present an overview of the current knowledge about the history, technological process, and biochemical and microbiological characteristics of the artisanal Coalho cheese variety. Additionally, a special focus is given to the Coalho cheese produced in the semi-arid region of Jaguaribe, in the state of Ceará, where studies for a protective Geographical Indication are in progress.

Historical and Geographical Aspects

The artisanal Coalho cheese manufactured from raw cow's milk is one of the most traditional products of the Northeast region of Brazil, and its methods of production in rural properties have been passed down through several generations, from father to son, since the cheese making process was introduced in Brazil as a heritage of the Portuguese colonization in the post-Columbian era.

The origin of cheese production in Brazil has been reported by several historians. Mello (2007) states that cheese production was a result of Portuguese colonization and a time line proposed by Dias (2010), indicates that the first cattle was brought to São Vicente, on the coast of the state of São Paulo, Brazil in 1532, during the first official colonization expedition of Portugal, commanded by Martim Afonso de Sousa. The first records of cheese making in Brazil dates back to 1581, where it was produced in a dairy facility at a Jesuit School located in Salvador, and at a sugar mill, both in the northeastern region of the state of Bahia. Other documents report cheese production in the southeast (São Paulo, Rio Grande do Sul, Minas Gerais and Piauí states, Brazil) around the same period.

Coalho cheese derives its name from the Portuguese word for rennet, which in turn is related to the use of strips or crude extracts of the stomach of, mainly, young goats (*Capra hircus*) or calves (*Bostaurus* and *Bosindicus*) for milk coagulation in the cheese-making process. Such strips have been in use from its origin until at least the 1970s. The use of the stomach of mocó (*Kerodon ruprestre*) and cavy (*Galea spixii*) for the production of Coalho cheese has also been reported. The traditional procedure employed for obtaining the coagulant enzymes was gradually abandoned, as it became easier to acquire commercial chymosin, in liquid or powder form.

The geographical comprehensiveness of the Coalho cheese production is wide, and nowadays a relevant volume is produced in at least nine states in the northeastern region of Brazil: Alagoas, Bahia, Ceará, Maranhão, Paraíba, Pernambuco, Piauí, Rio Grande do Norte and Sergipe (Araújo et al. 2012). Much of the cheese in this region is handmade on small rural properties by agricultural families, thanks to the simplicity of the employed technology. Thus, in addition to its cultural relevance, the artisanal Coalho cheese is an important source of income for many small family farms.

As a typical and popular dairy food of the region, Coalho cheese is sold in street and fairs, local markets and supermarkets. It is the most consumed variety of cheese, as the local population has integrated it into their traditional food. Raw, roasted or fried, it may be consumed as a side dish or as an ingredient in several northeastern typical dishes. It is served grilled in restaurants and also by street vendors, who are common on urban beaches in the northeast and southeast of Brazil.

Menezes (2011) points out that the Coalho cheese is a cultural heritage of the Northeastern population of Brazil that arouses the interest of producers, both in the public and private institutions. Recognizing its socio-economic importance, the key social influencers are playing an active and important role in the development of regulations compatible with the traditional small-scale production of this cheese.

Coalho Cheese Characteristics

The Brazilian Technical Regulations of Identity and Quality of Coalho cheese (Brazil 2001a) has set the minimal requirements for the product to be destined for national and international trade and consumption. The document defines the Coalho cheese as the product obtained through coagulation of milk by rennet action or other appropriate coagulating enzymes, supplemented or not by the action of selected LAB. As optional ingredients, the normative also provides for the use of calcium chloride, dairy solids, condiments and spices, and sodium chloride. The technical regulations require the pasteurization of milk as a process to be used in the cheese production, aiming to ensure product safety for consumers.

The same normative describes the main sensory characteristics of Coalho cheese: uniform yellowish white color, mild and slightly acidic flavor, sometimes salty, and presenting a slightly acidic aroma, resembling curd. Also according to the normative, the cheese has semi-hard and elastic consistency, a compact and tender texture, and its rind is characterized as thin, without cracks, having an unusually well-defined rind formation. Inside, the Coalho cheese small holes (eyes) may or may not be present.

Besides these typical features of Coalho cheese that are provided for by the regulations, relevant differences are observed in the physical, chemical and sensory characteristics of the handmade cheeses in each one of the northeastern states in Brazil, or even in geographically distinct areas of the same state. Such differences have been linked to local specificities in the manufacturing process, as well as to the composition and endogenous microbiota of the milk used as a raw material. Milk composition is influenced by environmental factors, including cattle feeding and management practices and the local climate (mostly temperature and humidity), which add to the regional details of the cheese making procedures, what is generally known as the *"terroir"*. The Brazilian regulations recognize the magnitude of the differences in artisanal Coalho cheese characteristics, thus categorizing it as either a medium to high moisture, semi-cooked or cooked curd cheese, with a fat content ranging between 35.0% and 60.0% (w/w) of the total solids (Brasil 2001a).

Differences in sensory attributes of the Coalho cheese produced in different places have been reported by consumers and are mainly attributed to the influence of pasteurization and other processing steps (Dias 2010; Benevides 2000b). The indigenous microbiota of milk is destroyed in the pasteurization process and its replacement with exotic LAB alters the typical characteristics of the cheese made from raw milk. In the state of Pernambuco, Brazil, the Coalho cheese produced from pasteurized or raw milk are differentiated by the classification type "A" and "B", respectively (Pernambuco 1999).

Recognizing the difficulty, or even the impossibility of replacing the unique microbiota characteristic of raw-milk cheeses, cheeses with Protected Designation of Origin (PDO), such as *Camembert, Roquefort, Pélardon, Saint-Nectaire* and *Saint Maure Touraine*, among others, continue to be made from raw milk in countries having a strong cheese making tradition, like France.

The size, weight and shape of the Coalho cheese can also vary according to the region where it is produced. There are round, square and rectangular Coalho cheeses, of different sizes. The cheeses produced in the state of Ceará, Brazil, can have a round, square or rectangular shape, and those produced in the Jaguaribe region of this state are characterized by the round format. Meanwhile, the Coalho cheese produced in the rural areas of the states of Pernambuco and Rio Grande do Norte (Brazil) are usually rectangular in shape.

Despite the observed differences, the artisanal Coalho cheese manufactured in each locality of the northeastern Brazil is greatly appreciated and consumed by the local population. Together, all these varieties of cheeses constitute the Coalho cheese renowned all over Brazil.

Coalho Cheese Manufacturing Process

The main steps of Coalho cheese production is schematically shown in Fig. 2.1, and described below. The sequence of the processing steps practiced in different areas of production is similar, however specificities in the procedures and process parameters are observed.

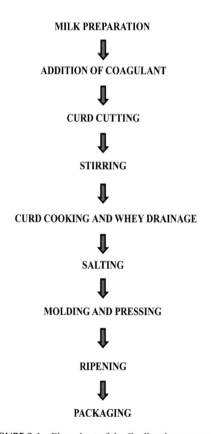

MILK PREPARATION

⇓

ADDITION OF COAGULANT

⇓

CURD CUTTING

⇓

STIRRING

⇓

CURD COOKING AND WHEY DRAINAGE

⇓

SALTING

⇓

MOLDING AND PRESSING

⇓

RIPENING

⇓

PACKAGING

FIGURE 2.1   Flow chart of the Coalho cheese processing.

It is important to mention the differences observed in the curd cooking and salt-ing process, as well as in the cheeses ripening conditions among the states in the northeastern region of Brazil, the micro-regions of the same state, and even among specific producers. These variations in the processes can add to the differences in the milk composition and microbiota, creating cheeses with unique and distinguished sensory, microbiological and physico-chemical characteristics.

Milk Preparation

Even though the Coalho cheese is traditionally made from raw milk, at the indus-trial level the milk employed for cheese production is pasteurized. Both rapid (72–75 °C /15 s) and slow (62–65 °C/30 min) pasteurization are employed in dairy factories for Coalho cheese production. After pasteurization, the starter culture, cal-cium chloride and the coagulant are added to the milk at 32-35 ºC, in this order. It is important to highlight that in the traditional production with raw milk, only rennet is used as an ingredient.

When pasteurized milk is employed commercial starter cultures are added to replace the autochthonous lactic acid bacteria and promote acid production before the addition of coagulant. In general, freeze-dried cultures composed by *Lactococcus lactis* subsp. *cremoris* and *Lactococcus lactis* subsp. *lactis* are utilized. Milk temper-ature is kept at 32–35 ºC for starter culture addition and action. Currently, indigenous LAB cultures for the production of Coalho cheese are not available in Brazil. In general, dairy products use imported lyophilized LAB cultures, with the consequent alteration of regional and traditional characteristics of the cheese.

The addition of calcium chloride also becomes necessary in the case of milk pas-teurization, aiming to restore the level of ionic calcium in milk, and to promote an adequate milk coagulation process. Calcium chloride is dissolved in filtered water before the addition occurs.

Liquid or powdered rennet, dissolved in a small volume of water, is added to the milk at a concentration strong enough to promote coagulation in a period of around 40–50 min, according to the manufacturer instructions. The addition of the rennet solution is followed by stirring the milk for a few seconds to obtain a homogeneous mixture. The milk then rests until coagulation.

Curd Cutting

Milk coagulation and proper curd formation occurs around 40–50 min after the addi-tion of the coagulant. In general, the cutting time is determined by empirical inspection of the gel firmness, visually or through manual testing. The curd is considered ready to be cut when it is bright, is easily removed from the vat edges, or breaks cleanly when a knife is inserted at an angle around 45° and slowly raised to the gel surface.

At the industrial level, and sometimes at the artisanal medium scale, the curd cut-ting step is done using cutting harps, to obtain cubes (curd grains) of 1.5 to 2 cm, in average. On rural properties, however, this step is usually done with a knife. After cutting, the curd grains rest for a period of 3 to 5 min, to start the syneresis and the whey releasing process.

Stirring

The cutting procedure is followed by a stirring step, which begins with mild and slow movements to preserve the integrity of the curd grains. Stirring is done intermittently, alternating periods of 10 to 20 min agitation with resting periods. The speed is gradually increased as the whey leaves the grains, which becomes firmer, and stops when the curd tends to deposit at the bottom of the vat.

Curd cooking and whey drainage

The curd particles are heated to favor whey release, reducing their size and moisture, to obtain a semi-hard cheese. The traditional procedures involve the drainage of around 50% of the whey from the vat, followed by the heating up of the whey to around 70 °C, and its return to the vat to elevate the curd temperature to 45–50 °C. In some regions, the curd cooking process is done in hot water, after the partial drainage of the whey. The cooking period is usually around 10–20 min, varying among producers, being stopped when the curd grains settle down to the bottom of the vat. At this point the curd grains are firm and tend to aggregate easily when pressed by hand. The complete drainage of the whey is then conducted, and the curd forms a block of cheese mass. Even though the cooking step is usually done in the Coalho cheese making process, it is not practiced in some regions of the state of Pernambuco, Brazil.

Salting

The salting process varies according to the production area of the Coalho cheese. In general, vat salting predominates among artisanal producers, in which the salt is added directly to the mass (dry salting) or diluted in whey. The amount of salt to be added is usually calculated as 1 to 2% of the initial milk volume.

Dry surface salting is also employed in some production areas, by means of spreading dry salt on the surface of the molded pressed cheeses, as observed by Silva et al. (2010) in the state of Alagoas, Brazil.

Molding and pressing

After the whey drainage, the curd mass is transferred to molds and subjected to a pressure operation that determines the cheese format. The shape and size of the mold employed in artisanal Coalho cheese manufacture varies according to the production region. The cheese format can be round, square or rectangular, in small to large size, weighing between 0.5 to 5.0 kg, although the usual weight is around 1 kg. In general, larger cheeses are manufactured in square or rectangular format. Traditionally, square or rectangular wooden molds covered with cotton cloths are used, which in recent years have been replaced by polyethylene shapes and synthetic cheese cloths.

The pressing step aims a further removal of the whey of the curd by compressing the cheese mass placed in molds through the application of mechanical pressure, to obtain a cheese of a harder consistency. At traditional small dairies, pressing is performed placing wooden planks covered with cloths at the mold surface, on which the weights are placed. At industrial dairies, pneumatic cheese presses are used to control the amount of pressure applied on the cheese. In general, pressing is performed for a period of approximately 12 to 16 h, varying according to the amount of pressure applied.

Coalho cheeses of a harder consistency are nowadays found in the market. These cheeses are usually obtained with additional pressing and drying time. In the northeastern Brazil, these cheeses are made to be grilled without melting, and are commonly sold in small pieces inserted in wooden sticks.

Ripening

The ripening time required to achieve the main sensory characteristics of Coalho cheese is around ten days, however, commercialization has shortened this period to less than one week. At the industrial level, after the pressing operation the cheeses are kept in cold storage (4 to 12 °C) for about 12 –24 h for surface drying, and then the ripening process follows or they are directly packaged.

Packaging

Traditionally, artisanal Coalho cheeses are sold unpacked. However, nowadays, the product is commonly found in markets packaged in polyethylene plastic bags (Fig. 2.2). The Coalho cheeses produced industrially are usually packaged in vacuum polyethylene sacks immediately after the surface drying period.

FIGURE 2.2   Coalho cheese exposed in a street market (A) and at a bus station shop (B) in the city of Tianguá, state of Ceará, Brazil.

Microbiological Aspects of Artisanal Coalho Cheese

The traditional Coalho cheeses are made from unpasteurized milk, and the artisanal process does not involve any deliberate addition of selected starter cultures. Therefore, the biodiversity of the microbiota naturally present in raw milk as adventitious contaminants, besides specific technological procedures, determines the initial microbial ecosystem associated with the cheese, which undergoes transformations during ripening and storage.

LAB, naturally present in milk or added as a starter, are responsible for the acidification required for whey expulsion from the coagulum during the cheese making process. The production of lactic acid throughout a fermentation process also generates an acidic environment that favors the control of pathogens and spoilage microorganisms.

The microbiota of traditional cheeses is also responsible for their typical organoleptic characteristics, defined by a complex set of biochemical processes. The ability of the autochthonous bacterial strains to generate flavor compounds throughout cheese ripening is well documented, and the unique textural qualities derive from specific proteolysis and lipolysis. The presence of adventitious LAB introduces a high variability in the cheese ripening process and may results in relevant differences in the final product characteristics (Randazzo et al. 2013).

Knowledge about the particular LAB microbiota of an artisanal product is important to ensure its quality, authenticity and traceability, and can also support protected geographical indications (PGI), and protected designations of origin (PDO) (Giannino et al. 2009; Silva et al. 2012).

In the case of Coalho cheese, even though the absence of systematic studies of its typical microbial ecosystem, there are some studies evidencing the LAB diversity of the cheese bacterial populations (Carvalho 2007; Freitas et al. 2013). Carvalho (2007) investigated the LAB microflora of the Coalho cheese produced from unpasteurized milk in the state of Ceará, Brazil, using traditional microbiological techniques for isolation and enumeration of LAB, with subsequent identification of the isolates using a set of biochemical tests. Among 643 isolates classified as LAB, *Enterococcus* (59.6%) was the predominant genus, followed by *Lactobacillus* (22.0%), in addition to a considerable proportion of *Streptococcus* (12.8%). Among the enterococci, the majority was identified as *Enterococcus faecium*. *Lactobacillus paracasei* species predominates among the lactobacilli, and *Streptococcus thermophilus* was the main species among the *Streptococcus* strains. *Lactococcus* and *Leuconostoc* were also identified among the strains. Carvalho (2007) also studied the lactic microbiota in raw milk samples, and observed the higher Enterococci population in Coalho cheese in comparison with the milk, evidencing the resistance and even the multiplication of these micro-organisms during the cheese making process. The prevalence of the thermophilic *Enterococcus, Lactobacillus* and *Streptococcus* in the Coalho cheese samples, according to Carvalho (2007), showed that the high temperatures employed in the curd cooking step promotes a selection of heat resistant strains.

Silva et al. (2012) and Dias (2014) studied the LAB profile of the artisanal Coalho cheese produced in the state of Pernambuco, Brazil. Silva et al. (2012) detected the presence of *Lactobacillus, Lactococcus, Streptococcus* and *Enteroccocus* genus in the cheese samples, and confirmed the presence of the species *Enterococcus faecium, Enterococcus faecalis, Streptococcus thermophilus* and *Lactococcus lactis* using molecular techniques. In the study performed by Dias (2014), the LAB isolated from artisanal Coalho cheese included *Enterococcus* (37.18%), *Lactococcus* (19.23%), *Streptococcus* (25.64%) and *Leuconostoc* (15.38%).

Freitas et al. (2013) investigated the LAB microbiota of Coalho cheese produced with traditional procedures in rural properties located in different regions in the state of Paraiba. Using traditional microbiological techniques to access the LAB profile of the cheese samples, the authors registered the predominance of *Enterococcus, Lactobacillus* and *Lactococcus* among the 49 isolated, characterized as LAB, although reporting some divergence among the employed biochemical methods of identification.

The elevated proportion of *Enterococcus* spp. found among the LAB isolates from artisanal Coalho cheese may indicate inadequate sanitary conditions during milking, milk handling and cheese processing, as highlighted by Carvalho (2007). In fact, several studies focusing on the microbiological quality and safety of Coalho cheese from different localities in the northeastern region revealed a high prevalence of total and thermotolerant coliform, *Escherichia coli*, and *Staphylococcus aureus*. Several studies detected populations of coliforms and *Staphylococcus aureus* exceeding the limits established by the Brazilian sanitary legislation (Azevêdo et al. 2014; Freitas et al. 2013; Santana et al. 2008; Feitosa et al. 2003; Nassu et al. 2000; Benevides et al. 2000a). A low detection rate for *Salmonella* spp. and, in general, an absence of *Listeria monocytogenes* in Coalho cheese was reported (Azevêdo et al. 2014; Borges et al. 2003; Silva et al. 2010).

Raw milk cheeses can be carriers of microbial pathogens. If these organisms are present in the milk and are able to survive the specific processing and ripening conditions, it raises concerns about product safety and the use of unpasteurized milk. Meanwhile, some researchers pointed out that the raw milk indigenous LAB population can produce bacteriocins and other antimicrobial agents capable of inhibiting or controlling the growth of pathogens and spoilage micro-organisms in the cheese environment (Azevêdo et al. 2014; Nero et al. 2008), acting as biopreservatives. According to Silva et al. (2010), the reported absence of *Listeria monocytogenes* and *Salmonella* spp. in Coalho cheese samples can be explained by the antimicrobial metabolites produced by the autochthonous LAB population. Moreover, some strains of *Enterococcus faecium* isolated from artisanal Coalho cheese manufactured in Ceará state, Brazil, were characterized as bacteriocin producers (Dos Santos et al. 2015a). Antimicrobial peptides have been found in Italian cheeses—*Pecorino Romano, Canestrato Pugliese, Caciocavallo, Crescenza,* and *Caprino del Piemonte*, showing inhibitory activity against various bacteria including *Escherichia coli, Bacillus megaterium, Listeria innocua, Staphylococcus aureus, Salmonella* spp. and *Yersinia enterocolitica* (Rizzello et al. 2005).

Total and thermotolerant coliforms are hygienic indicators in cheese production, since their presence is related to the sanitary conditions of raw milk production and storage and to inadequate handling during the manufacturing process (Azevêdo et al. 2014). The implementation of good agricultural and manufacturing practices can ensure a safe final product, as shown by many studies. Cheese matrix characteristics–water activity, lactic acid content and salt concentration, among others–are determinants of the growth and viability of the microorganisms in the final product. Altogether, intrinsic and extrinsic parameters are involved in cheese microbiological safety.

Fermented Foods of Latin America

TABLE 2.1 Average composition and physico-chemical parameters of Coalho cheeses from different dairies of northeastern Brazilian states.

State, Brazil	Moisture (%)	Fat (%)	Protein (%)	Ash (%)	pH	Titratable acidity (% lactic acid)	Reference
Alagoas	45.50–51.50	36.59–48.16	26.93–29.63	3.40–3.70	5.98–7.13	0.34–0.44	Silva et al. 2010
Paraíba	43.72–59.31	19.04–25.38	17.17–22.64	2.88–3.54	4.80–5.60	0.21–0.78	Freitas et al. 2013
Sergipe	21.00–29.38	–	–	–	5.18–5.68	0.74–1.01	Sousa et al. 2014
Ceará	44.34–44.52	24.83–26.88	20.70–22.49	4.10–4.68	5.84–6.33	0.42–1.29	Andrade 2006
Pernambuco	52.76–58.37	21.25–29.9	17.04–21.59	–	–	–	Silva 2012

Compositional and Functional Aspects of Coalho Cheese

Recent studies evaluated the composition and physico-chemical parameters of the Coalho cheeses manufactured in different states and areas of the northeast. A wide range of moisture, fat, and protein content, as well as of pH values and titratable acidity, have been reported in the available studies, as shown in Table 2.1.

Usually, the Coalho cheese manufactured in the state of Ceará, Brazil, is characterized by its medium moisture content, low acidity, and a salt content close to 3% (Andrade 2006; Carvalho 2007). On the other hand, the moisture content of the cheeses from Pernambuco State of Brazil, as reported by Silva (2012), were remarkable higher than those registered for the cheese from the other Brazilian states. The fat content of the Coalho cheese ranged from 19% in samples obtained in the Paraíba state, Brazil (Freitas et al. 2013) to around 48% in cheeses coming from the Alagoas state, Brazil (Silva et al. 2010). According to the available data and a cheese classification based on the fat content proposed by the Brazilian regulations, the Coalho cheese from the states of Ceará, Pernambuco, and Paraíba are classified as low fat or semi-fatty, and those from Alagoas state as a semi-fatty or fatty cheese.

The differences in cheese pH values and/or titratable acidity are also remarkable. The pH values reported for cheeses from Sergipe and Paraiba states (Brazil) tended to be lower than those observed from the other sampled states (Freitas et al. 2013; Sousa et al. 2014).

Silva et al. (2010) evaluated the influence of the manufacturing procedures on the physico-chemical, sensory and microbiological characteristics of the Coalho cheese produced in the Alagoas state, Brazil, and found that the steps of the heating of the curd and the salting processes differed from one dairy to another, resulting in products with different characteristics. The discrepancies in the composition of the cheeses from different localities have been attributed to differences in milk composition, climate, and especially to the technological process.

Since the pressing step in most dairies is manual, it is very difficult to control the final moisture content of the cheeses. This is not observed in industrial dairies that use a pneumatic press that controls the output of the liquid and consequently the moisture.

The proteolysis profile of the Coalho cheese from Jaguaribe valley, Ceará state, Brazil, was analyzed through Sodium dodecyl sulfate polyacrylamide gel electrophoresis (SDS-PAGE) electrophoresis on the first day after production. The results, shown in Fig. 2.3, revealed a profile typical of traditional cheeses produced by enzymatic coagulation, registering the presence of α_s and β-casein, and the formation

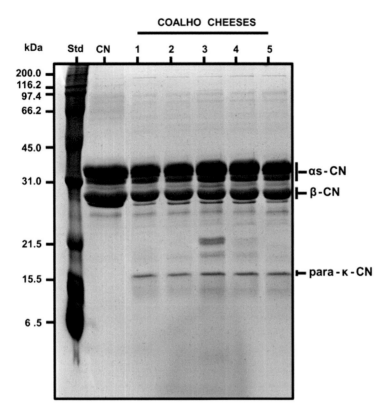

FIGURE 2.3   Sodium dodecyl sulfate polyacrylamide gel electrophoresis of Coalho cheeses made from cow's milk from Jaguaribe Valley, state of Ceará. Quantities of 80 μg of proteins were loaded in the gels. Proteins were stained by Coomassie blue. Std: molecular mass standards; CN: bovine sodium caseinate; 1–5: Coalho Cheese with 1 day of manufacturing; αs-CN: αs1- + αs2-caseins; β-CN: β-casein; para-κ-CN: para-κ-casein.

of peptides at the first day after production (Fontenele 2013). The para-κ-casein originated from the hydrolysis of κ-casein by chymosin also appears in the electrophoresis gel of the first day's sample of cheese. Chymosin is an aspartic protease (EC 3.4.23.4) that, during the cheese making process, initially cleaves to the bovine κ-casein in the Phe_{105}-Met_{106} (Jolles et al. 1963), promoting the destabilization of the casein micelles, which results in milk coagulation. This hydrolysis yields the para-κ-casein, composed by the N-terminal part of kappa-casein (f1-105), which remains associated with the casein in cheese, and a glycomacropeptide (GMP), which is the C-terminal part (f106–169) of κ-casein, released in the whey during the production of the cheese. According to Trujillo et al. (1997), the order of the hydrolysis of the milk caseins through chymosin occurs preferably in κ-casein followed by the demanding αs -casein and β-casein, being that the β-casein is more resistant than the others.

The process of proteolysis undergoing in caprine Coalho cheese was followed during 180 days of ripening through SDS-PAGE electrophoresis. It was possible to observe that the caseins hydrolysis occurred initially in κ-casein, due to the presence of para-κ-casein in the gel at the first day after manufacturing, and was followed by α_s caseins hydrolysis, that decreased in intensity according to the ripening time of the cheese, becoming very clear its position in the gel at the end of the period. At all times, it was possible to observe the presence of β-casein without major changes, because of its higher resistance to hydrolysis. The α_{s1}-casein is the casein fraction most prone to hydrolysis, as verified in a wide variety of cheeses, such as the Canestrato Pugliese, Fiore Sardo, and Pecorino Romano (Di Cagno et al. 2003), as well as Mozzarella (Feeney et al. 2002) and Caciocavallo Pugliese (Gobbetti et al. 2002), among others. The degradation of β-casein is always inferior to the other casein fractions (Di Cagno et al. 2003; Kalit et al. 2005). The proteolysis process of Coalho cheeses is therefore similar to those of other traditional European cheeses.

Functional Aspects of Coalho Cheese

Cheeses, like other dairy products, can be a source of biologically active peptides, generated as LAB metabolites and / or as a result of the hydrolysis of milk proteins by endogenous enzymes or enzymes added in the cheese making process. Along with the afore mentioned antimicrobial activity, several properties are attributed to bioactive peptides that are found in traditional cheeses, like the opioid activity of the β-casomorphin peptide isolated from Brie cheese (Jarmolowska et al. 1999), and the inhibition of the angiotensin converting enzyme (ACE) activity detected in Idiazabal, Manchego, Roncal, Mahon, and Cabrales cheeses (Gomez-Ruiz et al. 2006), Cheddar cheese (Ong et al. 2007), and Castello, Port Salut and Brie cheeses (Pripp et al. 2006).

Recent studies have shown that Coalho cheese can be also a source of bioactive peptides. Silva et al. (2012) found that water extracts of artisanal Coalho cheese from the state of Pernambuco contained soluble peptides presenting antimicrobial activity against *Enterococcus faecalis*, *Bacillus subtilis*, *Echerichia coli* and *Pseudomonas aeruginosa*, as well as antioxidant and zinc-binding activity.

In a similar way, hydrosoluble peptides isolated from Coalho cheese manufactured in Vale do Jaguaribe, in the Ceará state, Brazil, were analyzed by Fontenele (2013) with mass spectrometry, and compared to bioactive peptide sequences obtained from casein described in the literature. The study identified peptides derived from β-casein and casein-α_{s1} with immuno-modulatory activity, from β-casein, α_{s1}-casein, α_{s2}-casein and κ-casein with antihypertensive activity and antimicrobial activity originating from β-casein and casein-α_{s1}. Besides the potential health benefits related to the bioactive peptides of artisanal Coalho cheese, the antimicrobial activity of the can enhance the safety of this traditional food.

Geographical Indications for the Coalho Cheese

The Coalho cheese from specific production regions such as the Jaguaribe Valley, in the Ceará state, Brazil, and the Agreste Pernambucano in countryside of Pernambuco state, Brazil, is distinguished for its special sensory attributes, and enjoys a solid reputation among its consumers. The milk produced in these areas by animals fed on native pastures as well as the traditional know-how of local cheese producers makes the special characteristics of these cheeses. Producers, researchers, governmental and non-governmental organizations have been undertaking actions aiming towards the protection of these traditional cheeses, based on the geographical indication and legal framework provided by the Brazilian regulation (Brasil 1996).

The accurate mapping and demarcation of the geographical region, as well as the detailed description of the traditional cheese making process and characterization of the product through physical, chemical, microbiological, biochemical, and proteomic analysis, are some of the first steps to establish the distinguished quality of the Coalho cheese produced in these two regions.

The main characteristics and procedures employed for the production of the Coalho cheese from the Agreste Pernambucano and Jaguaribe region are discussed below, as well as (some of the actions done to obtain a protected geographical indication for these traditional food products).

The Coalho Cheese from the Agreste Pernambucano

The Agreste Pernambucano is a meso-region delimited between the coast and the interior/inland of the state. The geographical location gives the region a milder climate compared to the state semiarid area, and a higher pluviometric index, favoring the native pastures, and consequently influencing the quality of milk.

The artisanal Coalho cheese from the Agreste Pernambucano is produced from raw milk, following traditional procedures established from the time of the Portuguese colonization. Rectangular in shape, the artisanal Coalho cheese is usually sold in medium size pieces weighing around 1 kg, in fairs and small markets, at room temperature. It is generally consumed fresh, after a short ripening period (few days).

The cheese making process is distinguished from those of other areas mainly because it does not involve the cooking of the curd, and the milk is usually not heated prior rennet addition, thereby increasing the coagulation period up to 60 min. Another distinguishing factor is the salting step, done after the molding and pressing of the Coalho cheese through dry surface-salting.

Regarding compositional parameters, the Coalho cheese from this region has a high to very high moisture content, and can be classified as a low-fat or fatty cheese, according to the results obtained by Silva (2012) which were based on the analysis of cheeses samples from six municipalities (Table 2.1). The study showed the remarkable differences in the artisanal cheese composition, and the author highlights a need for procedures standardization among the local producers, aiming to preserve the identity of the Coalho cheese from the Agreste Pernambucano.

The Pernambuco state, Brazil, has specific regulations for the artisanal Coalho cheese production, which describes the main steps of the traditional cheese making process in the state (Pernambuco 2007). Resulting from the joint action of artisanal producers and health authorities, this resolution aimed towards the protection of the traditional artisanal technology. The normative specifies the use of milk not subjected to heat treatment, establishes that the cheese making process should start up to 2 h after the onset of milking, and points out the dry salting of the Coalho cheese. Fresh whole raw milk—obtained from milking cows, buffaloes or goats—and rennet are considered the mandatory ingredients.

The Coalho cheese from Jaguaribe

The Jaguaribe valley is a socio-economic region around the Jaguaribe river in the Ceará state, in northeast Brazil (Fig. 2.4-A). Semi-arid climate predominates in the region and the neighboring areas, with an open and dense shrub savanna vegetation, which is characteristic of the Caatinga Biome.

The artisanal Coalho cheese produced in the Jaguaribe region is renowned for its organoleptic qualities, having a significant economic importance to the Ceará state, Brazil. The influence of the animal feeding in native pastures on the milk quality, as well as the local cheese making procedures, may be the key factors related to the peculiarity of the Jaguaribe Coalho cheese. It is well known that technology incorporates the characteristics of the producing region, and the production process passed from generation to generation is an important factor in the features of the final product.

The specific production mode and characteristics of the Jaguaribe Coalho cheese were studied by the Brazilian Agricultural Research Corporation (EMBRAPA) and co-workers, with the aim of the preservation of traditional procedures and product quality. The research focused on a survey of the requirements for the construction of the geographical indication process, since the cheese of this region has a widely recognized reputation. Fig. 2.4 shows the area corresponding to the demarcated region in the Jaguaribe valley and marks the 10 dairies that participated in the EMBRAPA project.

For the product characterization, the study carried out the mapping of milk production, the geographical and environmental characteristics, and the location of cheese processing units, among other actions with the aim to gain the knowledge of the traditions associated with the Coalho cheese production.

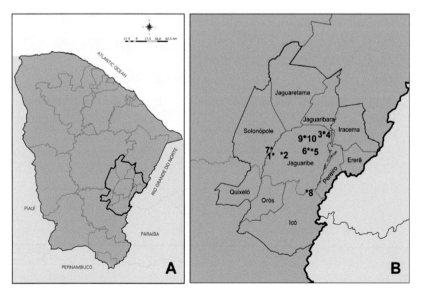

FIGURE 2.4   Location of dairies at the region of Jaguaribe valley, state of Ceará. A: Ceará map showing the demarcated region; B: Details of the demarcated area corresponding to the Jaguaribe region; *= Dairies that participated in the EMBRAPA project.

The main steps of cheese making process carried out by the dairies in the project area, based on the information from the producers association and project participants, are similar to those employed in other regions in the Ceará state, Brazil. This fact highlights the importance of the factors linked to milk quality in the distinguished characteristics of Jaguaribe Coalho cheese. However, there are notable differences regarding the procedures employed in other northeastern states.

In general, the Coalho cheese production in the Jaguaribe region involves heating the raw milk at 32-35 °C before the addition of rennet, and the coagulation time is around 35-40 min. After partial whey drainage, the curd is cooked at 45-60 °C, using the whey or added water as the heating medium. Salt is added to the curd after the last drainage step, before the molding step. The pressing period varies from 2-6 h.

Some differences were registered among the Jaguaribe producers regarding the curd stirring periods, the use of water or whey as the heating medium for curd cooking, and the number of drainage steps, among other parameters of the processing operations. These conditions can explain the observed differences in the composition and characteristics of the regional Coalho cheese.

The composition and physico-chemical parameters of Jaguaribe Coalho cheese from twenty producers were monitored over a period of two years, and the data is shown in Table 2.2.

TABLE 2.2 Composition and physico-chemical parameters of Jaguaribe Coalho cheese

Parameter	Minimum	Maximum	X ± S*	CV**
Titratable acidity (% lactic acid)	0.14	0.62	0.32 ± 0.27	84.87
pH	5.83	6.46	6.12 ± 0.26	4.26
Moisture (%)	38.00	46.31	42.94 ± 1.78	4.14
Ash (%)	3.50	4.79	4.11 ± 0.47	11.45
Fat (%)	28.03	32.72	30.55 ± 2.54	8.32

* average of the samples ± standard deviation ** = coefficient of variation (standard deviation / mean) × 100

The differences in lactic acid content observed among the cheese samples can be related to the microbiological quality of the milk employed, as well as to the manufacture and storage conditions. The acidity exerts importance in cheese quality, since it has a direct influence on the pH and the whey expulsion from the curd during the cheese production and in an early stage of ripening.

Jaguaribe Coalho cheese samples presented a medium moisture content, and a fat content around 30% (Table 2.2), being classified as a semi-hard and semi-fatty cheese according to Brazilian regulations (Brasil 2001a). Moisture content is related to water activity and cheese preservation, influencing the composition and metabolic activity of the cheese microbiota during storage, with consequences in texture, pH, flavor and aroma of the product (Ferreira and Freitas 2008). Moreover, the fat content has a direct influence on the cheese texture attributes, being related to the ability to slice deformity and to the melting behavior of the product. A rubbery texture is a desirable attribute for local consumers, as the Coalho cheese is widely consumed roasted.

Coalho cheese samples of different artisanal producers of the Jaguaribe region were analyzed for the identification and description of their main organoleptic features, with the aim of product characterization. Based on the results, the artisanal Coalho cheese from Jaguaribe was characterized as round, pale yellow, soft, with a rubbery and uniform texture, with a slight creak in mastication, a cooked milk / buttery aroma, pleasant acidity, and slightly salty. It has a diameter of around 12.5–13.0 cm, a height of 6.5–7.0 cm, and weight of 0.9–1.0 kg.

Innovative Varieties of Coalho Cheese

A number of new Coalho cheese varieties have been developed based on the traditional cheese making process and the additional use of spices, such as oregano and pepper, as well as other ingredients. Moreover, Coalho cheese made with milk from others species such as goats and buffaloes has been marketed in northeastern Brazil.

Coalho cheese made from buffalo milk is considered a promising product in the Northeastern Brazil market. Studies conducted by Costa (2007) of the bubaline Coalho cheese standardization, showed its distinguished physical, chemical and sensory features in comparison to those made from bovine milk. Besides presenting a remarkable high cheese yield, around 6.0 L / kg, and moisture content around 48% (w/w), the bubaline Coalho cheese has different physico-chemical and sensory features compared to the bovine cheese. Costa (2007) also reported the elevated sensory acceptability of this type of Coalho cheese among consumers of other cheeses made from buffalo milk.

Goat milk has also been employed to make Coalho cheese (Egito and Laguna 1999), in order to develop new varieties of caprine cheese by EMBRAPA Goats and Sheep (Sobral, Ceara state, Brazil). Among them, the matured and smoked Coalho cheese (Laguna and Egito 2008), the caprine Coalho cheese with aromatic herbs such as oregano (*Origanum vulgare* L.), cinnamon (*Cinnamomum zeylanicum*) and cloves of India (*Caryophilus aromaticus*) (Egito et al. 2007) stood out.

The use of the biodiversity of northeastern Brazil as a source of aromatic ingredients for incorporation to the Coalho cheese was also researched. In these studies, the use of pequi oil (*Caryocar brasilienses*) stood out. Pequi is a native fruit of Brazil, rich in oil, proteins and carotenoids, having a pleasant aroma and an attractive dark yellow coloring (Benevides et al. 2009). The bark of Cumaru tree *(Amburana cearenses)* was also used in the flavouring of caprine Coalho cheese. Native to Brazil, Cumaru has been used as a condiment in goat cheese in some regions of the Northeast, targeting a wider acceptability of these products or just to add the distinctive flavor of the product that comes from the Caatinga bioma (Benevides et al. 2010)

Queiroga et al. (2013) demonstrated the feasibility of producing a Coalho cheese variety using a mixture of caprine and bovine milk. The cheese proved to be a unique dairy product, with a mild caprine flavor, which improved consumer acceptance, and retained the positive nutritional properties of goat cheese. Additionally, this type of Coalho cheese can be a promising product for the northeastern region, which stands out as the main producer of goat milk and Coalho cheese in Brazil.

A study carried out by Egito et al. (2008) monitored the ripening process of the caprine Coalho cheese at a room temperature of around 30 °C, that is usual in the northern area of the Ceará state, Brazil. The microbiological analysis of the cheese at 20 days of ripening indicated the absence of microbial pathogens. Among the sensory attributes, the cheese presented a dry and firm consistency, and a unique pleasant flavor.

Santos et al. (2012) also obtained an innovative type of caprine Coalho cheese, naturally enriched in conjugated linoleic acid (CLA), and added with probiotic bacteria. An improved content of CLA in goat milk was achieved through animal nutrition strategies, mainly by the dietary supplementation with vegetable oil rich in polyunsaturated fatty acids, which was transferred to the cheese. The cheese was ripened for 60 days, during which the viability of the probiotic *Lactobacillus acidophilus* was maintained, and the product was well evaluated by potential consumers. The product sums up the benefits of viable probiotic bacteria and healthy fatty acids, and is considered a functional food.

In addition, *Lactobacillus rhamnosus* and *Lactobacillus plantarum* strains isolated from Coalho cheese samples collected at the Jaguaribe region were considered as probiotic candidates, based on beneficial properties detected by *in vitro* essays (Santos et al. 2015b). One of these strains, *Lactobacillus rhamnosus* EM1107 was successfully employed in Coalho cheese manufacture as a starter and a potentially probiotic culture, showing a good survival rate after exposure to simulated gastrointestinal conditions (Rolim et al. 2015). Furthermore, the strain demonstrated an inhibitory effect against intestinal pathogens through an *in vivo* essay.

Conclusion

Based on the reported studies, we can conclude that the artisanal Coalho cheese, nowadays, denotes a group of closely related cheeses differing in specific technological aspects, and presenting a wide range of moisture and fat content. Data from several studies have evidenced the need for continuous improvement of the sanitary conditions of Coalho cheese manufacturing at the artisanal level, with the aim to enhance its microbiological quality and guarantee product safety.

In this context, the artisanal Coalho cheese from delimitated areas, as the Jaguaribe valley and the Agreste Pernambucano, are distinguished for their unique sensory quality and reputation among consumers. As the cheese making process perpetuated in these areas originate since Portuguese colonization, both products deserved studies and actions to protect the existing traditional technology and to obtain indications of geographic origin.

Furthermore, in line with current food trends, the addition of spices chosen from Brazil's biodiverse produce has resulted in the emergence of innovative versions of the Coalho cheese, which are promising as products that incorporate new flavors and, as with the probiotic Coalho cheese, full of health attributes.

Acknowledgements

The authors acknowledge the financial support to the Jaguaribe Coalho Cheese Project from the Brazilian Agricultural Research Corporation (EMBRAPA), Conselho Nacional de Desenvolvimento Científico e Tecnológico (CNPq), and Serviço Brasileiro de Apoio às Micro e Pequenas Empresas (SEBRAE).

References

Araújo, J.B.C., J.C.M. Pimentel, F.F.A. Paiva, F. A. Marinho, P.F.A.P. Pessoa and H.E.M. Vasconcelos. (2012). Pesquisa participativa e o novo modelo de produção de queijo coalho artesanal da comunidade de Tiasol, em Tauá, CE. Cad Cienc Tecnol 29: 213–241.

Azevêdo, M.C., G.B. dos Santos, F. Zocche, M.I.C. Horta, F.S. Dias and M.M. da Costa. (2014). Microbiological evaluation of raw milk and coalho cheese commercialized in the semi-arid region of Pernambuco, Brazil. Afr J Microbiol Res 8: 222–229.

Andrade A.A. (2006). Estudo do perfil sensorial, físico-químico e aceitação de queijo de coalho produzido no Ceará. M.Sc. Thesis, Universidade Federal do Ceará, Fortaleza, Ceará.

Brasil (2001a). Instrução Normativa n. 30 de 26 de junho de 2001 do Departamento de Inspeção de produtos de origem animal do Ministério da Agricultura, Pecuária e Abastecimento. Aprova os Regulamentos Técnicos de Identidade e Qualidade de Manteiga da Terra ou Manteiga de Garrafa; Queijo de Coalho e Queijo de Manteiga. Diário Oficial [da] Republica Federativa do Brasil, Brasília, DF, 16 jul 2001. Seção I, p.13–5.

Brasil (2001b). Ministério da Saúde. Agência Nacional de Vigilância Sanitária. Resolução RDC n° 12, de 02/01/2001. Regulamento técnico sobre padrões microbiológicos para alimentos. Diário Oficial [da] República Federativa do Brasil,Brasília, 02/01/2001, p. 1–54.

Brasil (1996). Casa Civil. Regulamentação de direitos e obrigações relativos à proprie-dade industrial. Presidência da República, Subchefia para Assuntos Jurídicos. Lei N° 9.279, de 14 de maio de 1996.

Benevides S.D., F.J.S. Telles, A.C.L. Guimarães and A.N.M. Freitas. (2000a). Aspectos físicoquímicos e microbiológicos do queijo de Coalho produzido com leite cru e pasteurizado no estado do Ceará. Bol CEPPA 19: 139–153.

Benevides S.D., F.J.S. Telles, A.C.L. Guimarães and M.C.P. Rodrigues. (2000b). Estudo bioquímico e sensorial do queijo de Coalho produzido com leite cru e pasteurizado no estado do Ceará. Bol CEPPA 18: 193–206.

Benevides, S.D., K.M.O. Santos, A.D.S. Vieira and F.C.A. Buriti. (2009). Processamento de queijo de Coalho de leite de cabra adicionado de óleo de pequi. Comunicado Técnico 103. Embrapa Caprinos, Sobral, Ceará.

Benevides, S.D., K.M.O. Santos, F.C.A. Buriti, A.L.S. Junior, L.E. Laguna and A.S. Egito. (2010). Processamento de queijo tipo Coalho de leite de cabra adicionado de (Cumaru) Amburana cearenses. Comunicado Técnico 120. Embrapa Caprinos, Sobral, Ceará.

Borges, M.F., T. Feitosa, R.T. Nassu, C.R. Muniz, E.H.F. Azevedo, and E.A.T. Figueiredo. (2003). Microrganismos patogênicos e indicadores em queijo de Coalho produzido no Estado do Ceará, Brasil. Rev Bras CEPPA 21: 31–40.

Carvalho, J.D.G. (2007). Caracterização da microbiota lática isolada de queijo de Coalho aratesanal produzido no ceará e avaliação de suas propriedades tecnológicas. PhD Thesis, Universidade Estadual de Campinas, Campinas, São Paulo.

Costa, R.G.B. (2007). Tecnologia de fabricação e caracterização de queijo de Coalho obtido de leite de búfala. PhD Thesis, Universidade Federal de Lavras, Lavras, Minas Gerais.

Dias, J.C. (2010). Uma longa e deliciosa viagem. Barleus, São Paulo, São Paulo.

Dias, G.M.P. (2014). Potencial tecnológico de bactérias ácidas láticas isoladas de queijo de Coalho artesanal produzido no município de Venturosa—Pernambuco. PhD Thesis, Universidade Federal de Pernambuco, Recife, Pernambuco.

Di Cagno, R., J. Banks, L. Sheehan, P.F. Fox, and E.Y. Brechany. (2003). Comparison of the microbiological, compositional, biochemical, volatile profile and sensory characteristics of three Italian PDO. Int Dairy J 13: 961–972.

Egito, A.S., K.O. Santos, S.D. Benevides, S.C. Pereira and L.E. Laguna. (2008). Processamento artesanal de queijo do sertão fabricado com leite de cabra. Embrapa Caprinos. Comunicado Técnico 93. Embrapa Caprinos, Sobral, Ceará.

Egito, A.S., and L.E. Laguna. (1999). Fabricação de queijo Coalho com leite de cabra. Circular Técnica 16. Embrapa Caprinos, Sobral, Ceará.

Egito, A.S., K.O. Santos, L.E. Laguna, and S.D. Benevides. (2007) Processamento de queijo de cabra com ervas aromáticas. Comunicado Técnico 81. Embrapa Caprinos, Sobral, Ceará.

Feitosa, T., M.F. Borges, R.T. Nassu, E.H.F. Azevedo, and C.R. Muniz. (2003). Pesquisa de *Salmonella* sp., *Listeria* sp. e microrganismos indicadores higiênico-sanitário em queijo de Coalho produzido no Estado do Rio Grande do Norte. Cienc Tecnol Alim 23: 162–165.

Freitas, W.C., A.E.R. Travassos and J.F. Maciel. (2013). Avaliação microbiológica e físico-química de leite cru e queijo de Coalho produzida no Estado da Paraíba. Rev Bras Prod Agroind 15: 35–42.

Ferreira, W.L., and J.R. Freitas Filho. (2008). Avaliação da qualidade físico - químicos do queijo Coalho comercializado no município de Barreiros-PE. Rev Bras Prod Agroind 2: 127–133.

Fontenele, M.A. (2013). Caracterização físico-química, avaliação sensorial, proteômica e bioquímica do queijo Coalho do Jaguaribe-CE visando o processo de indicação geográfica. PhD Thesis, Universidade Estadual do Ceará, Fortaleza, Ceará.

Feeney, E.P., T.P. Guinee, and P.F. Fox. (2002). Effect of pH and calcium concentration on proteolysis in Mozzarella cheese. J Dairy Sci 85: 1646–1654.

Giannino, M.L., M. Marzoto, F. Dellaglio, and M. Feligini. (2009). Study of microbial diversity in raw milk and fresh curd used for Fontina cheese production by culture independent methods. Int J Food Microbiol 30: 188–195.

Gobbetti, M., M. Morea, F. Baruzzi, M.R. Corbo, A. Matarante, T. Considine, R. Di Cagno, T. Guinee, and P. Fox. (2002). Microbiological, compositional, biochemical and textural characterisation of Caciocavallo Pugliese cheese during ripening. Int Dairy J 12: 511–523.

Gómez-Ruiz J.Á., G. Taborda, L. Amigo, I. Recio and M. Ramos. (2006). Identification of ACE- inhibitory peptides in different Spanish cheeses by tandem mass spectrometry. Eur Food Res Technol 223: 595–601.

Jarmolowska, B., E. Kostyra, S. Krawczuk and H. Kostyra. (1999). β-casomorphin-7 isolated from Brie cheese. J Sci Food Agric 79: 1788–1792.

Jollès, P., C. Alais, and J. Jollès. (1963). Study of k-casein form cows. Characterization of the linkage sensitive to the action of rennin. Bioch Bioph Acta 69: 511–517.

Kalit, S., J. Lukac Havranek, M. Kaps, B. Perko, and V. Cubric Curik. (2005). Proteolysis and the optimal ripening time of Tounj cheese. Int Dairy J 15: 619–624.

Laguna, L.E., and A.S. Egito. (2008). Processamento do Queijo de Coalho Fabricado com Leite de Cabra Maturado e Defumado. Circular Técnica 90. Embrapa Caprinos, Sobral, Ceará.

Mello, E.C. (2007). Olinda restaurada: guerra e açúcar no Nordeste, 1630–1654. Editora 34, São Paulo, São Paulo.

Menezes, S.S.M. (2011). Queijo de coalho: tradição cultural e estratégia de reprodução social na região Nordeste. Rev Geogr 28: 40–56.

Nassu, R.T., R.S. Araújo, C.G.M. Guedes. and R.G.A. Rocha. (2003). Diagnóstico das condições de processamento e caracterização físco-química de queijos regionais e manteiga no Rio Grande do Norte. Boletim de Pesquisa e Desenvolvimento 11. Embrapa Agroindústria Tropical, Fortaleza, Ceará.

Nero, L.A., M.R. Mattos and M.B.T. Ortolani. (2008). *Listeria monocytogenes* and *Salmonella* spp. in raw milk produced in Brazil: occurrence and interference of indigenous microbiota in their isolation and development. Zoon Publ Health 55: 299–305.

Ong, L., A. Henriksson and N.P. Shah . (2007). Proteolytic pattern and organic acid profiles of probiotic cheddar cheese as influenced by probiotic strains of *Lactobacillus acidophilus, Lb. paracasei, Lb. casei* or *Bifidobacterium* sp. Int Dairy J17: 67–78.

Pernambuco (2007). Assembléia Legislativa. Processo de Produção do Queijo Artesanal. Lei no. 13.376, de 20 de dezembro de 2007.

Pernambuco (1999). Secretaria Produção de Rural e Reforma Agrária. Resolução n. 002 de 19 de abril de 1999. Estabelece a identidade e os requisitos mínimos de qualidade que deverá cumprir o Queijo Coalho produzido no Estado de Pernambuco e destinado ao consumo humano. Diário Oficial do Estado de Pernambuco, Recife, 20 de abril de 1999.

Pripp, A.H., R. Sorensen, L. Stepaniak, and T. Sorhaug. (2006). Relationship between proteolysis and angiotensin-I-converting enzyme inhibition in different cheeses. Food Sci Technol 39: 677–683.

Queiroga, R.C.R.E., B.M. Santos, A.M.P. Gomes, M.J. Monteiro, S.M. Teixeira, E.L. Souza, C.J.D. Pereira and M.M.E. Pintado. (2013). Nutritional, textural and sensory properties of Coalho cheese made of goats', cows' milk and their mixture. Food Sci Technol 50: 538–544.

Randazzo, C.L., I. Pitino, F. Licciardello, G. Muratore and G. Caggia. (2013). Survival of Lactobacillus rhamnosus probiotic strains in peach jam during storage at different Food Sci Technol 33: 652–659.

Rizzello, C.G., I. Losito, M. Gobbetti, T. Carbonara, M.D. De Bari and P.G. Zambonin. (2005). Antibacterial activities of peptides from the water-soluble extracts of Italian cheese varieties. J Dairy Sci 88: 2348–2360.

Rolim, F.R.L., K.M.O. Dos Santos, S.C. Barcelos, A.S. Egito, T.S. Ribeiro, M. L. Conceicao, M. Magnani, M.E.G. Oliveira and R.C.E. Queiroga. (2015). Survival of Lactobacillus rhamnosus EM1107 in simulated gastrointestinal conditions and its inhibitory effect against pathogenic bacteria in semi-hard goat cheese. LWT 63: 807–813,

Santana, R.F., D.M. Santos, A.C. Martinez and Á.S. Clima. (2008). Qualidade microbiológica de queijo-coalho comercializado em Aracaju, SE. Arq Bras Med Vet Zoot 60: 1517–1522.

Santos, K.M.O., M.A.D. Bomfim, A.D.S. Vieira, S.D. Benevides, S.M.I. Saad, F.C.A. Buriti and A.S. Egito. (2012). Probiotic caprine Coalho cheese naturally enriched in conjugated linoleic acid as a vehicle for *Lactobacillus acidophilus* and beneficial fatty acids. Int Dairt J 24: 107–112.

Dos Santos, K.M.O., A.D.S. Vieira, H.O. Salles, J.C.F. Nascimento, C.R.C. Rocha, M.F. Borges, L.M. Bruno, B.D.G.M. Franco and S.D. Todorov. (2015a). Safety, beneficial and technological properties of *Enterococcus faecium* isolated from Brazilian cheeses. Braz J Microbiol 46: 237–249.

Dos Santos, K.M.O., A.D.S. Vieira, F.C.A. Buriti, J.C.F. Nascimento, M.E.S. de Melo, L.M. Bruno, M. De F. Borges, C.R.C. Rocha, A.C. de S. Lopes, B.D.G.M. Franco and S.D. Todorov. (2015b). Artisanal Coalho cheeses as source of beneficial *Lactobacillus plantarum* and *Lactobacillus rhamnosus* strains. Dairy Sci Technol 95: 209–230.

Sousa, A.Z.B., M.R. Abrantes, S.M. Sakamoto, J. Silva, A. Berg, P.O. Lima, R.N. de Lima, M.O.C. Rocha and Y.D.B. Passos. (2014). Aspectos físico-químicos e microbiológicos do queijo tipo coalho comercializado em Estados do nordeste do Brasil. Arq Inst Biol 81: 30–35.

Silva, M.C.D., A.C.S. Ramos, I. Moreno and J.O. Moraes. (2010). Influência dos procedimentos de fabricação nas características físico-químicas, sensoriais e microbiológicas de queijo de coalho. Rev Inst Adolfo Lutz; 69: 214–221.

Silva, R.A., M.S.F. Lima, J.B.M. Viana, V.S. Bezerra, M.C.B. Pimentel, A.L.F. Porto, M.T.H. Cavalcanti and J.L. Lima Filho. (2012). Can artisanal "Coalho" cheese from Northeastern Brazil be used as a functional food? Food Chem 135: 1533–1538.

Silva, R.A. (2012). Caracterização microbiológica, físico-química, proteômica e bioativa de queijos de Coalho artesanal produzidos na Região Agreste do Estado de Pernambuco-Brasil. PhD Thesis, Universidade Federal Rural de Pernambuco, Recife, Pernambuco.

Trujillo, A.J., B. Guamis and C. Carretero. (1997). Proteolysis of goat β-casein by calf rennet. Int Dairy J 7: 579–588

3

Buffalo Mozzarella Cheese: Knowledge from the East and the West

Cristina Stewart Bogsan[a]* and*
Svestoslav Dimitrov Todorov[b]

Abstract

Buffalo mozzarella cheese was first produced in Italy in the 12th century. It is a fresh cheese - stretched curd or "pasta filata", produced after a short period of fermentation. Buffalo mozzarella cheese has a high consumer acceptance and is characterized with high value-add. In Latin America, it was firstly introduced to Brazil during the period of Italian emigration at the beginning of 20th century and today, almost all buffalo milk production in the country is used for the production of buffalo mozzarella cheese, especially as this cheese is increasingly participating in the international market.

Worldwide, the technological process of production of buffalo mozzarella cheese is still based on traditional Italian production technology. Briefly, the raw milk is heat-treated, followed by addition of rennet or acid to facilitate curd formation. The cutting process is made in two phases, followed by a rapid maturation period until pH reaches 4.9–5.1, then the scalding and stretching of the curd and, finally the shaping of the cheese.

The fermentation process could be conducted by commercial starter lactic acid bacteria (LAB) cultures or by the "back slopping method", a reintroduction of inoculums, as a part of the previous fermentation into new product. The "back slopping method" is cheaper; however, it does not guaranty constancy in production, due to changes in the natural microbiota of the whey, and the starter could also be a vehicle of microbial contaminations.

The physico-chemical parameters, chemical composition, rheological properties, applied technology and the interferences of indigenous microbiota with the added starter and probiotic cultures to buffalo mozzarella cheese production are explored in this chapter.

---

[a] Department of Biochemical and Pharmaceutical Technology, São Paulo University, 05508-000, Sao Paulo, SP, Brazil.
[b] Universidade Federal de Viçosa, Veterinary Department, Campus UFV, 36570-900, Viçosa, Minas Gerais, Brazil. E-mail: slavi310570@abv.bg
* Corresponding author: cris.bogsan@usp.br

Introduction

Mozzarela is a cheese of Italian origin which is, nowadays, commonly produced and commercialized in Brazil. The cheese is preserved in brine (saline solution) and consumed by Brazilians especially in Italian heritage cuisine, pizzas, salads, or appetizers (Silva et al. 2015). It is traditionally produced using raw or pasteurized milk and naturally present LAB in previously produced cheese as starters (back slopping method). However, influence of natural contaminants from the area of production and traditional tools used in production processes can influence its specific microbiota and organoleptic characteristics, including compromising of the entire production process (Losito et al. 2014; Silva et al. 2015).

The annual Brazilian production of buffalo milk reaches 92.3 million liters; however, the production of mozzarella cheese represents 70% of the total of buffalo's dairy product (Silva et al. 2015). Analysis of buffalo milk is normally characterized by 82.2% (w/100 g) of water, 4.8% (w/100 g) proteins, 7.5% (w/100 g) of fat, 4.7% (w/100 g) of lactose and 0.8% (w/100 g) of mineral salts in it's composition (Call and Relli 2015).

Buffalo mozzarella is a fresh stretched curd cheese, originally produced from Italian Mediterranean buffalo (*Bubalus bubalis*) milk. It is traditionally made from raw buffalo milk by adding natural whey cultures as starters obtained from previous day of manufacturing (back slopping method). After coagulation and curd fermentation until reaching pH values 4.9–5.1, the curd is sliced and placed in hot water and manually or mechanically stretched and shaped, and then placed in cold water to provide a first hardening, followed by immersion in a saline solution (Chapman and Sharpe 1981; Pisano et al. 2016).

In 1996, Italian Buffalo's Mozzarella cheese gained a Protected Designation of Origin (PDO) recognition under European Union (Commission Regulation EC # 1107/96) as cheese produced under conditions of transformation of buffalo's milk obtained from authochthonous animals bred in local farms (Pisano et al. 2016). For the mozzarella cheese to receive the PDO recognition, no additional milk (from other species of animals) is allowed and pasteurization is optional.

Additionally, the culturing should be made by adding starter resulting from spontaneously acidified whey from the previous day of cheese production. After coagulation and fermentation, the curd is cut, and cooked (Addeo et al. 2007).

The origin of buffalo mozzarella dates from the 12th century, when the monks of the abbey of St. Lawrence of Capua, in the province of Caserta in southern Italy, offered mozzarella cheese for peregrines, which was a less refined version of provolone and was not maintained for long time. By the 16th century, the buffalo mozzarella began to be served to Popes Pius IV and Pius V by Renaissance chef Bartolomeo S. (1500–1577). From the 18th century onwards, information about the production process of mozzarella and other buffalo milk products began spreading.

In 1895, Vicente Chermont de Miranda, imported the first buffalo herd of the Mediterranean breed to Brazil from Conde of Rospigliosi, the Italian mozzarella cheese producer (McManus et al. 2010). The size of the Brazilian buffalo herd in 2010 was estimated at 2.3 million heads, with an annual growth of 12.8% in the last 10 years (Andrighetto et al. 2005; McManus et al. 2010). The buffaloes are considered a good option for dairy farming, since their potential for milk production can be explored in various environmental conditions (Andrighetto et al. 2005). In Brazil, four races are recognized by the Brazilian Association of Buffalo Breeders (ABCB): Carabao, Mediterranean, Murrah and Jafarabadi (Maia 2014).

Due to the fact that physico-chemical properties of buffalo milk, the total solid content, fat and casein can be higher than that recorded for cow milk, buffalo milk is widely used for preparation of distinctive products, as mozzarella, provolone and ricotta that are commercialized in Brazilian marked (Andrighetto et al. 2005). However, In Brazil, only few dairy factories follow the typical Italian production process of buffalo's mozzarella cheese. The main production of mozzarella cheese is often by using pasteurized milk with addition of commercial starter cultures, composed of LAB which are responsible for the fermentation process of the curd (Silva et al. 2015).

Regulatory Affairs

In Brazil, the mozzarella cheese production is regulated according to the Regulation of Industrial and Sanitary Inspection of Animal Products - RIISPOA (Brazil 1997) and is made without distinction as to which milk may be used. There are no specific rules for buffalo mozzarella cheese. According to RIISPOA, "mozzarella type cheese is the product obtained from pasteurized milk, stretched curd and unpressed bulk, delivered to the consumer within five days of its manufacture, and should be presented with: variable format, between cylindrical and parallelepiped; weight from 15 g to 4 kg; thin crust, yellowish; semi-hard putty consistency; compact and closed texture; creamy white and homogeneous color and mild aroma and salty taste." (Brazil 1997; Maia 2014).

Physico-chemical Parameters

The buffalo mozzarella is characterized by significantly higher fat and protein content when compared to the mozzarella made with cow's milk. However, the pH, titratable acidity, moisture, soluble nitrogen and ash content do not show significant differences between the type of milk used. The buffalo milk mozzarella's stretching is relatively lower than that found in cow milk's mozzarella (Jana and Mandal 2011).

In Brazil, manufacture of mozzarella cheese on a commercial scale also uses both, buffalo and cows' milk. Buffalo milk contains approximately 16 to 20% total solids, whereas cow milk contains 11 to 13% total solids. In addition buffalo milk contains 40% to 60% more protein, fat and calcium than cows' milk, that contribute to a higher overall curd firmness, strength and quality to buffalo dairy products (Hussein et al. 2013; Hussein et al. 2012).

The mozzarella cheese produced in Brazil according to the standards required by the legislation, should present: 41.60% to 45.33% of moisture content, 23.43% to 24.66% of fat, pH between 5.06 to 5.09, and 54.66% to 58.40% of total solids (Mendes et al. 2015).

Buffalo's milk contains significantly more casein than cows' milk. It needs to be underlined that chemical composition and casein fractions play a vital role in curd formation. This is related to the higher protein, fat and total calcium content present in buffalo milk (Hussein et al. 2012).

The chemical composition analysis, made by using the gas chromatography mass spectrometer approach, indicated that the aqueous fraction of mozzarella, made by buffalo and cow milk, was composed of short-chain hydroxylated carboxylic acids, amino acids and saccharides. The main acids found were lactic acid, citric acid, succinic acid, fumaric acid, glyceric acid, gluconic acid and orotic acid. Also some amino acids, serine, lysine, valine, glycine, proline, threonine, phenylalanine, aspartic acid, and pyroglutamic acid were detected. Saccharides such as galactose, maltose, ribose, myoinositol and turanose were also identified, together with lactic acid dimer. The stearic acid, oleic acid and palmitic acid, belong to long chain fatty acids and are also present in mozzarella cheese (Pisano et al. 2016).

Buffalo mozzarella cheese presents superior organoleptic quality and nutritional values by the lower concentration of free fatty acids. Additionally, to minimize the bitter amino acids production, as leucine, phenylalanine, proline and valine, the calf rennet ratio is reduced (Jana and Mandal 2011).

Rheological Properties

The rheological characteristics of mozzarella, including texture of cheese, will depend on factors such as the chemical composition, microstructure (structural arrangement of its components), the effects of standardization and heat-treatment of milk, the physico-chemical state of the interactions of its components, such as type of acid and pH coagulation, whey draining acidity and proteolysis during storage, and its macrostructure (Jana and Mandal 2011; Maia 2014).

Mozzarella cheese is a visco-elastoplastic material at room temperature with visco-elastic characteristics at 60 °C. The yield stress texture gradually decreases with increase of temperature (Jana and Mandal 2011). During the curd formation, the buffalo milk presents more protein compared to the curd formation from cow milk. Furthermore, the gel network of buffalo mozzarella formation proceeds faster than the cow mozzarella because of the higher content of casein concentration that cause the gels to be stronger and faster forming. The buffalo's milk took less time to curd formation due to the casein micelles and fat globules be larger than that found in cow milk (Hussein et al. 2013). The final product shows a poor meltdown, fat leakage, acid flavor, free surface moisture and poor cohesiveness (Jana and Mandal 2011). The microstructure of mozzarella cheese exhibits large fat globules uniformly scatted trough the compact protein matrix with little aggregation (Jana and Mandal 2011).

Technology Process

The production of buffalo's mozzarella cheese is divided into two steps briefly described here. At the start of production, buffalo's milk is heated until it reaches 32–34 °C, then the natural whey starter, derived from the previous day's cheese manufacture (back slopping method) is added to it along with the calf rennet (providing microbial and enzymatic process of coagulation). Around of 45 min after the addition of chymosin, second stage begins when the curd is cut into 10–12 mm cubes and parcial drained to allow the fermentation. The curd-ripening phase is achieved when the curd reaches the pH between 4.9 to 5.1. This process takes about 4.0 to 4.5 h. Then the curd is sliced and manually or mechanically stretched and shaped in hot water at

90 °C to 95 °C. This step is followed by placing them in cold water (10 °C to 12 °C) for 30 min, followed by an immersion in a sodium chloride solution for 30 min (Jana and Mandal 2011; Silva et al. 2015; Pisano et al. 2016) (Fig. 3.1).

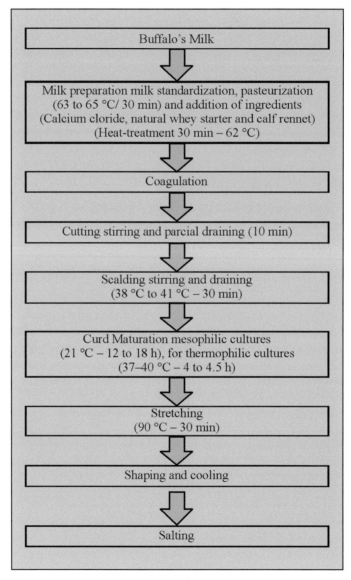

FIGURE 3.1   Flow chart of mozzarella production.

It is interesting to underline for the technological process, that the milk may be raw or heat-treated. However, the standardization and heat treatment of milk improves the cheese quality. It is recommended that milk be standardized to 3.0%–6.0% of its fat content, with the consideration that a lower fat content results in a harder texture and flavor (Jana and Mandal 2011). Aside from this technique, traditionally mozzarella was manufactured from raw milk; however it is recommended the pasteurization of milk for Mozzarella cheese that is to be consumed fresh, because the plasticizing process (stretching) do not always destroy potential pathogens. The homogenization of the milk could improve flavor and reduce fat losses in whey, and also increases the moisture content of the mozzarella cheese (Jana and Mandal 2011).

The coagulation process consists in transforming the milk into curd, which will be successively stretched until completely transformed into mozzarella cheese (Addeo et al. 2007). This process is done by rennet addition and can be improved by acidification with organic acids. The process might also be optimized by mixing both, rennet and organic acids (Callandrelli 2015). The higher content of solids in buffalo's milk results in better curd formation and curd moisture retention. The "back slopping method" employs thermophilic bacterial cultures present in a previous fermentation and is responsible for the reduction of the pH value of the pasteurized milk dropping from 6.5 to 4.9 in the mozzarella cheese (Van Vliet et al. 1997; Hussein et al. 2012). However, chemical acidification of milk is also a very common practice. Chemical acidification is achieved by the use of organic acids, normally lactic and/ or citric acid. Calf rennet or chymosin is used to produce the rearrangements of particle strands and lead to an increase in syneresis (Van Vliet et al. 1997; Hussein et al. 2012). The rennet employed is a mixture of various enzymes which perform a specific function on the milk proteins and the rennet could be of animal origin, such as bovine, or porcine pepsin or from vegetable origin as *Irpex lacteus* (Jana and Mandal 2011). The renin, natural rennet for the manufacture of cheese, is a proteolytic enzyme secreted by the gastric mucosa of the fourth stomach of ruminants, specially from calves and lambs prior to weaning. This secretion is produced as an inactive precursor, pro-renin, which while in the neutral range, has no enzymatic activity, however when introduced to an acid medium it rapidly becomes active renin. The calf rennet has two enzymes: pepsin and chymosin. Fungal rennet also could be used in cheese manufacture (Jana and Mandal 2011). Rennet can be industrial or farm-produced in origin, the ratio used in the coagulation process is 1:10000, rennet: milk (Callandrelli 2015). The retention of chymosin is affected by the curd's pH, the higher the pH the shorter the retention rate of chymosin is achieved. Consequently less casein proteolysis will occur and the residual coagulant still present in the final product that promotes a residual proteolysis and alteration of mozzarella cheese texture, even during brine and refrigerated storage (Addeo et al. 2007). Casein proteolysis has negative effects on curd stretching and mozzarella cheese properties, to limit the proteolysis process, using calf rennet with high levels of chymosin and very low of pepsin content was suggested. The pepsin has better activity in low pH, cleaving at aromatic amino acid residues and chymosin has limited proteolytic effect on casein at any pH value (Addeo et al. 2007).

The slow acid development of mozzarella related to the addition of organic acids or as result of the metabolism of bacterial cultures (LAB) decreases the rate of whey expulsion and could promote better moisture in the curd and in final product. This occurs by the stretching done at various final pH. A 0.2 pH variation does not affect the chemical composition nor functional properties of buffalo mozzarella cheese (Addeo et al. 2007). During the decrease of pH, ash, calcium and phosphorous concentrations also decrease and calcium ratio needs to be controlled to promote the best meltability of the cheese (Jana and Mandal 2011).

The cutting process transforms the curd into different small sized granules, the curd will then be scalded at 38 °C–41 °C to separate the whey expelled from the curd. High temperatures used in scalding decreases the proteolysis and the moisture of cheese (Jana and Mandal 2011). In mozzarella production, the cutting process is divided into two phases, in the first: the curd is cut into cubes with a knife, this is followed by a second cutting about 5 min later, the second cutting reduces the curd into nut-size granules (3–6 cm) with a curd knife. In next step, the whey is partially drained and the curd is put onto a table and the process of curd maturation starts. The progressive decrease in pH during curd ripening promotes the soluble colloidal calcium phosphate associated with the casein micelles to be released in the whey. Higher rates of calcium in cheese promotes an increased firmness in the cheese, making the cutting process of curd can be difficult with the decrease in softness of mozzarella cheese (Addeo et al. 2007). Usually, at a room temperature of 21 °C the curd reaches the right acidity after 12–18 hrs (Callandrelli 2015) a process promoted by the autochthonous microbiota, responsible for the transformation of dicalcium para-k-casein into monocalcium para-k-casein and this occurs during curd fermentation and affects the stretching of the cheese in hot water (Chapman and Sharpe 1981).

The mozzarella cheese is usually stored in a brine solution and during this storage period, the calcium content is crucial to its proper storage, due to micellar calcium decreases related to action of the acidity of the medium in which the mozzarella is, undergoing changes in its melting point and hindering the ideal melting point (Addeo et al. 2007).

The stretching process consists of placing the matured curd on a curd-draining table for 10 minutes, then cut into thin strips with a knife and transferred into a wooden or iron curd vat with boiling water (95 °C). The manual stretching is done with the help of appropriate sticks and tools, while the mechanical stretching is performed in a cheese stretcher-cooker which uses high-speed screws in oposite directions to shear the curd in hot water. The stretching is usually considered finished when the cheese becomes shiny and homogeneous. After that the shaping and addition of salt could be performed (Callandrelli 2015). During the stretching, the best titratable acidity is 0.8% lactic acid, and it can influence moisture and high melting quality (Jana and Mandal 2011).

The mozzarella cheese could be salted by direct or brine salting. The direct salting of mozzarella cheese is where salt is added to hot water during the stretching process. After draining the water, 1.6% of salt is sprinkled over the cheese during mixing. The brine salting occurs when the final product is immersed in a solution of 23% of NaCl at pH 5.1 in a trough for 30 min and stored in a solution of 1.0% of NaCl for 12 days or 4 weeks at a temperature of 4 °C. The brine salting promotes an increase in moisture and yield to the mozzarella cheese; however, the last process is very costly (Jana and Mandal 2011).

Autochthonous Microbiota

The fermented food matrices present a large range of metabolites produced by autochthonous and starter microbiota and could help in identifying the manufacturing procedures, ageing and detrimental processes. The microbial metabolites also contribute to the taste, flavour and rheological properties of final food products. The microbiota identification could indicate the original site of mozzarella manufacturing (Mazzei and Piccolo 2012; Pisano et al. 2016).

The composition of microbiota in mozzarella cheese is complex, variable and may be affected by the environmental conditions, cheese-making technology and the dairy's location site (Pisano et al. 2016). The microorganisms detected in mozzarella cheese are mainly composed of LAB and yeasts. The phenotypical identification of LAB is very difficult, because they present a lot of similar characteristics. On the other hand, omics technology could improve the knowledge about the autochthonous microbiota. Buffalo mozzarella shows highest microbial diversity of with regard to species when compared to cow mozzarella. Psychrotrophic bacteria could be found and were mainly represented by *Pseudomonas fluorescens*, *Moraxella* and *Serratia*, whereas *Bacillus licheniformis* was the most recovered psychrotrophic species. LAB strains accounted for 57.5% of total isolates in buffalo mozzarella compared to 24% in cow mozzarella samples. A higher diversity in yeast species was observed, in addition to *Rhodotorula mucilaginosa*, *Filobasidium globisporum* and *Candida sake* were recovered in both mozzarella cheese types, other species such as *Kluyveromyces marxianus*, *Candida lipolytica*, *Cattleya intermedia* and *Saccharomyces cerevisiae* were found only in buffalo's mozzarella. This large diversity, which has already been observed by other authors who reported *Kluyveromyces* and *Saccharomyces cerevisiae* as dominant yeasts in buffallo's mozzarella cheeses (Aponte et al. 2010; Romano et al. 2001), reflects the microbiological complexity of natural whey used as starters, the composition of which may be related to the production context and the dairy's location within the production district (Coppola et al. 1988; Pisano et al. 2016).

In Brazilian mozzarella cheese, buffalo and cow milk can be mixed or not for mozzarella production, generating higher autochthonous microbiota diversity, which is specific to the dairy's location and production methods.

Starter and Probiotic Culture

In last few decades, special attention has been given to the development of the new probiotic products. Since Metchnikoff gave attention to the LAB and yoghurt, focus on beneficial properties of fermented dairy products has increased. In the middle of 20th century, the beneficial contribution of LAB was well established scientifically and the consumption of functional dairy products that provide health benefits has increased significantly. The industry has begun looking for strains with probiotic characteristics for future application in functional foods. Probiotics are defined as live, non-pathogenic microorganisms that, when administrated in adequate amounts, confer a health benefit to the consumer (FAO/WHO 2002).

The commercial starter cultures as *Streptococcus thermophilus* and *Lactobacillus delbrueckii* subsp. *lactis* were found only in mozzarella produced by commercial mozzarella starters samples (Pisano et al. 2016). However, de Paula et al. (2014) reported on *Leuconostoc mesenteroides* subsp. *mesenteroides* and Jeronymo-Ceneviva et al. (2014) on *Lactobacillus casei*, *Leuconostoc citreum*, *Lactobacillus delbrueckii* subsp. *bulgaricus* and *Leuconostoc mesenteroides* subsp. *mesenteroides* isolated from Brazilian water buffalo mozzarella cheese with potential to be considered as probiotics. Role of *Lactobacillus* spp. as probiotics has been well established.

However, some epidemiological studies have shown the beneficial effects of *Leuconostoc* strains. The reduction of acute diarrhea in children was confirmed after the children were fed with Indian Dahi, traditional Indian fermented milk containing *Lactococcus lactis*, *Lactococcus lactis* subsp. *cremoris* and *Leuconostoc mesenteroides* subsp. *cremoris* (Agarwal and Bhasin 2002). Use of pre- and probiotics have been found to reduce bacterial infection rates after liver transplantation and in patients who have undergone other high-risk surgeries (Rayes et al. 2005, 2009).

The use of LAB as probiotic cultures by dairy industries has become more frequent, and it is acceptable to consumers due the production of flavor, and the aroma that is added by the culture (Shiby and Mishra 2013). Moreover, when present in dairy products, these microorganisms can produce lactic acid and other antimicrobial compounds, such as bacteriocins, which can inhibit undesirable microorganisms, thus extending the shelf-life of the products, and promoting therapeutic, sensory, and technological food benefits (Kos et al. 2007).

Some studies on the evaluation of *Leuconostoc mesenteroides* subsp. *mesenteroides* as a potential probiotic culture are still preliminary. This microorganism has important technological properties, such as production of dextran, acetaldehyde, diacetyl and acetoin, lipolytic and proteolytic enzymes, low production of acid and ability to grow under stress conditions (acid, high salt content and elevated temperature) (Nieto-Arriba et al. 2010). A few research studies have observed the probiotic characteristics of this microorganism (Agarwal and Bhasin 2002; Aswathy et al. 2008; Tamang et al. 2009; Shobharani and Agrawal 2011; Allameh et al. 2012; Seo et al. 2012). These characteristics draw our attention towards a better understanding of the potential probiotic properties of *Leuconostoc mesenteroides* subsp. *mesenteroides.*

In the study of de Paula et al. (2014) and Jeronymo-Ceneviva et al. (2014) special attention was given to the evaluation of basic characteristics in the selection of the potential probiotic strains. Additionally, the authors have paid special attention to the safety features of the studied LAB. They evaluated the survival of the LAB under study in the presence of ox-bile, pH, aggregation properties, hydrophobicity, reduction of cholesterol concentrations in human blood through bile salt hydrolases, halotolerance, ability to adhere to the Caco-2 cells, and production of β–galactosidase etc. (de Paula et al. 2014; Jeronymo-Ceneviva et al. 2014).

Nowadays, lifestyle, stress and an inadequate food intake are raising the consumption of different groups of medications for pain and other kinds of illness. However, many consumers undergoing these therapies are not aware of the side effects of these compounds. In study by de Paula et al. (2014), the *in vitro* tests showed that different medications can inhibit the *Leuconostoc mesenteroides* subsp. *mesenteroides* SJRP55, strain isolated from mozzarella cheese. Most of the tested analgesic, antipyretic and anti-inflammatory medicaments are commonly used by people of different ages ranging from infants to the aged. Most of these medications are freely available without the need for a prescription (de Paula et al. 2014; Jeronymo-Ceneviva et al. 2014). The negative effect of these drugs on potential probiotic LAB seems to be common, and has been observed in other studies (Todorov et al. 2011; Todorov et al. 2012). In addition to previous reports, de Paula et al. (2014) and Jeronymo-Ceneviva et al. (2014) investigated the interaction between LAB (*Lactobacillus casei,*

Leuconostoc citreum, Lactobacillus delbrueckii subsp. *bulgaricus* and *Leuconostoc mesenteroides* subsp. *mesenteroides*) isolated from the Brazilian water buffalo mozzarella cheese with commonly used, non-antibiotic drugs from different generic groups, with special attention to the MIC (minimal inhibition concentrations). Such information is important for the optimal application of conventional western medicine and using the beneficial properties of the probiotics.

Conclusion

A large amount of studies have been done on mozzarella cheese, and the regional specificity is important for the quality of the final product. A purity seal, similar to Italy was created by the Brazilian Association of Buffalo Breeders (ABCB) which, after the use of an electrophoresis test, is granted to dairies that exclusively use buffalo milk in their products (Buzi et al. 2009).

Production of mozzarella cheese is a complex process and quality of final products depends on the raw material, applied technology and specificity of the microbial starter cultures. Besides the high nutritional value, mozzarella cheese is highly appreciated by consumers due to its gastronomical characteristics and possible contribution to health which is provided by beneficial LAB within the cheese.

References

Addeo, F., V. Alloisio, L. Chianese and V. Alloisio. (2007). Tradition and innovation in the water buffalo dairy products. Italian Journal of Animal Sciences 6(2): 51–57.

Agarwal, K.N. and S.K. Bhasin. (2002). Feasibility studies to control acute diarrhoea in children by feeding fermented milk preparations Actimel and Indian Dahi. European Journal of Clinical Nutrition 56: S56–S59.

Allameh, S.K., H. Daud, F.M. Yusoff, C.R. Saad and A. Ideris. (2012). Isolation, identification and characterization of *Leuconostoc mesenteroides* as a new probiotic from intestine of snakehead fish (*Channa striatus*). African Journal of Biotechnology 11: 3810–3816.

Andrighetto, C., A.M. Jorge, M.I.F.V. Gomes, A. Hoch and A. Piccinin. (2005). Efeito da monensina sódica sobre a produção e composição do leite, a produção de mozzarela e o escore de condição corporal de búfalas murrah. Revista Brasileira de Zootecnia 34(2): 641 – 649.

Anonymous. (2012). A Mussarela - O queijo das pizzas. Pizzas &Massas 3: 28–42. http://www.insumos.com.br/pizzas_e_massas/materias/113.pdf. Last access 4 September 2015

Aponte, M., O. Pepe, and G. Blaiotta. (2010). Short communication: Identification and technological characterization of yeast strains isolated from samples of water buffalo mozzarella cheese. Journal of Dairy Science 93: 2358–2361.

Aswathy, R.G., B. Ismail, R.P. John and K.M. Nampoothiri. (2008). Evaluation of the probiotic characteristics of newly isolated lactic acid bacteria. Applied Biochemistry and Biotechnology 151: 244–255.

Brazil (1997). Regulamento da Inspeção Industrial e Sanitária de Produtos de Origem Animal - RIISPOA. Brasília: Ministério da Agricultura Pecuária e Abastecimento. 154.

Buzi, K.A., J.P.A.N. Pinto, P.R.R. Ramos, and G.F. Biondi. (2009). Microbiological analysis and electrophoretic characterization of mozzarella cheese made from buffalo milk. Ciência e Tecnologia de Alimentos 29(1): 7–11.

Callandrelli, M. (2015) Manual on the production of traditional buffalo mozzarella cheese. http://www.fao.org/ag/againfo/themes/documents/milk/mozzarella.pdf last acess: 30 August, 2015.

Chapman, H.R. and M.E. Sharpe. (1981). Microbiology of cheese. In: Robinson, R.K. (ed.), Dairy Microbiology. Applied Sciences, London, pp. 157–243.

Coppola, S., E. Parente, S. Dumontet and A. La Pecerella. (1988). The microflora of natural whey cultures utilized as starters in the manufacture of Mozzarella cheese from water-buffalo milk. Le Lait 68: 295–309.

FAO/WHO. (2002). Guidelines for the evaluation of probiotics in food. Report of a joint Food and Agriculture Organization (FAO) of the United Nations/World Health Organization (WHO) working group on drafting guidelines for the evaluation for the probiotics in food.

Hussein , I., J. Yan, A.S. Grandison and A.E. Bell. (2012). Effects of coagulation temperature on Mozzarella-type curd made from buffalo and cows' milk: 2. Curd yield, overall quality and casein fractions. Food Chemistry 135: 1404–1410.

Hussein, I., A.E. Bell and A.S. Grandison. (2013). Mozzarella-type curd made from buffalo, cows' and ultrafiltered cows' milk. 1. Rheology and Microstructure. Food and Bioprocess Technology 6(7): 1729–1740.

Jana, A.H., and P.K. Mandal. (2011). Manufacturing and quality of Mozzarella cheese: A review. International Journal of Dairy Science 6(4): 199–226.

Jeronymo-Ceneviva, A.B., A.T. de Paula, L.F. Silva, S.D. Todorov, B.D.G.M. Franco and A.L.B. Penna. (2014). Potential beneficial properties of lactic acid bacteria isolated from water-buffalo mozzarella cheese. Probiotics and Antimicrobial Proteins 6(3-4): 141–156.

Kos, B., J. Šušković, J. Beganović, K. Gjuračić, J. Frece, C. Iannaccone and F. Canganella. (2007). Characterization of the three selected probiotic strains for the application in food industry. World Journal of Microbiology and Biotechnology 24: 699–607.

Maia, L.R. (2014). Estudo do comportamento mecânico da muçarela de búfala obtida de massa fermentada congelada. M.S. Thesis. State University of Southwest Bahia, Itapetinga, Bahia.

Mazzei, P. and A. Piccolo. (2012). H HRMAS-NMR metabolomic to assess quality and traceability of mozzarella cheese from Campania buffalo milk. Food Chemistry 132: 1620–1627.

McManus, C., S. Paiva, H. Louvandini, C. Melo, L. Cassiano, J.R.F. Marques and L. Seixas. (2010). Búfalos no Brasil. INCT: Informação genético-sanitária da Pecuária Brasileira. http://inctpecuaria.com.br/images/informacoes-tecnicas/serie_tecnica_bufalos.pdf. Last acess: 30 August, 2015.

Mendes, B.G., K.A. Castro, K.A.L. Silva, A.I.A. Pereira and J.V.C. Orsine. (2015). Quality and performance of musssarela in times of acidified mass under refrigerated storage. Revista Brasileira de Tecnologia Agroindustrial 9(1): 1744–1756.

Nieto-Arribas, P., S. Sesena, J.M. Poveda, L. Palop and L. Cabezas. (2010). Genotypic and technological characterization of *Leuconostoc* isolates to be used as adjunct starters in Manchego cheese manufacture. Food Microbiology 27: 85–93.

dePaula, A.T., A.B. Jeronymo-Ceneviva, L.F. Silva, S.D. Todorov, B.D.G.M. Franco, Y. Choiset, J.-M. Chobert, T. Haertlé, D. Xavier and A.L.B. Penna. (2014). *Leuconostoc mesenteroides*: bacteriocinogenic strain isolated from Brazilian water-buffalo mozzarella cheese.Probiotics and Antimicrobial Proteins 6(3–4): 186–197.

Pisano, M.B., P. Scano, A. Muriga, S. Consentino and P. Caboni. (2016). Metabolomics and microbiological profile of Italian mozzarella cheese produced with buffalo and cow milk. Food Chemistry 192: 618–624.

Rayes, N., D. Seehofer and P. Neuhaus. (2009). Prebiotics, probiotics, synbiotics in surgery--are they only trendy, truly effective or even dangerous? Langenbecks Archives in Surgery / Deutsche Gesellschaft fur Chirurgie 394: 547–555.

Rayes, N., D. Seehofer, T. Theruvath, R.A. Schiller, J.M. Langrehr, S. Jonas, S. Bengmark, and P. Neuhaus. (2005) Supply of pre- and probiotics reduces bacterial infection rates after liver transplantation - a randomized, double-blind trial. American Journal of Transplantation 5: 125–130.

Romano, P., A. Ricciardi, G. Salzano and G. Suzzi. (2001). Yeasts from water buffalo Mozzarella, a traditional cheese of the Mediterranean area. International Journal of Food Microbiology 69: 45–51.

Seo, B.J., I.A. Rather, V.J. Kumar, U.H. Choi, M.R. Moon, J.H. Lim and Y.H. Park. (2012). Evaluation of *Leuconostoc mesenteroides* YML003 as a probiotic against low-pathogenic avian influenza (H9N2) virus in chickens. Journal of Applied Microbiology 113: 163–171.

Shiby, V.K. and H.N. Mishra. (2013). Fermented milks and milk products as functional foods - a review. Critical Reviews in Food Science and Nutrition 53:482–496.

Shobharani, P. and R. Agrawal. (2011). A potent probiotic strain from cheddar cheese. Indian Journal of Microbiology 51: 251–258.

Silva, L.F., T. Casella, E.S. Gomes, M.C.L. Nogueira, J.D. Lindner and A.L.B. Penna. (2015). Diversity of lactic acid bacteria isolated from Brazilian water buffalo mozzarella cheese. Journal of Food Science 80(2): M411-M417.

Tamang, J.P., B. Tamang, U. Schillinger, C. Guigas and W.H. Holzapfel. (2009). Functional properties of lactic acid bacteria isolated from ethnic fermented vegetables of the Himalayas. International Journal of Food Microbiology 135: 28–33.

Todorov, S.D., D.N. Furtado, S.M. Saad, E. Tome and B.D. G.M. Franco. (2011). Potential beneficial properties of bacteriocin-producing lactic acid bacteria isolated from smoked salmon. Journal of Applied Microbiology 110: 971–986.

Todorov, S.D., J.G. Leblanc and B.D. G.M. Franco. (2012). Evaluation of the probiotic potential and effect of encapsulation on survival for *Lactobacillus plantarum* ST16Pa isolated from papaya. World Journal of Microbiology and Biotechnology 28: 973–984.

Van Vliet, T., J.A. Lucey, K. Grolle and P. Walstra. (1997). Rearrangements in acid induced casein gels during and after gel formation. In: E. Dickison and B. Bergentahl (eds.), Food colloids: protein, lipid and poly saccharides. Royal Society of Chemistry, Cambridge, pp. 335–345.

4

Cheese Production in Uruguay

*Silvana Carro**

Abstract

In Uruguay the cheese industry produced 66,992 tons of cheese in 2014, most of which is generally for export. The artisanal cheese (AC) is, however, predominantly produced for the domestic market. Cheese making in Uruguay started with European immigration, in the zones of Colonia and San José. Regarding AC in Uruguay, studies have been conducted in order to assess their quality. In this chapter, the production of milk and the typical artisanal cheeses in Uruguay is described, as well as their characteristics and legislation requirements. We are currently working on a line of research, in which we have isolated Lactic Acid Bacteria (LAB) from artisanal cheese, raw milk and non-commercial starter cultures in samples from a group of farms at Nueva Helvecia, Colonia, Uruguay. The initial goal was to generate a collection of native LAB, and make it available and standardize the elaboration of AC. In turn, to improve these standards native LAB could be used to inhibit pathogenic or spoilage microorganisms. The study also aimed to evaluate the potential production of bacteriocin and to select those with promissory potential for cheese elaboration. The results of the study are promising, and progressing towards being able to use these native strains for cheese manufacture and to obtain safe products. Currently, this line of work is being developed and evaluating technological properties of some of these LAB and its applications in cheese making and to test its antimicrobial effect against *Listeria* spp. Moreover, livestock production in Uruguay has been characterized by sheep and cattle production although producers have sought new alternatives including the production of goat milk. In this chapter, some studies and results about the quality of goat milk and conservation are presented.

Introduction

Uruguay's land mass covers an area of 176,215 km^2 and has a population of about 3 million people. It´s located between Argentina and Brazil and bordered in the south east by the De la Plata River and the Atlantic Ocean. Its economy is based mainly on its agricultural trade, where milk production has an important role. In fact, Uruguay produces 430 L/yr of milk per capita, and has a high milk consumption of 250 L/yr/person (Fresco 2004; DIEA 2014 a). Uruguay has had a sustained growth

---

Las Places 1620–Montevideo–Uruguay-Z.P. 11600.
* Corresponding author: scarro@fvet.edu.uy

of milk production which has increased at an average annual rate of about 3% (DIEA 2014 b). At present, the dairy exports represent 9.9% of trade which corresponds to 900 million dollars. Around, 70% of the cheese produced is exported, which represents 30% of all dairy exports, semi hard cheese is the cheese in highest demand. In this chapter we will try to introduce the dairy history of Uruguay until today, milk quality, artisan and industrial cheese, types of cheese produced and its starter culturess.

Brief History of Milk Production in Uruguay

Cheese making in Uruguay started with European immigration. From 1881, the first Swiss settlers began to locate in Nueva Helvecia, Department of Colonia (Fig. 4.1). They brought their traditional practices: agriculture, animal domesticating for milk production, milking, cheese making and products marketing. Their main customers came from Montevideo (Borbonet 2001). Swiss immigration has been characterized by an appreciation of quality. This new way of doing things is also evident in the social organization, based on a strong sense of the community participation. Today, the city of Nueva Helvecia (Department of Colonia) is cleaner and more orderly than other cities and it is the regional epicenter for artisan cheese production (Arocena 2009). The cheese region was subsequently spread to Ecilda Paullier (Department of San José), Puntas del Rosario, Colla, Quevedo and Tarariras (Department of Colonia) (Fig. 4.1).

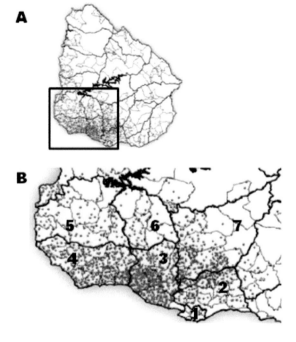

FIGURE 4.1 Distribution of commercial milk production in Uruguay (2010–2011). A) The image shows the Uruguayan politic map, divided into departments (wide black traces). B) Enlargement of remarked part in A. The numbers indicate the department names in which the main concentration of producers is located. 1-Montevideo, 2-Canelones, 3-San José, 4-Colonia, 5-Soriano, 6-Flores, 7-Florida. Each point represent two producers (DIEA 2014).

During those years there was immigration of other European origins: Spanish, Italian, etc. allowing the establishment of different types of cheese production (Borbonet 2001).

Cheese making regions have expanded to the South-West Coast and there institutions linked to milk and cheese have been created. For example, in 1930, the Society of Rural Development of Colonia Suiza founded the Colonia Suiza Dairy School, which nowadays instructs technicians for dairy industries. Today, in this region, there are other educational institutions: Faculty of Veterinary, University of the Republic (UdelaR) and Technologic University (UTEC Institute Tertiary Superior, created in 2011), both involved in teaching of dairy and dairy products at degree level.

Until 1910, the sanitary veterinarian control services of meat and milk for consumption depended on the **Experimental Institute of Hygiene** of Montevideo. With relation to human health, the main problem was bovine tuberculosis, due to poor hygiene practices in the place where cows were milked and milk distribution (Bertino and Tajam 2000).

In April 1910, the law about sanitary conditions of animals was created, and policed by the **Livestock Division of the Ministry of Industry, Employment and Public Instruction**. The law stated the obligation of owners and veterinarians to report the existence of sick animals. The government, could then declare the property, department area or corresponding department infected, and isolate, vaccinate etc., the animals in the infected zone, as well as prohibit the transport of these animals and animal fairs, disinfect the properties and euthanise animals in certain cases, within the established indemnifications (Bertino and Tajam 2000).

In 1911, the Uruguayan government created regulation rules linked to dairy production, which included considerations about animal welfare, building regulations and milk quality concept (Casaux 2005).

In the years 1911 and 1915 milk sale in the capital city was regulated, sales could only be conducted in tanks with faucets, closed with sealed tops, and sale was prohibited in open containers. The maximum price of milk then was established daily by the Supply Direction and Tablada (Bertino and Tajam 2000).

In 1922, the first milk processing companies were established in Montevideo. In 1925 the first pasteurizing company, the Central Dairy Uruguayan Kasdorf S.A. was set up, and in 1927, the activity was regulated. Later, in 1930, the Cooperative Milk S.A. was founded when pasteurization became obligatory. In April 1932, the Dairy Cooperative Melo (COLEME) was founded and in December 1936, the commerce of pasteurized milk started. However, on 25 April 1933 was stated that only in the city of Montevideo pasteurization needs to be effectuated immediately and this process started from 1 of January 1934. In 1934, apart from the previously mentioned companies, there were others in nearby Montevideo: "Mercado Cooperativo S.A", "La Palma S.A", "Alianza de tamberos y Lecheros de la Unión" and "La Nena" (Bertino and Tajam 2000; Borbonet 2001).

In 1935, during the dairy crisis, CONAPROLE (National Cooperative of Milk Producers) started operations. Interesting, the creation of cooperatives was supported by State and established by law (law nº 9526), as a way to overcome the dairy crisis. This was a key contribution to further develop a strong dairy industry in Uruguay The main purpose of CONAPROLE was to organize the marketing, processing and distribution of milk in the city of Montevideo. Most industrial plants had to transfer their activity to CONAPROLE (Bertino and Tajam 2000; Casaux 2005; Martí 2013; Viera et al. 2013).

With the creation of CONAPROLE, the state fixed the price of milk: to the producers, to the consumers and for intermediate stages of commercialization (Bertino and Tajam 2000). Thus, differentiated prices for the raw materials were regulated, stipulating the amount of liters of milk (quota) devoted to consumption. In the case of CONAPROLE, the milk "quota" was an average of the volume the company accepted from each of the producers. This average was calculated based on the quantity of milk sent during the winter months, where milk that was not going to be sold for consumption was proportionally deduced. This function required a continuous acquisition of milk throughout the year and ended in an instrumentation of a double system of prices, named "quota" and "industry". The price of "quota" milk was fixed with the objective of warranting milk supply (and for this reason it was always higher), and compensating the producer for the higher costs of production during the winter. The "industry" milk was devoted to the production of dairy products (cheese, yogurt, etc.).

Milk to the consumers is fixed with the objective of warranting its availability and accessibility. However, in the middle of the 70s the price of "industry" milk was set free and at the beginning of the year 2008 the price of "quota" milk as well was set free. Today, the price of milk for consumers is the only one still regulated (Viera et al. 2013).

CONAPROLE has been the most important industry of Uruguay. In addition to this, it has been the largest dairy company in the country along its entire existence, and the main exporter of the country for long periods (Viera et al. 2013).

From 1935 to 1950, milk production increased further towards self-sufficiency in Uruguay. In the period 1950–1975, milk production was stable without significant changes. Since the early 70´s our country was self-sufficient in milk production and became a strong exporter as well to the extent that approximately 60% of the milk industries are entering international market today, reaching 65 markets across the continents in 2011 (CINVE 1987; Uruguay XXI 2012).

Local dairy areas developed around the most populated cities, many with pasteurizing plants. There was a geographical expansion of watersheds while a technological renewal occurred, resulting in an increased referral of industrial plants during 1975–1985. All of which would lead for a greater impact into foreign markets by Uruguayan milk (CINVE 1987).

Viera et al. (2013) states that since 1975 to the present day, there has been a new spread of milk production which is based mainly on the technological transformation of the places where cows are milked.

The dairy industry as a result of its strong export profile has achieved internationally accepted quality standards and subsequently a presence in world markets. Taking into account the national market, the main market for milk is the capital city of Montevideo, where more than half of the population of the country live. For this reason, an important milk catchment area has been developed which has the production concentrated in Departments (administrative division of Uruguay) Colonia, San José, Soriano and Florida (Fig. 4.1) (DIEA 2014b).

On the other hand, in relation to the quality of milk in Uruguay, the first actions were taken in 1963 for qualified milk, taking into account only animal sanitation and the infrastructure of the cowsheds where they were milked. In 1976, another important step was taken by the Departments; who paid an addition of 10% for milk to be tested for redutase and lactofiltration; this incentive resulted in radically decreased the bacteria count. For over twenty years the Department continued to pay for the mentioned tests. It was not until 1993 that studies were started to pay to measure for the hygienic quality of the milk (Mesa Tecnológica de la Cadena Láctea 2005). In this context, the industry instrumented registration and classification of milk through parameters, as total aerobic mesophilic bacteria count in milk to evaluate its hygienic quality; somatic cells count (SCC) with the objective of verifying the sanity of the mammary gland and the detection of antibiotic residues (Ibarra 1997).

The characterization of milk through certain parameters of raw milk quality made it possible for the National System of Quality Milk to improve the quality of milk sent to dairy plants; this was implemented by decree in 1995 (MGAP 1995).

At the same time based on this decree, the **Ministry of Livestock, Agriculture and Fisheries (MGAP)** made a resolution which put the Veterinary Laboratories Division (DiLaVe-MGAP) in charge of the training the private laboratories to determine the milk quality parameters (MGAP 2001).

Milk quality has been improved immensely since the implementation of the National Quality System (1997), which classifies milk into categories according the bacteria and the somatic cells count, to the extent that the fixed limits designated at the beginning have been modified as part of the quality improvement (MGAP 1995; MGAP 2013).

According to data from MGAP, in Uruguay the averages of total aerobic mesophilic bacteria count decreases immensely (almost 90%) since the setting up of The National Quality System until 2003 (DIEA 2003). With the new system, in which the total bacteria count and the SCC taking into account, a fundamental stage in the improvement of milk quality (milk to be processed in dairy plants) was accomplished.

As a consequence, almost 90% of milk sent to dairies by producers has less than 50,000 UFC/mL of somatic cells and nearly the 80% of milk sent to dairies plants fulfills the international standard of having less than 400,000 somatic cells/mL (INALE 2015). At the same time, a technological revolution and management of herds occurred, which has consequently generated an improvement in the quality of raw milk. With the addition of the legislation, an important boost in quality was added to an industry already looking for quality.

In fact, the Uruguayan government has recently approved Decree 359/13 (MGAP 2013) which has 14 articles, the sixth article established the maximum total bacterial count and how it will be adjusted in future years (Table 4.1) (MGAP 2013).

TABLE 4.1 Maximum values accepted for bacteria and somatic cells count according to Decree 359/13 (MGAP 2013)

Date	Bacteria Count (UFC/mL)	Somatic Cell (cells/mL)
November 2013	500,000	800,000
November 2014	300,000	600,000
November 2016	100,000	400,000

Finally, it is important to consider that during the last decade of the 21st century, the milk sector suffered a series of setbacks for milk production, such as strong variations in climate conditions and prices for milk producing. However, the growth that characterized milk industry since the first half of the 20th century was maintained (Zorrilla de San Martín 2013).

Nowadays, 70% of produced milk is exported to multiple markets and the technology and organization of the primary production model is one of the most efficient at the world level.

Cheesemaking in Uruguay

In Uruguay, the cheeses produced in a greater volume are those of the semi-hard variety (39274 tons), soft variety (14678 tons) and hard variety (4092 tons) (DIEA 2014 a). Cheese exportation is related to industrial production, while the artisanal cheese is for the domestic market, either by the producer via direct sale or in urban centers through sales by intermediaries dedicated to the cheese industry. The most common cheese types are Sbrinz (grana cheese or semi-hard cheese), Danbo and Colonia (PACPYMES 2007). Around 54% of Uruguayan families consume different types of cheese weekly, which contributes to the demand for artisanal cheese. In the dairy zone par excellence (Colonia and San José) and in the main consumer market (Montevideo), higher cheese consumption levels have been observed. Although the national demand grows, the consumption of cheese per person in small cities and rural areas without tradition for cheese, is still about 15% below the national average. The estimated volume is about 20 kg/yr/household and 6.5 kg/person/yr (FONADEP 2008).

It is estimated that over 1,500 producers devote to the artisan cheese making tradition (DINAPYME 2013), as previously mentioned, artisanal production began as a tradition and also as a technique of preserving milk, and it has subsequently turned into a profitable business that has been adopted as a way of life for many families in the areas where immigrants arrived. As a result of this, producers are identified with the activity of cheese making and played a very important role in the development of artisanal cheese making, without losing its relation to the dairy industry at a more general level. In addition, there are approximately 5,000 people directly involved with the production of artisanal cheese, often the defining feature being the family workforce. There is a strong presence of women in the development of these products along with a significant number of young people working with other cheese related tasks in this endeavor. The presence of women and young people in this industry implies the families roots to the rural area (Jeruselmi et al. 2008).

Since 1990, the artisanal cheese making industry accompanied the growth of the dairy industry in Uruguay, and became a sub sector of agricultural production, which was slowly showing growth nationwide. Among the difficulties found in the artisanal cheese industry, there were also problems concerning livestock health, infrastructure of the farms ("tambos") and dairies; and hygiene habits employed in cheese making. This situation affects the desirable safety of a high protein value product as cheese (Anchieri et al. 2007).

According to Decree 315/994 of the National Bromatological Regulation (RBN) (MSP 2009), *cheese* is "the fresh or matured product obtained by partial separation of the whey or reconstituted milk (whole, partially or totally skimmed) or whey, that is coagulated by action of rennet, specific enzymes, specific bacteria, organic acids alone or combined, all of them with food grade quality".

This RBN decree, states "the artisanal cheese is cheese made in artisanal conditions in individual, family or associative manner, except for mass production involving industrial facilities and processes". In turn, Decree 65/003 of MGAP 2003 (MSP 2009), states artisanal cheese is that "which is made from raw, pasteurized or thermized milk, produced on the farm, exclusively". It is evident from the existing differences in these definitions, that in the first one there exists the possibility of association (milk from several farms) while in the latter one is restricted to milk produced in the same establishment or farm. However, MGAP Decree 65/003 indicates that the artisanal cheese producer is "any natural or legal person that produces artisan cheese in an individual, family or associative manner". In turn, this decree defines as cheese producer or middleman "any natural or legal person processing cheese in artisanal or industrial amount for commercialization. This activity includes the maturation or packing of cheeses. The decree also defines as carrier "any natural or legal person transporting artisanal cheeses or cheeses from industrial plants from the place of its manufacture or collection, to the site of collection or marketing" Finally, it defines as cheese processors "any natural or legal person applying one or more of the transformations or processes to the artisanal cheese, or cheese from another origin, such as splitting, shredding or melting."

The artisanal cheese (AC) making should preferably be made with pasteurized milk, using autochthonous lactic acid cultures, prepared by the producer themselves or by any other competent body or company. Pasteurization can be excluded when ripening times exceed 60 days and when the use of raw milk and/or shorter ripening times are needed to obtain the intrinsic product features. Cheeses made from unpasteurized milk or with shorter ripening times should be evaluated at an official or renowned laboratory (MSP 2009).

During the preparation of artisanal cheese, it is essential to respect optimum hygienic and sanitary conditions in order to ensure the safety of the product. In this regard, in Uruguay, there are programs created some years ago to support the AC process. They focussed on that required help in activities for AC licensing, as well as the implementation and application of Good Manufacturing Practices (GMP), these programs help in all aspects of quality improvement to be appreciated in the domestic market (with added value) and assist in being able to gain space in the export market.

AC Licensing Decree 65/003 (DIEA, MGAP, 2003) regarding the "tambo" or dairy farm producers of bovine, ovine, caprine or buffalo milk and artisanal cheese makers should provide among other information, the health attestation of livestock, health certification of the staff members, report of the physical, chemical and microbiological analysis of water and the name of the person responsible for the cheese making process. As for the cheese making facilities, physical, chemical and microbiological analysis of water from the property, information on the production process performed, pest control and water control program is also required. Regarding the product control, Decree 274/004 (MSP 2009, MGAP 2010), amends the scope of aerobic mesophilic bacteria in raw milk (1×10^6 CFU/mL), total coliforms (10^4 CFU/mL) and *Staphylococcus aureus* (10^3 CFU/mL), the decree also sets new requirements for collectors, processors, artisan cheese makers, as well as the microbiological requirements for AC made with raw milk (Table 4.2). The presence of enteropathogenic *Escherichia coli* and staphylococcal enterotoxin in these products is not allowed.

TABLE 4.2 Microbiological requirements established by Decree 274/004, MGAP for AC

Microorganism	Requirements
Staphylococcus aureus	n* = 5
	c* = 2
	m* = 1000 CFU/g
	M* = 10000 CFU/g
Escherichia coli	n = 5
	c = 2
	m = 10000 CFU/g
	M = 10000 CFU/g
Listeria monocytogenes	n = 5
	c = 2
	m = absence in 25 g

* n: number of sample units that are to be drawn independently and randomly from the lot;
  c: marginally acceptable units with microbiological concentrations between m and M;
  m: marginally acceptable quality;
  M: unacceptable microbiological limit.

The values presented in Table 4.2 are higher than those established for cheeses made by the industries. Therefore Decree 315/994 (MSP 2009), specifies the values according to cheese moisture content, e.g. coagulase-positive *Staphylococcus* in low, medium and high moisture is specified as: n: 5, c: 2, m: 100 CFU/g and M: 1000 CFU/ g. This might be because the conditions of some of these producers are covered in the Decree, therefore regulations towards them have been more flexible.

In addition, the identification of artisan cheese is made by marking on one side of the cheese the registration number, which has to be printed immediately after the cheese elaboration (this registration will be awarded at the time of the licensing and it consists of a capital letter in print, for the Department where the property is located, followed by the serial number which corresponds to the business license number). The cheese will also clearly display the date of manufacture. In the case of cheese or products where the condition or classification is required, it will also carry an expiration date. Cheeses made from sheep, goat, buffalo or mixed milk should include the species from which the milk come from by indicating their respective proportions in descending order. Cheeses made in facilities exclusively authorized to produce cheese for processing will also be identified allowing traceability of the finished product.

The processor must also keep records of the raw material used in cheese making, in which establish the origin of the product. To maintain the license, the hygiene and sanitary documentation is required, and it is revised annually. Every two years, the collectors, processors of cheese, or the professional responsible for the establishment must submit documentation that ensures hygienic-sanitary conditions are kept. In this same Decree 65/003 of MGAP (MGAP 2003), the conditions of the facilities for processing AC and establishments of collection were described.

Chapter 3 of the decree states that the collectors and processors of cheese must have a GMC Handbook and comply with the requirements of the Health Authority Officer (ASO), designed by the company (Cleanup Program, Pest Control, etc.). This handbook will form the basis of operations for the establishment, on which the authorization shall be served and shall be permanently available in the company to inspectors accredited by the ASO. All operations to be performed within the facility must be described in the GMP Handbook. Operations should be described and conducted in accordance with the provisions of current regulations.

Regarding the conditions during the manufacture of artisan cheese, among others, it states all artisan cheese producer must use ingredients, additives and materials accepted as safe for public health. Hygiene conditions must be described in the GMP Handbook. Methods of cleaning and disinfection used in the establishment should also be detailed, including the products used and the frequency in which cleaning is done. All cleaning and disinfection products must be approved and registered for use in the food industry. In addition, all establishments shall implement a pest control program; this program should be described in the GMP Handbook. All products for pest control must also be approved and registered by the ASO.

The road to licensing has been possible for many of the AC by the support they have had over the years until now by different institutions, such as groups formed by the cheese makers. The association of artisan cheese makers had its beginning in the 1980s with the "Union of Artisanal Cheese Producers. In turn, the Society for the Promotion of Colonia Valdense, the Technological Laboratory of Uruguay (LATU) and the Cooperative Veterinary Laboratory of Colonia (COLAVECO) worked together with the aim of identifying artisan cheese makers and putting that information on a map. Then in 2000 the "Group of 30" was created, which brought together 30 producers belonging to the artisanal cheese making activity with initiative in various areas. This partnership in this sector has not been limited to links between producers" (Jerusalmi et al. 2008).

In 2003, it was founded the "*Mesa del Queso*", whose official name is "Regional Development Program of Artisanal Cheese in the East of the Department of Colonia and the West of the Department of San José, integrated by the National Cheese Makers Association of Uruguay members, Artisanal Cheese Makers Union, the Artisanal Cheese Makers of San Jose Association, the Artisanal Cheese Makers of Ecilda Paullier and Cheese Makers Association of Ismael Cortinas, collectors, processors and Municipal Governments of San José, Colonia, Soriano and Flores" (Jerusalmi et al. 2008; PACPYMES 2007)

The Board's objectives are: to formalize the artisanal cheese making, to integrate the different players, to better position their products in the local market, working to export quality products, to increase the technical and productive capacity of small farmers, to consolidate a valid interlocutor that represents the artisanal cheese production, and to support improvements in physical infrastructure of the farms, "tambos" and dairies.

With this in mind, in 2006, they began a cluster initiative in the community supporting the development of artisanal cheese making in areas of San José and Colonia. It had the approval of the Program for the Support to the Competitiveness and Export Promotion (PACPYMES), which originated from a bilateral cooperation agreement between the European Union and the Uruguayan Government, and subordinated to the Ministry of Industry, Energy and Mining (MIEM) (PACPYMES 2007).

Also in 2006, the Board of cheese of the littoral was created, which was a "promotion stage" intended to consolidate the Programme for the Development of Artisanal

Cheese. This is part of the Development Plan of the Dairy Industry of the Littoral and is one of the strategic lines of local governments (Government of Paysandú), so funds from FONADEP are being requested to enhance the cheese industry in the region. The project *"100 Queserías"* of LATU aims to give technical assistance to the dairies of the area (FONADEP 2008).

Likewise, with creation of the National Dairy Institute (INALE) in 2008, where artisanal cheese makers are also represented, supported the associative matter (Jeruselmi et al. 2008). Moreover, the document of FONADEP (2008) "the Development of Artisan Cheese of the Dairy Littoral Basin, Departments of Paysandú, Salto and Rio Negro" reports the existence of a wide variety of producers that are also developed in other areas, including the west coastal region of Uruguay (Rio Negro, Salto and Paysandú). The latter has been enhanced by the presence of 77 producers of cheeses, of which 74% are located in Paysandú, and 18% in Río Negro and the rest in Salto.

In general, there are two types of products in the Uruguayan Artisanal Cheese Making industry, those who make a basic cheese or curd which is sold to small local industries for "processed cheese" and those who produce artisan cheese itself. The former differs from the latter by having lower costs in terms of infrastructure (storage rooms, etc.). The limitations in this sector are factors such as the low production volume (cheese kg by dairy or "tambos"), variable production quality and sometimes poor handling with regard to product safety. In addition, one of the critical points of the artisanal cheese is milk pasteurization. One of the conditions that determines the quality of artisan cheese is in fact the use of raw milk, given that pasteurized milk may lack bacteria that provides characteristics of aroma and texture to artisanal cheeses (PACPYMES 2007).

The following sections describe the main characteristics of Grana, Colonia and Danbo cheese types.

Grana-type Cheese (Sbrinz Cheese or Reggianito Uruguay)

It is a low moisture cheese or hard cheese it which is originally from Italy and Switzerland. Adaptations have been incorporated in Uruguay (Borbonet et al. 2010). It is characterized by 35.9% maximum moisture, and high fat content of 45 to 59.9% fat in dry matter (FDM). It is made with raw, thermized or pasteurized cow milk with 2.7 to 2.8% of fat, natural whey culture, calcium chloride, genetic or animal coagulant, vegetable dye and sodium chloride. It is cylindrical, weights 7 to 9 kg, and presents smooth rind, clean, straw colour, closed texture without eyes, brittle and grainy, hard consistency, salty and slightly spicy, and pleasant aroma. The minimum ripening time is 6 months (Fig. 4.2).

It is important to consider that the starter or culture applied in many of the AC in Uruguay is called whey culture, which is used in a similar manner in other countries (Italy, France, Switzerland, Greece). Whey-culture is, a blend of native lactic acid bacteria (LAB) contained in whey of previous lots of cheese that replicates as inoculum for the new batch processing. Hence, the "whey culture" of the cheese making process for the day is incubated at a high temperature (45–52°C) for a period of 4.5 to 18 hrs, depending on the crop, to be used in the preparation of the following day.

Cheeses are generally made from raw milk, and therefore, these cultures depend on the presence of LAB (Fox et al. 2000).

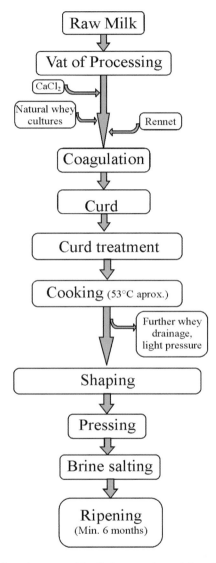

FIGURE 4.2   Manufacturing protocol for Sbrinz type cheese (adapted from Fox et al. 2000).

Danbo Cheese

This cheese, whose origin is Denmark, has been adapted to the conditions of Uruguay. It is made with cow's milk and belongs to the "washed-rind" cheeses because during its manufacturing process after cutting the curd, it is "washed" so part of the whey is removed and hot water (70 °C) is added. A semi-cooked mass is obtained and after partial removal of whey, pre-pressed under whey, molded, pressed, salted and ripened. This cheese is soft and semi firm. It has small eyes, well dispersed or no eyes. It has a yellowish white color and soft flavor, typical of the variety (Borbonet et al. 2010).

Colonia Cheese

Colonia cheese is a typical Uruguayan cheese, made according to the quality criteria and to the Swiss cheese tradition of immigrants residing in the Department of Colonia. It is a cheese made from raw, thermistor pasteurized cow's milk, with 2.6 to 2.8% fat, generally cylindrical in shape, with slightly convex sides, weighing 7 to 10 kg, clean, flexible smooth rind, and a straw yellow smell. The cheese is soft, semi firm with low acidity. Its texture is opened with spherical and bright eyes of 6–8 mm in diameter, presents 36% to 46% of humidity, 45–55% fat in dry matter, and ripening time is 20 to 30 days. It has natural external characteristics or plasticized red rind (Borbonet et al. 2010). Its process has not yet presented any origin denomination and in general the *starter* bacteria are *Lactococcus lactis*, *Lactococcus lactis* subsp. *diacetylactis, Leuconostoc* subsp. *mesenteroides* and *Propionibacterium* (responsible for the formation of eyes) (Crosa et al. 2009).

At present, in the Group of 30 includes the fourth generation of artisan cheese makers, descendants of European immigrants who founded the Swiss Colony. Many of them have inherited the art from their great-grandparents and grandparents who had arrived in the nineteenth century in Colonia and the surrounding area, becoming the quintessential cheese zone. In this area the typical cheese "Colonia" was generated, which remains as the best-known product of the region. Although Parmesan, Danbo, Cuartirolo and Parrillero cheeses are also produced, Colonia cheese is a product of the particular conditions which artisan Swiss cheese makers found in the country. The existing pastures and animals resulted in milk with different characteristics, so they had to adapt their procedure of cheese making; therefore, the flavour and texture properties were different, and the Colonia cheese became a typical product of Uruguay (Bielli 2009).

In general, the technology involved in the production of milk has long been characterized as being inferior for delivering producers to industrial plants, although there are those who are group leaders. Settlement of the family on the farm and the tradition and experience of the players involved in the system stands out. Because of their small size, most of the producers present weaknesses in infrastructure and technology, which affects production and health aspects, and impacts negatively in the access to credit and training (PACPYMES 2007).

Informal aspects of production and marketing and lack of traceability are some examples in which some producers must work to achieve quality cheeses in order to conquer foreign markets. However, there are producers called leaders who have a production capacity in line with international requirements for quality and safety, and can have a significant volume for exportation, in fact, they are exporting the artisanal cheese directly. These are the producers who have led the generations of various associations of producers of artisan cheese, and often a producer belongs to more than one organization. Their obstacle was the difficulty in marketing their products and the low margin obtained individually. These producers created an organization called "*Unión de Queseros Artesanales*" that included: Producers, buyers and input suppliers (PACPYMES 2007).

At the same time, there is a major boost in the industry by the Government; the Interior Development Fund participated in projects with departmental governments (Colonia, San José, Flores, Soriano) and also with non-state public entities (Technological Laboratory of Uruguay (LATU), Agricultural Plan Institute, the Program for the Support to the Competitiveness and Export Promotion (PACPYMES) and International Institutions. In this network public policy institutions also have an influence that regulates and defines the framework for the sector. LATU began its support in 1990 advising producers of AC to improve their facilities in line with the international standards of quality and safety.

The MGAP (Ministry of Livestock, Agriculture and Fisheries) took over some of the functions from LATU regarding the certification of health and quality of exportation type cheeses. This is more consistent with the demands of governments in the international markets that require Ministries to assume the roles of Official Health Authority.

The Colonia Suiza Dairy School is unique in the country, it depends on the public education and it's located in the cheese cluster region, and trains dairy technicians to work in the dairy sector. COLAVECO (Cooperative Veterinary Laboratory of Colonia) organization provides clinical services in animals, milk, cheese, water and others. It is a major player in the extension activities of animal production, linked to the quality assurance processes and products (PACPYMES 2007).

Sheep milk Cheese

In Uruguay, sheep production is traditionally oriented to wool, meat being a by-product of the system. In 1987, due to the low international price of wool, the first attempts of sheep milking were started, mainly the Corriedale breed. The milk produced in those years has been mainly destined to the production of ripened (Manchego, Sardo, Pecorino type) and fresh cheese (Feta type).

In 1991, some practical experiences started to be implemented not only in obtaining products derived from sheep milk but also some goat milk with producers interested in a plant in the Rural Development Society of Durazno, and their intention was to continue with this line of work. In Uruguay, these kind of products had been a growing interest and there were two companies producing sheep's milk cheeses for the foreign markets. One company had already been authorized to export sheep cheeses to Brazil. In 1992, with funding from INIA a sheep dairy was set up in the Experimental Field N° 1 (Migues, Canelones Department) of the Faculty of Veterinary, UdelaR. In the dairy, they performed activities intended to evaluate management, health, feeding, breeding and genetic alternatives to familiarise Uruguay in this non-traditional production system. First, it was by the milking of the Corriedale breed and, since 1993, it began the insemination with the dairy breed Milchschaf (Kremer 1995). Corriedale breed was developed in New Zealand, and was obtained by crossing Merinos with long-wool rams. It is well known for being a widespread breed, it is estimated that it ranks second in global stocks, after the Merinos, it is a dual-purpose animal (meat and wool), with a good body and robust development, and a milk production of 0.7 liters per day (Kremer 2005). From the milk of the flock of sheep, the Sardo cheese type which is a hard, fatty, prepared with whole milk, acidified by lactic bacteria culture and coagulated by rennet and/ or specific enzymes was developed. It has a minimum aging of 90 days, is cooked, molded, pressed, salted and matured curd cheese; it is hard, consistent, crisp and grainy. It has a characteristic spicy flavor, pleasant, clean, well-developed flavour and white yellowish colour (Fig. 4.3).

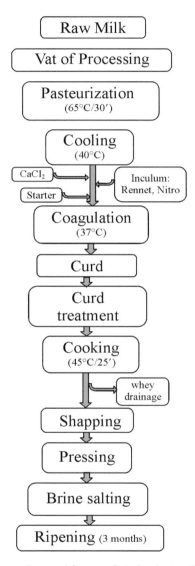

FIGURE 4.3 Proposed protocol for manufacturing the type Sardo sheep cheese.

Other cheeses produced and currently available in the Uruguayan market are acidic fresh cheeses (Feta and Ricotta) and ripened cheeses (Pecorino as an example).

Sheep milk has a number of advantages compared to cow milk, it has a greater cheese making ability, which is evidenced in its higher yield (1 kg of cheese per 5 liters of sheep milk, whereas 10 liters of bovine milk are needed for equal yield);

regarding technological parameters, sheep milk responds differently by coagulating faster, due to its higher protein content.

The two main components that affect cheese yield are casein and fat. Casein is the key component in the process of curd formation which traps fat globules; in the casein we pay attention to the relationship with the other components of the milk to predict the potential quality and cheese yield. The casein/fat (C/F) ratio is critical in controlling the final fat in dry matter (FDM) of the finished cheese. The specifications of FDM are defined for many cheeses with identity standards that already exist in several countries (Wendorff 2003).

Currently, the Uruguayan consumer has no habit of consumption of cheeses made from sheep's milk, because the taste is "stronger" than cheeses made with cow's milk. Moreover, domestic production of sheep cheeses does not reach volumes that allow exporting the product, so the production is oriented on a small segment of the population, preventing the sheep dairy from developing as a profitable activity. From the total volume, sheep cheeses represent about 0.4% of cheese production. It has been evidenced that increasing of production of this type of cheese is related to the demand behavior of the current consumer which presents an increment in demand. Among them are also included niche products, also known as "delicatessen", whose market position is based on its sales margin and not in its sales volume. These products are derivatives made from sheep's milk, among which there are a variety of highly valued, accepted and growing demand products (Kremer 2005). The sheep's milk is an alternative category in Uruguay and the development of cheese of this species has been an interesting possibility in the domestic and other markets. Certainly the types for elaboration should be standardized to keep this cheese on the inside of the market.

On the other hand, Bermudez and Reginensi (2014) indicate that sheep's milk is a productive alternative for small farms in Uruguay, because they normally require less investment for implementation. They also make reference to cheese making in particular; there are two main approaches, on one side imitation of products from other countries, mainly European and on the other, the development of authentic products (genuine) based on the re-engineering of specific biological aspects associated with the "territoriality". This includes the specific regional product (Paxon, 2010, cited by Bermúdez and Reginensi, 2014). These authors have developed a culture for use in the manufacture of cheese with sheep milk, characterizing lactic acid native bacteria (Reginensi et al. 2013).

Goat Milk Cheese

Livestock production in Uruguay has been characterized by sheep and cattle production although producers have sought new alternatives including the production of goat milk. In 1987 started the importation of specialized dairy breeds for raise them as purebred and also to make crosses with native goats (chivas) which were in some regions in semi-wild state.

The main imported breeds were: Anglo Nubian, Brown Alpine or French Alpine, Saanen and Toggenburg. The goat milk producers are concentrated in the southwest region of the country (approximately 16,000 km^2), this region has a long dairy tradition. The milk is usually transported frozen to other facilities either for direct sales or for raw materials, and so it was possible to mantain the production of artisan cheeses continuously throughout the year (Ciappesoni, 2006).

In Uruguay, most producers of goat milk have small size dairy farms where most of the production goes to making cheese (70%) and the remaining part goes to direct consumption and goat feeding (Barberis 2002 cited by Grille et al. 2013). The goat farm is centralized in milk production, which is used mostly for making yogurt (natural, with fruits and low-fat) and various types of cheese (pure or mixed with cow's milk, smoked, hard for grating, in olive oil, matured in wine, cream spreads, etc.). The meat is a by-product (Ciappesoni 2006).

The goat milk stands out because it has hypoallergenic properties. It also contains higher concentration of nutrients in relation to cow's milk and it is affected by several factors, which include breed, stage of lactation, production level, management, environment and feeding (Haeinlen 2002; Grille et al. 2013). In a work of Ceballos et al. (2009), values of 3.36% capric acid in cow's milk and 11.07% in goat's milk under the same production conditions were obtained. While the proportion of fatty acids (FA) can be modified by feeding, animals (individual and lactating) and/or the environment.

In general studies focus on feeding seek for changes to a higher percentage of unsaturated FA.

It is recognized that milk has a pivotal role in cheese characteristics, as the fatty matter composition gives properties to cheeses (yield, texture, taste). It contains more FA that are involved in cheese flavor, with higher levels of acid butyric (C4), caproic (C6) caprylic (C8) and capric (C10) than cow's milk (Oliszewski et al. 2002). Also towards human health, there are countries proposing methods to determine, for example atherogenic and thrombogenic indexes, which is useful in the application of nutritional value (Núnez-Sánchez et al. 2016), since consumers are increasingly interested in these products. Even some European countries (e.g. France and The Netherlands) have introduced FA composition among the parameters considered to determine milk price and supported producers in the adaptation of compliant systems for feeding their animals (Coppa et al. 2014; Núñes-Sánchez et al. 2016).

The medium chain FA have different properties from those of the long chain when metabolized by humans, especially caprylic and capric acids. This is mainly due to the tendency of these acids to provide energy and not to contribute to the formation of adipose tissue, as well as its ability to limit and dissolve deposits of serum cholesterol, which is related to a decrease in heart disease, cystic fibrosis and gallstones (Haenlein 2002; Grille et al. 2013).

The values obtained in the work of Grille et al. (2013), in goat milk from a herd of Saanen breed, were to capric acid (8.4%) 2 to 3 times greater than bovine milk. In this work it was confirmed that milk from Saanen breed goat has higher amounts in acids caproic (1.3%), caprylic (2%) and capric (8.4%) and low amounts of trans fatty acids (3.04 %) compared to those reported in cow's milk, which results in health benefits for this milk compared with bovine.

For some time there has been a program of the Park of Agro industrial Activities (PAGRO) of the Municipality of Montevideo (IMM) that promotes goat farming in families whose young population is at health risk conditions in the rural area of

Montevideo. Currently in the PAGRO there is a herd of goats and it also operates a processing plant called Caprino Alto.

In the Uruguayan market are also available cheeses made with goat milk and acidic coagulation similar to Quark and other types (pure or mixed with cow milk, smoked, hard for grating, in olive oil, matured in wine, cream spreads, etc.).

Conservation of goat milk

Conservation of milk is achievable, among other methods, by means of cooling or freezing, which is one of the methods of conservation of goat milk used in certain establishments in order to get a longer life and greater availability of milk in different seasons (Grille et al. 2013).

Our research group (Faculty of Veterinary, UdelaR) together with PAGRO (IMM) and LATU, conducted a study in order to guarantee the preservation of goat milk by freezing at $-18°C$ for six months, verifying oxidative stability, compositional and microbiological quality. To guarantee the milk preservation the following specific objectives were defined: 1. to evaluate and compare, with respect to raw goat milk, how heat treatment (pasteurization) affects prior to freezing, the various parameters of quality and stability. 2. To evaluate the effect of two methods of thawing on the parameters of quality and stability of goat milk. 3. To evaluate how it affects the freezing times values regarding quality and stability of raw and pasteurized goat milk. The conclusions of this study were that the freezing of goat milk for six months did not affect the quality in terms of the parameters studied, so this method of conservation could be an alternative in rural establishments to maintain a steady volume of goat milk through the year. It is considered important to continue studying the effects of the freezing of the milk used as raw material for cheese making, given that 70% of goat milk produced in the country goes to the production of these products (Grille et al. 2013).

It is noteworthy that this study was developed by producers demand (milk was usually frozen as a preservation method, but this process was not officially approved), so these results were fundamental to draft the "Licensing and counter-signature handbook of dairy farms and Artisan dairies" MGAP (2010) (*Manual para la Habilitación y Refrendación de los Establecimientos Productores de Leche y Queserías Artesanales*). In which, it is specified that the goat milk producer can choose cooling or freezing as a method of milk storage. Regardless of the method used, the time between milking and refrigeration or freezing should not exceed two hours. In the case of refrigerated milk, it must meet the same requirements described for bovine milk handling. The freezing of milk, only allowed for goat and sheep milk, should also comply with the following: The milk must be filtered immediately after milking and collected in labelled containers (for traceability) used for first time in freezing. Freezing and / or maintenance of frozen milk should be performed using equipment with storage capacity and cold production compatible with the volume of production and product storage period at the processing establishment. Frozen milk should be kept at lower temperature then $-18\ °C$ for a maximum period of 5 months. The equipment used for freezing (freezer) should be used exclusively for milk. It must have a device for temperature control and should be maintained in adequate hygienic conditions (MGAP 2010).

Small Ruminants Milk Cheese

The official specifications for licensing procedures and sanitary control of establishments producing both sheep and goat milk are determined by Decree 164/004 (MGAP 2010). According to Albenzio and Santillo (2011), in relation to milk from sheep and goats, the proteins and native enzymes influence directly in its ability to be processed and in the quality of dairy products. The characteristics of casein from sheep and goat milk are of particular interest due to the high degree of polymorphism which is linked to the properties of cheese making.

Of course, it is essential the compositional quality of raw milk and its biochemical characteristics; the technological treatment of milk, rennet process, curd production, ripening (in case required) represent a multiplicity of factors that affect, directly or indirectly, on the quality of these products. Each factor of this complex system is affected by the activity of enzymes originated from milk (i.e., native enzymes and microbiota), coagulants, and starter and no starter microbiota. It is important to recognize the proteolytic role of native and lipolytic enzymes in the capacity of cheese making and its effects during ripening. So there is a relationship with the characteristics achieved in terms of flavour, rheology, texture and functionality for each variety of cheese. All of them are involved in defining the quality of cheese and are related to the intensity of the ripening process in terms of proteolysis, lipolysis and glycolysis (Albenzio and Santillo 2011).

Artisanal Cheeses: Quality

Regarding artisanal cheeses in Uruguay specific studies have been conducted in order to assess their quality. During July 2007–July 2008, it was proposed to evaluate the hygienic-sanitary quality of artisanal hard cheeses from the dairies in the area of Colonia, Uruguay. The quality of the raw milk, raw material for the production of cheese was also evaluated. Samplings were conducted for a year, in 10 establishments, in total 83 samples of raw milk and its respective cheese. These were analyzed in the Laboratory of Milk Science and Technology in the Faculty of Veterinary, UdelaR Montevideo, Uruguay. The results were compared with existing regulations at national level. The dairies of this study have different characteristics in relation to infrastructure, equipment and processing practices that are reflected in the quality of their products. The process of licensing in dairies and cheese companies is fundamental to ensure the quality of the product. The raw milk samples in general had high counts of aerobic mesophilic bacteria, but not high counts of total coliforms and *Staphylococcus* coagulase positive being its composition acceptable. Regarding the hygienic and sanitary quality of the cheeses of this study, positive *Staphylococcus* coagulase was the most common contamination of samples, which is not suitable for consumption. A 22.9% (n–19) of cheese samples had high counts for this microorganism according to current regulations. The results indicated the absence of *Listeria monocytogenes* and *Salmonella* spp. in all the samples analyzed. Additionally, eight of the dairies studied did not have potable water for use in the dairy and artisanal cheese making. In summary, in the study 69.9% of the samples (n–58) of artisanal cheese were suitable for consumption (Barneche and Villagrán 2012).

On the other hand during 2012, another study was conducted in artisan dairies located on the west coast of Uruguay. The aim of the study was also to evaluate the hygienic-sanitary quality of cheeses from 10 establishments located in the Departments of Salto, Paysandú and Rio Negro. The quality of the raw material used in the cheese making was also evaluated and it was characterized by the type of cheese produced according to its composition (fatty in dry matter and moisture). Two sample collections were conducted at each site, obtaining samples of milk, water, brine and cheeses. These samples were analyzed in the Laboratory of Milk Science and Technology in the Faculty of Veterinary, UdelaR. The results were compared with current nation regulations. Milk samples from two dairies were suitable as well as a sample from another dairy after the heat treatment was performed. The most important factor in milk quality was the count of *Staphylococcus* coagulase positive followed by total coliforms. Regarding artisanal cheeses according to the classification established in current regulations, eight were are classified as "fatty, medium moisture or semi-hard" and two "fatty, high moisture or soft". In addition, 90% of cheese samples of the dairies were not suitable for consumption, which corresponds to exceeding values for *Staphylococcus* coagulase positive. Regarding *Escherichia coli*, two samples exceeded the allowed limit and *Listeria monocytogenes* and *Salmonella* spp. were absent in samples. Only 1 of the 10 artisanal cheeses analyzed meet the hygienic-sanitary requirements for quality according to current regulations. Possible sources of contamination are: the raw milk used as the raw material, the dairy itself and the cheese making environment, the people involved and the water used in the process of cheese making. It was also observed that establishments producing the analyzed artisan cheese did not apply good manufacturing practices in the process of making cheese (Barca 2012).

This latest study was conducted in the west coastal region which is not a traditional area for cheese making unlike the previous zone of Colonia, so the path travelled in terms of the qualification or licensing, conditions of the dairies and others are well reflected in their products. The study was conducted by a specific demand to establish a diagnosis of the situation by the Faculty of Veterinary, UdelaR and the researches continue working with the producers of the region.

During the last decade, there is a growing demand for artisanal cheeses, in which consumers require those specific products and even traditional procedures strictly related to their territory (Aquilanti et al 2013 and Di Cagno et al. 2007 cited by Velčovská and Sadilek 2015). The interest appears also associated with health and its origin identity. This last aspect unlike other countries is not developed in Uruguay, but it is heading in that direction.

We are currently working on a line of research in which we have isolated LAB from artisanal cheese, raw milk and non-commercial starter cultures from a group of farms at Nueva Helvecia, Colonia, Uruguay. The goal was to initially generate a collection of native LAB, to make it available and standardize the elaboration of AC. In turn, to improve these standards, native LAB could be used to inhibit pathogenic or spoilage microorganisms.

Thus the "bioconservation" is defined as the prolongation of life and increment of food health safety through natural microbiota or its metabolites (Devlieghere et al. 2004). LAB have been extensively studied as microorganisms and as preservation agents since they can exert antimicrobial activity against pathogens and survive for long periods of time without altering the quality or organoleptic properties of the product. Availability of these strains could be useful for the production of good quality artisan cheese (Devlieghere et al. 2004; Leroy and De Vuyst 2004; García et al. 2010). Antibacterial compounds produced by these bacteria, such as bacteriocins, may inhibit pathogens such as *Listeria monocytogenes*, *Staphylococcus aureus* or *Escherichia coli*, and could also have a role in food preservation and are considered extremely effective tool for food safety applications (Deegan et al. 2006; Galvez et al. 2007; Fraga et al. 2013).

The study conducted by Fraga et al. (2013), also stated as a goal to evaluate potential bacteriocin production, and to select those with an promissory potential for cheese elaboration. A culture collection of 509 isolates was obtained and five isolates were bacteriocin-producers and identified as *Enterococcus durans,* two strains of *Lactobacillus casei* and two strains of *Lactococcus lactis.* No evidence of potential virulence factors were found in *Enterococcus durans* strains. These results are promising in order to use these native strains for cheese manufacture and in obtaining safe products (Fraga et al. 2013). Currently, this line of work is still being developed evaluating technological properties of some of these LAB and its application in cheese making to test the antimicrobial effect against *Listeria* spp.

Conclusion

In Uruguay, the milk of small ruminant production is still incipient with an audience that has a deeply rooted preference for cow's milk and cheese. That is why a permanent disclosure should be made. The differences in milk composition of small ruminants compared to cow's imply that dairy products will developed different sensory characteristics, those widening the spectrum of consumers.

Studies of artisan cheeses from cow's milk presented here have been to the point, with producers who in some cases were not qualified, therefore did not in the past have control of their products. However, programs to support the artisan cheese making have succeeded in the application of GMP with their handbooks and of course the appropriate license.

Finally, it is worth mentioning that the Uruguayan dairy industry is going through a critical time, due to various factors including: the drought, the international market prices have fallen so much Uruguay has also had to recently close two plants in the southwest region. In this sense the MGAP instructed the INALE (where it exist representation of producers, industry and government), the study of funding alternatives that will allow dairy farmers to face this crisis.

Acknowledgements

The author grateful for the financial support from *"Comisión Dedicación Total"* UdelaR, Uruguay. The author would like to thank Lic. Karina Cal, Dra. Verónica Merletti and Dra.Mara Olmos for their contribution with suggestions and layout for chapter. Also to the teachers Catherine Morris and Elizabeth Carro for translation into English.

References

Albenzio, M. and A. Santillo. (2011). Biochemical characteristics of ewe and goat milk: Effect on the quality of dairy products. Small Ruminant Research 101: 33–40.

Anchieri, D., D. Carrera, P. Lagarmilla and E. Aguirre. (2007). Prácticas de Higiene en la Quesería Artesanal. Proceedings of Programa de desarrollo tecnológico Ministerio de Educación y Cultura and Facultad de Veterinaria. UdelaR, Uruguay: pp. 1–29.

Arocena, F. (2009). La contribución de los inmigrantes en Uruguay. CEIC ISSNN: 1695–6494 Papers, 2: 1–42.Available in: http://www.identidadcolectiva.es/pdf/47.pdf Accessed on: 01 June 2015

Bagnato, D. (2004). Quesería artesanal. Situación actual y desafíos para Uruguay. 20p. Available in: http//www.iica.org.uy/data/documentos/5050.doc Accessed on: 04 May 2015.

Barca, J. (2012). Evaluación higiénico-sanitaria de quesos artesanales producidos en la zona litoral oeste, Uruguay. Tesis de grado, Universidad de la República (Uruguay). Facultad de Veterinaria: 50 pp Available in: https://www.colibri.udelar.edu.uy/bitstream/123456789/2712/1/FV-29948.pdf Accessed on: 01 July 2015.

Barneche, M. and M. Villagrán. (2012). Evaluación de la calidad higiénico-sanitaria de quesos artesanales de pasta dura elaborados en la zona de Colonia, Uruguay". Tesis de grado, Universidad de la República (Uruguay). Facultad de Veterinaria: 51 ppAvailable in: https://www.colibri.udelar.edu.uy/bitstream/123456789/2723/1/FV-29988.pdf Accessed on: 01 July 2015.

Bermúdez, J. and S. Reginensi. (2014). Alternativas para la producción de quesos ovinos diferenciados en Latinoamérica In: Guía práctica de producción ovina en pequeña escala en Iberoamérica [CITED] : pp. 148–154, ISBN 978-9974-99-696-0.

Bertino, M. and H. Tajam. (2000). La agroindustria láctea en el Uruguay 1911–1943. IECON (FCEA, UdelaR) DT 04/00. Available in: http://www.iecon.ccee.edu.uy/dt-04-00-la-agroindustria-lactea-en-el-uruguay-1911-1943/publicacion/87/es/ Accessed on: 04 May 2015.

Bielli, A. (2009) LATU y las MI Pymes Experiencias de transferencia tecnológica para el desarrollo 24- 53 ppISBN 978-9974-8213-0-9 Available in: http://catalogo.latu.org.uy/doc_num.php?explnum_id=762 Accessed on: 04 May 2015.

Borbonet, S., P. Urrestarazú and R. Pelaggio. (2010). Quesos Artesanales. Conceptos generales y recomendaciones prácticas y productivas. Proyecto de Cohesión Social Social "Colonia Integra" Laboratorio Tecnológico del Uruguay: 6–7 Available in: http://www.latu.org.uy/docs/Publicacion-Quesos-Artesanales-12072011.pdf Accessed on: 04 May 2015.

Borbonet, S. (2001). Historia de la quesería en Uruguay. MEDEA S.A., Montevideo, Uruguay.

Casaux, G. (2005). Temas de Legislación Sanitaria Especial. Tomo I: Tambos. Oficina de Publicaciones de la Facultad de Veterinaria (UdelaR), Montevideo.

Ciappesoni, C.G. (2006). La producción caprina en Uruguay y Latinoamérica. Department of Tropical and Subtropical Animal Production, kamycka 129, Suchdol 165 21, Praga 6, Republica Checa. Available in: htttp://www.caprahispana.com/mundo/uruguay/uruguay.htm Accessed on: 05 May 2015.

[CINVE] Centro de Investigaciones Económicas. (1987). Una década de cambio en la lechería uruguaya. Ediciones de la Banda Oriental, Montevideo: pp. 18–19.

Crosa, M. J., R. Harispe, P. Mussio, R. Pelaggio, L. Repiso, C. Silvera. (2009). Comparación de los cambios químicos y microbiológicos en la maduración del queso Colonia salado tradicionalmente y por impregnación en vacío –Revista del Laboratorio Tecnológico del Uruguay INNOTEC 4: 22–27.

Deegan, L.H., P.D. Cotter, C. Hill and P. Ross. (2006). Bacteriocins: Biological tools for biopreservation and shelf-life extension. International Dairy Journal 16:1058-1071.

[DIEA] División de Estadística Agropecuaria, Ministerio de Ganadería Agricultura y Pesca (MGAP). (2003). Estadísticas del sector lácteo 2002, Montevideo, 44p. Available in: http://www.mgap.gub.uy/portal/page.aspx?2, diea, diea-pub-lecheria, O, es, 0, Accessed on: 28 April 2015.

[DIEA] División de Estadística Agropecuaria, Ministerio de Ganadería Agricultura y Pesca (MGAP). (2014a). Anuario Estadístico Agropecuario 2014, Montevideo: 56–59 pp. Available in: http://www.mgap.gub.uy/Dieaanterior/Anuario2014/Diea-Anuario%202014-Digital01.pdf. Accessed on: 28 April 2015. Consultation date: 04 May 2015.

[DIEA] División de Estadística Agropecuaria, Ministerio de Ganadería Agricultura y Pesca (MGAP) (2014b). Estadísticas del sector lácteo. (2013). Montevideo, 7, 14 pp. Available in:http://www.mgap.gub.uy/portal/page.aspx?2,diea,diea-pub-lecheria,O,es,0. Accessed on: 28 April 2015.

[DINAPYME] Dirección Nacional de Artesanías, Pequeñas y Medianas Empresas (2013) Con apoyo de Dinapyme, queseros artesanales implementan buenas prácticas para la elaboración Adapted for Inale.Available in:

http://www.portalechero.com/innovaportal/v/3640/1/innova.front/con-apoyo-de-dina-pyme-queseros-artesanales-implementan-buenas-practicas-para-la-elaboracion.html Accessed on: 05 June 2015.

Devlieghere F, L. Vermeiren, J. Debevere. (2004). New preservation technologies: Possibilities and limitations. International Dairy Journal 14: 273–285.

García P, L. Rodriguez, A. Rodriguez and B. Martinez. (2010). Food biopreservation: promising strategies using bacteriocins, bacteriophages and endolysins. Trends in Food Science & Technology 21: 373–382.

[FONADEP]. Fondo Nacional de Preinversión. (2008). Primer Informe de Avance. Desarrollo de la Quesería Artesanal de la Cuenca Lechera del Litoral, Departamentos de Salto Paysandú y Río Negro. Consultoría financiada con recursos de FONADEP-Salto-Paysandú-Río Negro- Uruguay: pp. 1–72

Fox, P., T. Guinee, T. Cogan and P. McSweeney. (2000). Fundamentals of Cheese Science *In*: Chapter 5: Starter Cultures and Chapter 17: Principals Families of Cheese. Aspen Publishers, Inc. Gaithersburg, Mary land, pp. 54–96 and 388–428

Fraga, M., K. Perelmuter, S. Giacaman, P. Zunino, S. Carro. (2013). Antimicrobial properties of lactic acid bacteria isolated from uruguayan artisian cheese. Ciência e Tecnologia de Alimentos 33: 801–804. Fresco, E. (2004). Situación y perspectivas del mercado mundial de lácteos-Desafíos para el Uruguay.Available in: http://www.iica.int/Esp/regiones/sur/uruguay/Documentos%20de%20la%20Oficina/CoyunturaAgropecuaria/coy-diciembre2004.pdf Accessed on: 02 June 2015

Ibarra, A. (2007). Sistemas de Pago de Leche. Proceedings de Seminario Regional de Calidad de Leche. Instituto Plan Agropecuario. Uruguay: 38–52 pp. Available in: https://www.google.com.uy/#q=Aldo+Ibarrra+Calidad+de+Leche+Uruguay Accessed on: 01 June 2015.

[INALE] Instituto Nacional de la Leche. (2015). Uruguay Lechero, Uruguay. Available in: http://www.inale.org/innovaportal/v/3204/4/innova.front/uruguay-lechero.html Accessed on: 01 June 2015.

Grille, L., S. Carro, D. Escobar, C. Fros, G. Cousillas, F. Lazzarini, A. Borges, and S. González. (2013). Effect of goat milk freezing on oxidative stability, hygienic sanitary and composition quality in a Saanen breed herd. INNOTEC 8: 60–66 - ISSN 1688-3691

Jeruselmi, C., M. Camacho and M. Mortorio. (2008). Estudio de Caso Cluster Quesería Artesanal en San José y Colonia. Instituto de Competitividad, Universidad Católica del Uruguay: X-XX. Available in: http://ucu.edu.uy/sites/default/files/facultad/fce/i_competitividad/cluster%20queseria. PDF Accessed on: 01 June 2015.

Kremer, R. (1995). Experiencias sobre un sistema de producción ovina orientado a la producción de leche. Proceedings of Jornadas de actualización en reproducción y producción de leche ovina y caprina. Facultad de Veterinaria, UdelaR, Uruguay: pp. 1–5.

Kremer, R. (2005). Leche Ovina en Uruguay. Resultados de Composición y Parámetros Tecnológicos. Proceedings of Seminario Leche y Productos Lácteos: Aspectos moleculares y Tecnológicos. Facultad de Veterinaria, UdelaR, Uruguay: pp. 19–23

Leroy F. and L. De Vuyst. (2004). Lactic acid bacteria as functional starter culture for the food fermentation industry. Trends in Food Science & Technology 15: 67–78.

Martí, J. P. (2013). National Cooperative of Dairy Producers. Its Creation is Analyzed from the Perspective of Public Policies. América Latina en la Historia Económica 20: 90–113.

[Mesa Tecnológica de la Cadena Láctea] (2005). Temas seleccionados para orientar las investigaciones en las áreas: calidad de la leche y tecnología para el sector industrial. Available at http://www.fagro.edu.uy/investigacion/MESAS/lacteos/prioridades2. html Accessed on: 01 June 2015.

[MGAP] Ministerio de Ganadería, Agricultura y Pesca. (2013). Decreto 359/13, Uruguay. Available in: http://www.presidencia.gub.uy/normativa/decretos/decretos-11-2013 Accessed on: 01 June 2015.

[MGAP] Ministerio de Ganadería, Agricultura y Pesca (2010). Manual para la habilitación y refrendación de establecimientos productores de leche y queserías artesanales: 33 pp. Available in:http://www.mgap.gub.uy/dgsg/Resoluciones/ Res_27_011_Manuales_habilitaci%C3%B3n_refrendaci%C3%B3n_leche_ artesanal/I_Manual%20habilitaci%C3%B3n%20y%20refrendaci%C3%B3n%20 tambos%20y%20queser%C3%ADas%20artesanales_v01m.pdf Accessed on: 30 June 2015.

[MGAP] Ministerio de Ganadería, Agricultura y Pesca (1995). Decreto 90/995, Uruguay. Available in: http://www.aduanas.gub.uy/innovaportal/v/6967/3/innova.front/ decreto_n%C3%82%C2%B0_90_995.html Accessed on: 01 June 2015.

[MGAP] Ministerio de Ganadería, Agricultura y Pesca (2001). Legislación Sanitaria Animal. Tomo 1. Dirección General de Servicios Ganaderos, División Sanidad Animal, Programas Servicios Agropecuarios, Montevideo: 188pp. Available in: http://www.mgap.gub.uy/dgsg/legislacion/Cap2_Sanidades_Especiales.pdf Accessed on: 01 June 2015.

[MSP] Ministerio de Salud Pública (2009). Reglamento Bromatológico Nacional. Decreto Nº 315/994. Con Apéndice Normativo 3ra Edición Dirección Nacioonal de Impresiones y Publicaciones y Publicaciones Oficiales IMPO. Montevideo Uruguay Chapter 16: 134-145 Apéndice Normativo: 405, 410.

Núñez-Sánchez, N., A. Martínez-Marín, O. Polvillo, V. Fernández-Cabanás, J. Carrizosa, B. Urrutia, J. Serradilla. (2016) Near Infrared Spectroscopy (NIRS) forthedetermination of the milk fat fatty acid profile of goats. Food Chemistry: 244–252

Oliszewski, R., A. Rabasa, J. Fernández, M. Poli, M. Núñez. (2002). Composición química y rendimiento quesero de leche de cabra criolla serrana del noroeste argentino. Zootecnia Trop. 20: 179 –189.

[PACPYMES] Programa de Apoyo a la Competitividad y Promoción de Exportaciones de la Pequeña y Mediana Empresa Cooperación Unión Europea. (2006–2007). Uruguay Cluster de Quesería artesanal.Available in: http://www.industrianaval.com.uy/c/document_library/get_file?p_l_id=6392&folderId=9652&name=Diagnostico+Queseria.pdf Accessed on: 01 June 2015.

Reginensi, S., M. González, J. Bermúdez. (2013). Phenotypic and genotypic characterization of lactic acid bacteria isolated from cow, ewe and goat dairy artisanal farmhouses. Brazilian Journal of Microbiology 44: 427–430.

Uruguay XXI. (2012). Promoción de Inversiones y Exportaciones. Sector Lácteo Oportunidades de inversión en Uruguay. Available in: http://www.uruguayxxi.gub.uy/inversiones/wp-content/uploads/sites/3/2014/09/Sector-Lacteo-Uruguayy-XXI-Julio-2012. pdf Accessed on: 01 June 2015.

Viera, E., F. Bengoa, G. Bagnato, I. Arboleya, (2013). El Sector Lechero Uruguayo. XX Reunión Especializada de Agricultura Familiar (REAF). Seminario sobre producción, comercialización y políticas públicas para la seguridad alimentaria. Contribuciones de las Políticas Públicas y la Institucionalidad Sectorial a su desarrollo. Proceedings of Programa Regional FIDA Mercosur – CLAEH, Uruguay. Available in: http://fidamercosur.org/site/images/BIBLIOTECA/2013/Publicaciones/El_sector_lechero_uruguayo.pdf Accessed on: 01 June 2015.

Velovská, Š., T. Sadilek. (2015). Certification of cheeses and cheese products origin by EU countries, British Food Journal 117: 12 pp. Available in: http://dx.doi.org/10.1108/BFJ-10-2014-0350

Wendorff, B. (2003) Milk composition and cheese yield Proceedings 7th Great Lakes Dairy Sheep Symposium. Ithaca, Department of Food Science University of Wisconsin-Madison. Available in: http://ansci.wisc.edu/Extension-New%20copy/sheep/Publications_and_Proceedings/Pdf/Dairy/Milk%20Composition%20and%20cheese%20yield.pdf Accessed on: 01 July 2015.

Zorrilla de San Martin, D. (2013). Lechería en el Uruguay: una década de fuertes impactos. Proceedings of Instituto Interamericano de Cooperación para la Agricultura, Uruguay. Available in: http://www.iica.int/Esp/regiones/sur/uruguay/Documentos%20de%20la%20Oficina/CoyunturaAgropecuaria/coy-lecheria.pdf Accessed on: 01 June 2015.

5

Brazilian Kefir: From Old to the New World

Cristina Stewart Bogsan[a]* and Svestolav Dimitrov Todorov*[b]

Abstract

Traditional kefir originated from the Caucasus region of Asia. It is produced by inoculation of the "kefir grain" into the fresh milk. The final product is a mildly alcoholic fermented dairy beverage with a refreshing taste, and is lightly carbonated and slightly acidic. Kefir grains and preparation of kefir were introduced to Latin America by lebanese and syrian immigrants at the end of 19th century. Brazilian kefir could be described as two types—milk kefir and sugary water kefir. In both products, the kefir grains are washed with water and stored at a low temperature, prior to serving as inoculums for a subsequent fermentation in milk or sugary water, respectively.

The complex kefir matrix comprises of a mixture of various bacteria and yeasts. The traditional kefir beverage is designated as fermented milk by an intricate symbiotic matrix formed by acid lactic bacteria (LAB) as *Lactobacillus*, *Lactococcus*, *Leuconostoc* and *Streptococcus* and yeasts, *Kluyveromyces* and *Saccharomyces*. This complex matrix could also be fermented in a raw sugar and water and is called sugary kefir which is mostly consumed in Mexico and Brazil.

This chapter discusses the differences in microbial population, physico-chemical characteristics, technological specificity and structure of Brazilian kefir.

Introduction

Kefir is a fermented milk product produced by incubation of milk which has been inoculated with kefir grains. The final product is distinguished by a slight acidic taste, natural carbonation, and specific dairy aroma (Otles and Cagindi 2003; Tamime 2006; Kabak and Dobson 2011; Ahmed et al. 2013; Leite et al. 2013; Nielsen et al. 2014). This type of fermented milk is characterized by specific symbiotic microbiota which resides in a carbohydrate polysaccharide matrix (Ahmed et al. 2013).

---

[a] Department of Biochemical and Pharmaceutical Technology, São Paulo University, 05508-000, Sao Paulo, SP, Brazil.
[b] Universidade Federal de Viçosa, Veterinary Department, Campus UFV, 36570-900, Viçosa, Minas Gerais, Brazil. E-mail: slavi310570@abv.bg
* Corresponding author: cris.bogsan@usp.br

The meaning of kefir is "feeling good", as one is supposed to feel after its ingestion (Tamime 2006; Leite et al. 2013; Ferreira 2014). According to popular believes and religious folklore, Mohammed gave kefir grains to the Orthodox people as "Grains of the Prophet" and taught them how to cultivate it and it was believed that "kefir grains have curative power, and need be kept in secret, passed on generation to generation as part of tribe's wealth" (Yemoos 2015; Serpa 2015, Tamime 2006; Ferreira 2014).

History

Knowledge of the beneficial properties of kefir can be traced back from 2000 B.C. and until the early 1900's it was consumed as fermented dairy beverage mainly in Caucasus Mountains (Tamime 2006; Leite et al. 2013; Ferreira 2014). Russian medical doctors recommended consumption of kefir as alternative for the treatment of symptoms of tuberculosis epidemiologic cases and gastrointestinal disorders. However, during that period, it was very difficult to obtain the kefir grains as the commercial production of kefir had yet to begin. Kefir was produced as an artisanal product in small quantity with no consistent microbial and organoleptic characteristics.

The small scale semi-industrial production of kefir manufacturing process can be dated to the beginning of 20th century in Russia (and later Soviet Union). Although by the 1930's, the large-scale kefir manufacturing had begun in Soviet Union, kefir was still produced using traditional methods and it was this approach that influenced microbial composition of the final product. Microbial characteristics of the grains and employed basic technology significantly affected quality of the final product. As a result the production of kefir by conventional (traditional) methods on a commercial scale did not maintain the same heath and taste properties observed in kefir made by the artisanal method (Yemoos 2015; Otles and Cagindi 2003). By the 1950's, in order to improve the quality of industrially produced kefir, the stirred method was developed by the All-Union Dairy Research Institute (VNIMI). This method comprises of fermentation, coagulation, agitation, ripening and cooling. It could produce kefir similar to that obtained by artisanal methods. Beside Russia, kefir is consumed in Hungary, Poland, Sweden, Norway, Finland, Germany, Greece, Austria, Canada, Brazil and Israel, and the popularity is growing in the United States and Japan (Otles and Cagindi 2003).

Unlike that of milk kefir grains, the origin of sugary kefir grain remains uncertain. The first reference to sugary kefir dates back to the 1855 Crimean war, when English soldiers brought back the product from the Crimean region. According to Pidoux (1989), the sugary kefir grains can be linked to the ginger beer plants. The kefir grains cultivated in sugary solution contain the symbiotic microbiota and the matrix structure very similar to the milk kefir grains (Magalhães et al. 2010). Nowadays, this beverage is consumed mainly in Mexico and Brazil (Miguel et al. 2011).

In Brazil, the consumption of kefir began in the beginning of the 20th century, and the scientific study and characterization of the entire process of preparation of sugary water and milk kefir started only at the end of the 20th century. However, til date, the manufacturing of kefir beverages in Brazil remains exclusively artisanal for personal consumption and is not very widespread (Magalhães et al. 2010; Leite et al. 2013).

Microbiota and Structure of Kefir Grains

Kefir grains can be cultivated in milk or in a sugary solution. However, the micro-biological structure of both, milk kefir and sugary kefir is similar (Magalhães et al. 2010; Miguel et al. 2011).

Traditional kefir grain is a complex symbiotic matrix, containing several distinct bacteria and yeasts. However, the essential role of insoluble polysaccharide and protein matrix that could contain small clusters of bacteria and yeasts in the structure of kefir grains need to be acknowledge. Kefir grains are gelatinous, yellowish, and vary in size from 0.3 to 3.5 cm in diameter (Kabak and Dobson 2011; Nielsen et al. 2014; Likotrafiti et al. 2015). During fermentation, the grains increase about 5 to 7% in weight and the additional biomass is harvested for new fermentations (Likotrafiti et al. 2015). The kefir grains are washed with water, and stored at a low temperature, prior to serving as inoculums in a subsequent fermentation, and their activity can be maintained for years when carefully preserved (Lopitz-Otsoa 2006; Nielsen et al. 2014).

The quantification of the specific microbial groups in symbiotic kefir grains were dependent on the fermentation time and food matrix used and also the grain's origin. Studies on the identification of this complex microbiota were based on application of selective growth media, morphological and biochemical characterization, as well quantitative real time PCR bio-molecular methods. The three major microbial groups that previously have been reported in kefir are LAB, acetic acid bacteria and yeast (Kim et al. 2014).

The complex matrix comprises microorganisms as LAB and acetic acid bacteria as *Lactobacillus parakefir*, *Lactobacillus kefir*, *Lactobacillus kefiranofaciens* subsp. *kefiranofaciens*, *Lactobacillus kefiranofaciens* subsp. *kefirgranum*, *Lactococcus lactis* subsp. *lactis*, *Lactococcus lactis* subsp. *cremoris*, *Lactococcus lactis* subsp. *diacetylactis*, *Leuconostoc mesenteroides* subsp. *cremoris*, and yeasts *Candida kefyr*, *Saccharomyces unisprous*, *Saccharomyces turicensis*, *Kluyveromyces marxianus*, *Candida inconspicua*, *Debaryomyces hansenii*, *Torulaspora delbrueckii*, *Zygosaccharomyces rouxii*, *Candida maris*, *Torulopsis kefir*, *Saccharomyces cerevisiae* and *Klyveromyces lactis* (Simova et al. 2002; De Moreno de LeBlanc et al. 2006; Vinderola et al. 2006; De Moreno de LeBlanc et al. 2007; Powell et al. 2007; Miguel et al. 2011; Ahmed et al. 2013).

The microbiota of the kefir grains differ from those encountered in the final fermented beverage. This is due to the complex symbiotic interactions between the organisms in the kefir grain during the production of kefir (Nielsen et al. 2014). The LAB represents 83–90% of the microbial counts in the grains while yeasts represent about 10–17% (Simova et al. 2002). Despite intense research and many attempts made to produce kefir grains from pure or mixed cultures that are normally present in the grain, still no positive result has been reported (Carneiro 2010).

The interactions between yeasts and bacteria in kefir grains and their interdependence are not completely understood, and even when they are grown seperately, the capacity of vitamin B production to metabolize lactose or produce CO_2 are compromised. This can probably be attributed to the fact that very little is known about the mechanism of the formation of the grains (Carneiro 2010; Rattray and O'Connel 2011; Leite et al. 2013). These bacteria and yeasts exhibit different role during the periods of activity throughout the fermentation process of kefir (Ferreira 2014). The *Lactococcus* spp. tends to grow faster than yeasts in milk, hydrolyzing the lactose and producing a suitable environment for yeast development (Tamime 2006; Leite et al. 2013). The yeasts were able to synthesize vitamins from B complex group and to hydrolyze milk proteins, to produce CO_2 and ethanol as results from their metabolite capacity (Lopitz-Otsoa et al. 2006; Tamime 2006). Such symbioses between LAB and yeasts result in the development of metabolites that can present inhibitory or stimulating properties and can maintain the grains stability (Lopitz-Otsoa et al. 2006; Leite et al. 2013). Some yeasts involved in the composition of kefir grains could present proteolytic or lipolytic activity and can be involved in the formation of bioactive peptides and fatty acids (Rattray and O'Connel 2011; Leite et al. 2013; Ferreira 2014). Some yeasts can assimilate lactic acid produced by LAB, can raise the environmental pH and, in this way, can stimulate the bacterial growth (Leite et al. 2013).

The structure of the kefir grains can present a wide variation in microbial population which normally is associated to the origin of the grain (Leite et al. 2013). In addition, Brazilian kefir grains present a relative variation in microorganisms' distribution. In general, rod-shaped bacteria were observed both in the inner and the outer grain; cocci were present mainly in the outer portion while yeasts were most frequent in the inner portion of the grains (Leite et al. 2013; Ferreira 2014). The microorganisms on the surface of the grains cause a greater impact on the kefir fermentation process and it could also be observed that coagulated milk adhered to the surface of the grains and the polysaccharide spread over all parts of the grains (Magalhães et al. 2011; Leite et al. 2013).

Kefir beverage can be prepared from cow, goat, sheep, camel, and buffalo milk. It also can be made from milk substitutes such as soy milk, rice milk, and coconut milk, and this promotes a different composition of the grains and final product (Nielsen et al. 2014).

Chemical Characteristic

The physico-chemical characteristics of the kefir change for many reasons, such as the age of the grains, the food matrix used, including type and composition and quality of the milk, the origin of the grains and the technology employed to production of the kefir beverage (Carneiro 2010; Ferreira 2014).

All kefir grains produce a beverage that is characterized with pH 4.2 to 4.6, presence of lactic acid of 0.8% (w/w), alcohol of 0.1% to 2.0% (w/v), a refreshing, slightly carbonated and alcoholic fermentation, with a sour and tart flavor (Tamime 2006; Carneiro 2010; Kabak and Dobson 2011; Ebner et al. 2015). About 89% to 90% (v/v) of kefir is water, as lipids are around 0.2% (w/w), protein around 3.0% (w/w), sugar around 6.0% (w/w) and ash around 0.7% (w/w) (Carneiro 2010; Kabak and Dobson 2011). However, the CO_2 responsible for the refreshing and slightly effervescent taste and the ethanol are altered by the technology employed to produce the fermentation process varying the value about 0.01% to 0.1% (w/w) (Beskova et al. 2002; Carneiro 2010). Aromatic compounds as acetaldehyde, dyacetil and acetoin, pyruvic, acetic, propionic and butyric acid also ethanol decreases during kefir storage (Carneiro 2010; Santos et al. 2013).

Nutritional Characteristic

The nutritional characteristics of kefir beverage change in accordance to the food matrix used in the fermentation process. The balanced chemical substances promote a higher nutritional value. Kefir is rich in calcium, vitamins, essential amino acids, that has an essential role in the treatment and maintenance of body functions (Carneiro 2010; Ahmed et al. 2013).

Vitamins

Vitamins are micronutrients that play an essential role for the metabolism of all living organisms. They are necessary for the regulation of vital biochemical reactions in the cell as intracellular coenzymes (Le Blanc et al. 2011). In kefir, vitamins production depends on the food matrix and is also related to the microorganism present in the kefir grains. Milk kefir is considered a great source of folic acid, biotin, pyridoxine, pantothenic acid and other vitamins of B complex (Kneifel and Mayer 1991; Carneiro 2010). The numerous benefits of B complex vitamins include regulation of the kidneys functions, liver and nervous system, have a positive role in the treatment of skin, increase energy and promote longevity (Carneiro 2010).

Proteins

The milk fermentation by kefir grains is related to production of highly digestible proteins, essential amino acids and bioactive peptides (Carneiro 2010; Ebner et al. 2015). Kefir fermentation shows a higher rate of proteolysis when compared with other fermented milks. The main bioactive peptides found in fermented milks were related to angiotensin-converting enzyme (ACE)–inhibitory peptide (Ebner et al. 2015).

During the fermentation process, the proteolysis of milk also releases essential amino acids as tryptophan, threonine, serine, alanine, lysine and ammonia. The presence of valine, isoleucine, methionine, lysine, threonine, phenylalanine and tryptophan were also found in kefir (Carneiro 2010). Amino acids play a role in maintenance of the nervous system functions (Otles and Cagindi 2003).

Minerals

Some macroelements found in the kefir beverage, as potassium, calcium, magnesium, phosphorus and trace elements such as copper, zinc, iron, manganese, cobalt and molybdenum in kefir contribute to cellular growth and the maintenance of cellular energy (Carneiro 2010; Ahmed et al. 2015).

Carbohydrates

The total carbohydrates in kefir are about 6% (w/w). However, the main carbohydrate is known as kefiran, a water-soluble glucogalactan classified as heteropolysaccharide. Kefiran improves viscose-elastic properties and the gel formation and also presents some function properties that have been described above (Ahmed et al. 2013). The kefir beverage contains a range of polyssacharides derived from linear chains of glucose, galactose, manose, glucuronic acid, rhamnose, among others. The polysaccharides in kefir contribute to the texture, viscosity and elasticity of the beverage (Botelho et al. 2014).

Functional Characteristic

Antimicrobial properties

Kefir has been shown to possess antimicrobial activity against different variety of Gram-positive and Gram-negative bacteria, as well to some fungi (Saloff-Coste 1996; Powell et al. 2007; Garrote et al. 2000) and *Mycobacterium tuberculosum* (Powell 2006). Growth reduction and even total repression of some coliforms and pathogenic bacteria (*Shigella* spp. and *Salmonella* spp.) was documented in kefir products (Koroleva 1988). Metabolites produced by *Lactobacillus acidophilus* isolated from kefir have been shown to be responsible for the inhibition of several Gram-positive and Gram-negative bacteria (Gilliland and Speck 1977; Apella et al. 1992; Gupta et al. 1996). However, acetic acid bacteria and microphilic homofermentative lactococci are most active against coliforms. According to Van Wyk (2001), kefir possesses an inhibitory activity against *Staphylococcus aureus, Bacillus cereus, Escherichia coli, Clostridium tyrobutyricum* and *Listeria monocytogenes*. Koroleva (1988) and Naidu et al. (1999) showed that some yeasts isolated from kefir, including some *Torulaspora* spp. have antimicrobial activity against coliforms.

Most probably inhibitory activity is a complex action of different antimicrobial compounds produced by microorganism incorporated in the kefir grains. However, Gibson et al. (1997), Naidu et al. (1999) and Powell (2006) pointed out that role of LAB incorporated in the kefir grains, plays an essential role in the inhibitory relations. LAB are also capable of preventing the adherence, establishment, replication, and pathogenic action of some enteropathogenic bacteria (Saavedra 1995). Different antimicrobial metabolites are, most probably, involved in this process, including lactic acid or other volatile acids, hydrogen peroxide, carbone dioxide, acetaldehyde and diacetyl, bacteriocins and other antimicrobial peptides, and polysaccharides (Shahani and Chandan 1979; Juven et al. 1992; Helander et al. 1997; Powell et al. 2007).

In last few decades, antimicrobial proteins (bacteriocins) produced by LAB have been intensively studied. According to Cotter et al. (2005), bacteriocins are bacterial proteins or peptides with bactericidal or bacteriostatic activity against closely related species. Bacteriocins generally vary with regards to their specific mode of action, genetic origin, biochemical properties, and spectrum of activity. Generally bacteriocins are primary metabolites and can be expressed via ribosomal machinery of the bacterial cells spontaneously; however, some of them need to be induced. Genetic determinants for production and expression of bacteriocins can be located both, on bacterial chromosome and/or plasmid/s. Bacteriocins are normally released from the producer cells depending of the specific bacteriocin-release proteins and the presence of detergent resistant phospholipase A in the bacterial membrane (Naidu et al. 1999).

Many classifications have already been proposed (Cotter et al. 2005), but according to the most recent (Heng et al. 2007), bacteriocins of Gram-positive bacteria are grouped into four classes, based on their structure and function. Class I: lantibiotic peptides, class II: small non-modified peptides with <10 kDa, class III, large proteins with >10 kDa and class IV: cyclic proteins.

A common mode of activity of bacteriocins produced by LAB is due to pore formation in the cytoplasmatic membrane, and it has been suggested that it occurs through the "barrel-stave" mechanisms (Ennahar et al. 1999). According to the this model, bacteriocins adhere to the surface of the membrane, through ionic interaction with phospholipid groups or recognition of the specific receptors (lipid II). Presence of bacteriocin then causes displacement of the phospholipids and results in the thinning and destruction of the membrane (McAuliffe et al. 2001). It has been suggested that bacteriocin with an narrow antibacterial spectrum require specific receptor molecules, while those exhibiting a wide host-range do not (Sablon et al. 2000).

Antimicrobial activity of kefir against different pathogenic bacteria been related to the bacteriocins produced from LAB isolated form kefir by Rodrigues et al. (2005) with activity against *Candida albicans* and *Streptococcus pyogenes*; Yuksekdag et al. (2004) against *Staphylococcus aureus, Escherichia coli* and *Pseudomonas aeruginosa;* Atanassova et al. (1999) against *Listeria innocua*; Santos et al. (2003) agaist *Salmonella* Typhimurium; Powell et al. (2007) agaist *Enteroccoccus* spp. and *Listeria* spp.

Study of *Lactobacillus plantarum* ST8KF isolated from kefir (Todorov and Dicks 2008) has shown that this strain has a potential to be applied as a probiotic. In addition to the bacteriocin production by *Lactobacillus plantarum* ST8KF (Powell et al. 2007), this strain has a good survival and growth rate in presence of 0.3% ox-bile, low levels of pH, and high levels of co-aggregation with *Enterococcus faecium, Listeria innocua* and *Listeria ivanovii* subsp. *ivanovii*. Growth of *Lactobacillus plantarum* ST8KF was inhibited by several antibiotics and anti-inflammatory medicaments containing ibuprofen, hydrochlorothiaziden, thioridazine hydrochlorid, sodium diclofenac and dimenhydrinate. Based on PCR analysis, *Lactobacillus plantarum* ST8KF contain the *Mub, MapA* and *EF-Tu* genes, related to the adhesion properties (Todorov and Dicks 2008). *Lactobacillus plantarum* ST8KF was re-incorporated to kefir grains and was able to reduce growth of *Enterococcus* spp. as observed by *in situ* hybridisation (FISH) (Powell et al. 2006).

Exopolysaccharides

Exopolysaccharides are carbohydrate polymers that can be produced by several microorganism, consisting of a long linear or branched chains of monosaccharides as glucose, galactose, manose, glucuronic acid, rhamnose; among others (Welman and Maddox 2003; Botelho et al. 2014). Kefiran is the most abundant exo-polysaccharide found in kefir, and is present in the cell free fraction (De Moreno de LeBlanc 2006).

Organic acids

Organic acids as orotic, citric, pyruvic, lactic, uric, acetic, propionic, butyric, and hippuric acids can be found in kefir beverage (Guzel-Seydim et al. 2000; Ferreira 2014). These acids directly influence the flavor properties and directly affect the sensory characteristics. The organic acids are natural preservatives, associated to the inhibition of certain food pathogens.

The power of microorganism inhibition is related to different mechanisms of mode of action of organic acids. The undissociated forms of organic acids improve the inhibitory power against pathogens by the kefir grains (Guzel-Seydim et al. 2000). The undissociated forms of lactic acid and acetic acid founded in kefir beverage implies a modification of permeability of cell membranes through the acidification of cytoplasm and degradation of enzymes inhibiting the growth of some pathogenic species like *Escherichia coli* and *Bacillus cereus* (Guzel-Seydim et al. 2000; Garrote et al. 2000; Ferreira 2014).

The down grade of the pH promoted by organic acids produced during kefir fermentation influence an increase in inhibition of pathogenic bacteria. However, the organic acids, except for lactic acid, changes very little during kefir storage (Guzel-Seydim et al. 2000).

Hydrogen peroxide

The hydrogen peroxide presents bactericidal activity that promotes the peroxidation of lipid membranes and destruction of the molecular structure of proteins of the cells. Some of the LAB presented in kefir grains can produce hydrogen peroxide as metabolite as a protection mechanism for oxygen by the action of NADH peroxidases. Because it is catalase negative, the LAB accumulate inside this metabolite and thus possess bactericidal action to pathogenic organisms (Santos et al. 2013; Ferreira 2014).

Proteins

The knowledge of protein contents of kefir grains and kefir beverage are still under investigation. The food matrix where the kefir grain is fermented significantly influences the protein profile of the final product. The fermentation of kefir grains produces a high rate of proteolysis of milk proteins mainly as result of the metabolism of the microorganisms from *Lactococcus* genera (Otlers and Cagindi 2003), the most abundant fragments are derived from β-casein protein with average molecular weight of 5000 Da (Ebner et al. 2015). Ebner et al. (2015) reported that the fermentation process with kefir grains increase 2 fold the release of peptides than unfermented milk. It was also found that among the 257 peptides released, 43% were derived from β-casein, 22% from α_{s1}-casein, 21% from κ-casein, and 14% from α_{s2}-casein.

Special attention was granted to the formed bioactive peptides which influenced reduction of blood pressure. At least 12 sequences of peptides are described as ACE-inhibitors. Peptides with functions as antimicrobial properties, immunomodulating activity and those peptides that bind opioid receptors are also present in kefir beverage (Hayes et al. 2007; Ebner et al. 2015).

The casein phosphopeptides could bind minerals and improve the dental remineralization and can play essential role in reducing calcium resorption. These peptides are present in milk and are highly increased in kefir beverage (Cross et al. 2007; Ebner et al. 2015).

The most abundant peptide found in kefir is a multifunctional peptide casecidin 17 that promotes antimicrobial, ACE-inhibitory, immunomodulating, antioxidant, and antithrombotic activities (Rojas-Ronquillo et al. 2012; Eisele et al. 2013; Ebner et al. 2015).

Health Benefits

Antitumoral effect

In 1982, Shiomi et al. (1982) reported for the first time that kefir metabolites could promote antitumoral effect. They found a polysaccharide that can stimulate the immune function and play a role in the antitumor processes. After that, further studies on cancer treatment have brought more information about bioactive compounds present in kefir beverage that could act as adjuvant in cancer treatment. The protein, especially sulfur containing amino acids, is also an important nutrient involve in anticarcinogenic activity (Guzel-Seydim et al. 2003; Ahmed et al. 2013).

The anticarcinogenic mechanism of kefir are not fully understood, some studies had shown that kefir promotes a reduction of tumor growth and cancer prevention through the delay conversion of pro-carcinogenic to carcinogenic activity of enzymes in early stages of tumor formation (De Moreno de LeBlanc 2006; Ahmed et al. 2013; Leite et al. 2013; Ferreira 2014).

In most of studies, the anticarcinogenic effects are associated to the cell free fraction. In order to kefir beverage promotes antitumoral effects, it should be consumed twice a day for at least 7 days as shown in mice model (De Moreno de LeBlanc 2006). Epidemiological studies had shown that intake of fermented milk products may reduce the risk of breast cancer in women. The inhibition of tumor growth also is associated to the induction of apoptotic cell lysis in tumors and high levels of IgA, suggesting the interface of intestinal mucosa site to promote the immune system activation (Ahmed et al. 2013; Leite et al. 2013). In breast cancer, De Moreno de LeBlanc et al. (2006) showed that the modulation of mucosal immunity and endocrine systems occurs due to the consumption of bioactives metabolites of kefir, and it is associated with the decrease of IL-6 expression in oestrogen-dependent tumors.

Immune system activation

Studies have shown that innate immunity protects against pathogens through action of cytokines, antimicrobial peptides, chemokines, macrophages, dendritic cells and natural killer cells in the tissues and circulation (Vinderola et al. 2006). The probiotic functional food, such as kefir beverage, acts in the immune system through activation of innate immunity. The functionality of probiotics could be identifying in oral mucosa until final portion of the gut and could exert influence in distant immune sites (De Moreno de LeBlanc 2006). The kefir induce a typical Th2 response by increasing IgA B cells and production of IL-4, IL-6, IL-10 cytokines, however, pasteurized kefir can also stimulate cells of the innate immune system, promoting a basal inflammatory state, increasing Th1 phenotype response or promoting cell-mediated immune responses against tumors and intracellular pathogen infections without tissue damage (Perdigon et al. 2001; Liu et al. 2002; Vinderola et al. 2006; Leite et al. 2013).

The cells of Peyer's patches, as B cells, T cell, natural killer cells, and dendritic cells activated by kefir intake could migrate from the small intestine, to the respiratory, gastrointestinal, and genitourinary tract, as well as to exocrine glands such as the lachrymal, salivary, mammary, and prostatic glands (Brandtzaeg and Pabst 2004; De Moreno de LeBlanc 2006). The number of IgA B cells is also increased and could

De Moreno de LeBlanc 2006). The number of IgA B cells is also increased and could be found in other distant mucosal sites such as the bronchus and mammary gland (De Moreno de LeBlanc 2007).

The cytokines profile promoted by kefir intake could stimulate the production of proinflamatory cytokines as IL-6 and TNF-α. The activation of cytokines production is via TLR2 (Leite et al. 2013).

Anti-hypertensive effect

The kefir intake has been associated with having an anti-hypertensive effect. This effect is promoted by the exopolysaccharides and peptides produced during the kefir fermentation. The exopolysacharide kefiran was the most studied and showed the property to reduce blood pressure by inhibiting the ACE activity (Maeda et al. 2004; Punaro et al. 2014).

More than 250 bioactive peptides produced during kefir fermentation are described. And at least 12 sequences of these peptides are described as ACE-inhibitors that influence the blood pressure by inhibiting the conversion of angiotensin I to the vasoconstrictive angiotensin II and the degradation of the vasodilator bradykinin to its inactive fragments. The peptides with this properties identified are KAVPYPQ, NLHLPLP, SKVLPVPQ, LNVPGEIVE, YQKFPQY, RDMPIQAF, NIPPLTQTPV, SQSKVLPVPQ, VYPFPGPIPN, HKEMPFPKYPVEPF, YQEPVLGPVRGPFPIIV, and LLYQEPVLGPVRGPFPIIV (Ebner et al. 2015).

Anti-diabetes effect

Diabetes Mellitus is a global heath problem, characterized by a number of metabolic disorders, associated to microvascular and neurologic complications, affecting the eyes, kidneys, heart and nervous system. It is identified by the presence of hyperglycemia (Punaro et al. 2014). Now researchers have begun to bring forward scientific evidences that kefir could be used as a complementary adjuvant in drug therapy of diabetes. Experiments in animals revealed that kefir treatment in diabetic animal reduced the fasting blood glucose, promoting a glycemic control with reduced polyuria, polydipsia and polyphagia, and as such improving quality of life. Kefir intake also promotes a partial improvement in renal function, increased nitric oxide (NO) excretion and reduction of superoxide anion, lipid peroxidation and inflammation (Punaro et al. 2014; Ostadrahimi et al. 2015).

The mechanism of how kefir modulate's hyperglycemia is still not fully understood. One of possible explanation is that microorganisms of kefir affect the gut microbiota to produce insulinotropic polypetides and glucagon-like peptides to induce uptake of glucose by the muscle, they also promote the consumption the glucogen by the liver which reduces the blood glucose (Punaro et al. 2014; Ostadrahimi et al. 2015).

High blood glucose causes the formation of advanced glycation end products (AGEs) that triggers the production of reactive oxygen species (ROS) and increase auto-oxidation in various tissues with subsequent impaired nitric oxide (NO) bioavailability. High production of ROS participates in lipid peroxidation (LPO) of cellular membranes. The LPO produce malondialdehyde (MDA), and its secondary products such as thiobarbituric acid reactive substances (TBARS), which is utilized as a marker of LPO and is a highly toxic molecule. ROS are also responsible for the activation of nuclear factor-kappa B (NF-κB), which increases the expression of pro-inflammatory biomarkers, such as tumor necrosis factor-alpha (TNF-α) and inter-leukin-6 (IL-6) that augments the expression of C-reactive protein (CRP) (Punaro et al. 2014).

Gastrointestinal tract

The microbiota related to kefir intake, as occurs with other probiotics, are a combination of many factors as actions of bacteriocins and organic acids that direct inhibit the pathogens present in microbiota in the intestinal mucosa (Rattray and O'Connel 2011). These beneficial components have also shown to present bacteriostatic effect on Gram-negative organisms, but a better bactericidal effect against Gram-positive organisms (Czamanski et al. 2004). The main pathogens, *Enterobacteria* and *Clostridium* can be affected by kefir consumption. Activity against *Yersinia enterocolitica*, *Escherichia coli*, *Listeria innocua*, *Listeria monocytogenes, Staphylococcus aureus*, *Bacillus cereus* and *Salmonella Enteritidis,* were also related (Czamanski et al. 2004). In Russia, kefir is used as coadjuvant in treatment of peptic ulcers in the stomach and duodenum of human patients (Leite et al. 2013).

Hypercholesterolemic

The consumption of kefir beverage has a positive effect on the control of cholesterol level and fraction rate. With regular consumption of kefir, the total cholesterol, low density lipoproteins (LDL) and triglycerides decreases and high density lipoprotein (HDL) and protein C reactive (CRP) increase (Punaro et al. 2014; Ostadrahimi et al. 2015).

Possible mechanisms proposed for the hypocholesterolemic activity may involve the inhibition of the exogenous cholesterol absorption in the small intestine, by the binding and incorporation of cholesterol to bacterial cells and cholesterol uptake by probiotic bacteria used to their own metabolism (Ostadrahimi et al. 2015). In addition, the deconjugation of cholesterol to the bile salts, suppress the bile acid re-absorption (Leite et al. 2013; Ostadrahimi et al. 2015).

Lactose intolerance

About 70% of the Brazilian population has some level of lactose intolerance due to the Asian and African genetic heritage of the population. Lactose is the main carbohydrate disaccharide present in milk and the lactose intolerance is a disorder associated with the inability to digest lactose (Ahmed et al. 2013; Leite et al. 2013).

The β-galactosidase is an enzyme that hydrolyses the lactose and is present in an active form in kefir grains after the fermentation process. This enzyme present in kefir beverage has the property to promote a decrease in the hydrogen expired and flatulence level in lactose intolerant subjects (Ahmed et al. 2013; Leite et al. 2013).

Manufacturing Technology

Kefir could be produced in large scale by artisanal methodology, commercial process by Russian methodology or by industrial methodology using pure cultures (Otles and Cagindi 2003; Rattray and O'Connel 2011; Leite et al. 2013).

In artisanal methodology (Fig. 5.1), about 5% of the kefir grains are inoculated in milk for 18 to 24 h at room temperature (20 °C to 25 °C) (Sarkar 2008; Leite et al. 2013; Rocha et al. 2014). At the end of the fermentation process the grains are washed and re-inoculated in fresh milk and the kefir beverage is ready to be consumed and can also be be stored at 4 °C (Beskova et al. 2002; Otles and Cagindi 2003; Leite et al. 2013).

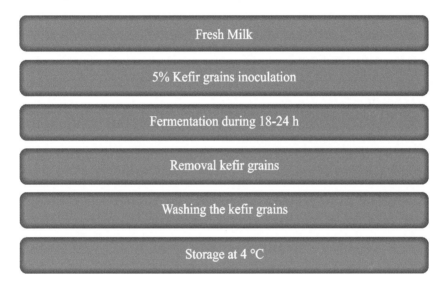

FIGURE. 5.1 Kefir production by artisanal methodology.

In artisanal methodology, some problems could occur due to agitation during fermentation which could influence the kefir microbial composition, increasing the fermentation by the Lactococci and yeast (Tamime 2006; Rattray and O'Connel 2011; Leite et al. 2013). Incubation above 30 °C stimulates thermophillic LAB in detriment of the yeast and the mesophillic LAB (Leite et al. 2013). Also the initial inoculums could be more than 5% which in turn could influence the pH, viscosity, lactose concentration and the microbial profile of the kefir beverage (Simova et al. 2002; Leite et al. 2013).

The Russian method of kefir production consists on a larger scale. The fermentation occurs in two steps: first of all, the mother culture is prepared of using 3 to 5% of kefir grains are inoculated into pasteurized milk which is allowed to ferment for up to 18 to 24 h at room temperature under stirring, then the grains are removed. The second step consists in fermenting 1 to 3% of the mother culture in fresh pasteurized milk during 12 to 18 h at room temperature to acquire the kefir beverage. The shear forces created by the stirrer can mechanically damage the kefir grain structure and can influence their growth dynamics (Kołakowski and Ozimkiewicz 2011). This product could or not be maturated for 24 h at 4 °C (Rattray and O'Connel 2011; Leite el al. 2013; Ôzer and Kirmaci 2014).

The industrial methodology to produce kefir beverage is the same process used in the artisanal method—using the kefir grains associated to pure cultures (Tamime 2006; Beskova et al. 2002; Rattray and O'Connel 2011; Leite et al. 2013). After fermentation, the kefir beverage is maintained 8 °C to 10 °C in a trough 24 h to maturation with the propose to enhance flavor produced by the yeasts (Behkova et al. 2002; Leite et al. 2013). During the storage, the yeast and heterofermentative LAB produce CO_2 that promote a bloating in the product package (Leite et al. 2013).

Commercial kefir beverages are found in many countries. In Brazil kefir is produced only by the artisanal method as an effect of kefir grains development. The commercial shelf life of kefir beverage is about 28 days under refrigeration; however, the artisanal kefir beverage is recommended to be consumed between 3 to 12 days after fermentation and is to be maintained under refrigeration (Rattray and O'Connel 2011; Leite et al. 2013).

Conclusion

Kefir is a functional food, which has being best characterized for its health benefits. It has been given different names in countries across the globe—Snow lotus, Talai, Tibico, kefiras, and Tibi. However, the main physico-chemical characteristics and technology employed to produce this beverage is the similar.

The nutritional and functional benefits of kefir intake are attributed to the presence of proteins, vitamins, antioxidants, minerals and biogenic compounds which have a therapeutic potential to alleviate gastrointestinal disorder and lactose intolerance. In addition, kefir could be used as coadjutant in treatment of diabetes, some kinds of cancer, high cholesterol and high blood pressure and promote a boost in the immune system.

Reference

Ahmed, Z., Y. Wang, A. Ahmad, S.T. Khan, M. Nisa, H. Ahmad and A. Afreen. (2013). Kefir and health: A contemporary perspective. Critical Reviews in Food Science and Nutrition 53: 422–434.

Apella, M.C., S.N. Gonzales, M.E. Nader de Macias, N. Romero and G. Oliver. (1992). *In vitro* Studies on the inhibition of the growth of *Shigella sonnei* by *Lactobacillus casei* and *Lactobacillus acidophilus.* Journal of Applied Bacteriology 73:480–483.

Atanassova, M., X. Dousset, P. Montcheva, I. Ivanova and T. Haertle. (1999). Microbiological study of Kefir grains. Isolation and identification of high activity bacteriocin producing strain. Biotechnology and Biotechnological Equipment 13: 55–60.

Beskova, D.M., E.D. Simova, Z.I. Simov, G.I. Frengova and Z.N. Spazov. (2002). Pure cultures for making kefir. Food Microbiology 19: 537–544.

Botelho, P.S., M.L.S. Maciel, L.A. Bueno, M.F.F. Marques, D.N. Marques and T. M. Sarmento Silva. (2014). Characterisation of a new exopolysaccharide obtained from of fermented kefir grains in soy milk. Carbohydrate Polymers 107: 1–6.

Brandtzaeg, P., and R. Pabst. (2004). Let's go mucosal: communication on slippery ground. Trends in Immunology 25(11): 570–577.

Carneiro, R.P. (2010) Desenvolvimento de uma cultura iniciadora para produção de kefir. M.Sc. Dissertation. UFMG, Belo Horizonte, MG.

Cotter, P.D., C. Hill, and R.P. Ross. (2005). Bacteriocins: developing innate immunity for food. Nature Reviews in Microbiology 3: 777–788.

Cross, K., N. Huq, and E. Reynolds. (2007). Casein phosphopeptides in oral health—chemistry and clinical applications. Current Pharmacology and Diseases 13(8): 793–800.

Czamanski, R.T., D.P. Greco and J.M. Wiest. (2004). Evaluation of antibiotic activity in filtrates of traditional kefir. Higiene Alimentar 18: 75–77.

De Moreno de LeBlanc, A., C. Matar, E. Farnworth and G. Perdigon, (2007). Study of immune cells involved in the antitumor effect of kefir in a murine breast cancer model. Journal of Dairy Science 90: 1920–1928.

De Moreno de LeBlanc, A., C. Matar, E. Farnworth and G. Perdigon (2006). Study of cytokines involved in the prevention of a murine experimental breast cancer by kefir. Cytokine 34: 1–8.

Ebner, J, A.A. Arslan, M. Fedorova, R. Hoffmann, A. Küketin and M. Pischetsrieder. (2015). Peptide profiling of bovine kefir reveals 236 unique peptides released from caseins during its production by starter culture or kefir grains. Journal of Proteomics 117: 41–57.

Eisele, T., T. Stressler, B. Kranz and L. Fischer. (2013). Bioactive peptides generated in an enzyme membrane reactor using *Bacillus lentus* alkaline peptidase. European Food Research and Technology 236(3): 483–490.

Ennahar, E.R., T. Sashihara, K. Sonomoto and A. Ishizaki. (1999). Class IIa bacteriocins: biosyntheis, structure and activity. FEMS Microbiology Reviews 24: 85–106.

Ferreira, C.L.L.F. (2014). Quefir como alimento functional–componentes bioativos e efeitos fisiológicos. In: Costa, N.M.B. and C.O.B. Rosa (eds.), Alimentos Funcionais. Rubio Ltda, Rio de Janeiro.

Garrote, G.L., A.G. Abraham and G.L. De Antoni. (2000). Inhibitory power of kefir: the ratio of organic acids. Journal of Food Protection 63: 364–369.

Gibson, G.R., J.M. Saavedra, S. MacFarlene and G.T. MacFarlene. (1997). Probiotics and intestinal infections. In: Probioitcs 2: Applications and Practical Aspects, R. Fuler (ed.), Chapman & Hall, New York.

Gilliland, S.E., and M.L. Speck (1977). Antagonistic action of *Lactobacillus acidophilus* toward intestinal and foodborne pathogens in associative culture. Journal of Food Protection 40: 820–823.

Gupta, P.K., B.K. Mital and S.K. Garg (1996). Inhibitory activity of *Lactobacillus acidophilus* against different pathogens in milk. Journal of Food Science and Technology 33: 147–149.

Guzel-Seydim, Z., A.C. Seydim and A.K. Greene. (2000). Organic acids and volatile flavor components evolved during refrigerated storage of kefir. Jounal Dairy Science 83: 275–277.

Hayes, M., C. Stanton, G.F. Fitzgerald and R.P. Ross. (2007). Putting microbes to work: dairy fermentation, cell factories and bioactive peptides. Part II: bioactive peptide functions. Biotechnol Journal 2(4): 435–449.

Helander, I.M., A. von Wright, and T.M. Mattila-Sandholm. (1997). Potential of lactic acid bacteria and novel antimicrobials agaist Gram-positive bacteria. Trends in Food Science and Technology 8: 146–150.

Heng, N.C.K., P.A. Wescombe, J.P. Burton, R.W. Jack, and J.R. Tagg. (2007). The diversity of bacteriocins in Gram-positive bacteria. In: M.A. Riley and M.A. Chavan (eds.), Bacteriocins Ecology and Evolution. Springer, Berlin.

Juven, B.J., F. Schved and P. Linder. (1992). Antagonistic compounds produced by a chicken intestinal strain *Lactobacillus acidophilus*. Journal of Food Protection 55: 157–161.

Kabak, B., and A.D.W. Dobson. (2011). An introduction to the traditional fermented foods and beverages of Turkey. Critical Reviews in Food Science and Nutrition 51: 248–260.

Kim, D.H., J.W. Chon, H. Kim, H.S. Kim, D. Choi, D.G. Hwang and K.H. Seo. (2014). Detection and enumeration of lactic acid bacteria, acetic acid bacteria and yeast in kefir grain and milk using quantitative Real-Time PCR. Journal of Food Safety 35: 102–107.

Kneifel, W., and H.K. Mayer. (1991).Vitamin profiles of kefir made from milks of different species. International Journal of Food Science Technology 66: 423–428.

Kołakowski, P., and M. Ozimkiewicz. (2011). Restoration of kefir grains subjected to different treatments. International Journal of Dairy Technology 6(1): 140–145.

Koroleva, N.S. (1988). Technology of kefir and kumys. Buletin of the International Dairy Federation 277, Chapter VII, 96-100.

Le Blanc, J.G., J.E. Laiño, M. Juarez del Valle, V. Vannini, D. van Sinderen, M.P. Taranto, G. Font, G. de Valdez, Savoy de Giori and F. Sesma. (2011). B-Group vitamin production by lactic acid bacteria–current knowledge and potential applications. Journal of Applied Microbiology 111: 1297–1309.

Leite, A.M.O., M.A.L. Miguel, R.S. Peixoto, A.S. Rosado, J.T. Silva and V.M.F. Paschoalin. (2013). Microbiological, technological and therapeutic properties of kefir: a natural probiotic beverage. Brazilian Journal of Microbiology 44(2): 341–349.

Likotrafiti, E., P. Valavani, A. Argiriou and J. Rhoades, (2015). *In vitro* evaluation of potential antimicrobial synbiotics using *Lactobacillus kefiri* isolated from kefir grains. International Dairy Journal 45: 23–30.

Liu, J.R., S.Y. Wang, Y.Y. Lin and C.W. Lin. (2002). Antitumour activity of milk kefir and soya milk kefir in tumour-bearing mice. Nutrition and Cancer 44: 183–187.

Lopitz-Otsoa, F., A. Rementeria, N. Elguezabal and J. Garaizar. (2006). Kefir: a symbiotic yeasts-bacteria community with alleged healthy capabilities. Revista Iberoamericana de Microbiologia Veterinaria 23(2): 67–74.

Maeda, H., X. Zhu, K. Omura, S. Suzuki and S. Kitamura. (2004). Effects of an exopolysaccharide (kefiran) on lipids, blood pressure, blood glucose, and constipation. BioFactors 22: 197–200.

Magalhães, K.T., G.V.M. Pereira, D.R. Dias and R.F. Schwan. (2010). Microbial communities and chemical changes during fermentation of sugary Brazilian kefir. World Journal of Microbiology and Biotechnology 33: 1–10.

McAuliffe, O., R.P. Ross and C. Hill. (2001). Lantibiotics: structure, biosynthesis and mode of action. FEMS Microbiology Reviews 25: 285–308.

Miguel, M.G.C.P., P.G. Cardoso, K.T. Magalhães and R.F. Schwan. (2011). Profile of microbial communities present in tibico (sugary kefir) grains from different Brazilian States. World Journal of Microbiology and Biotechnology 27: 1875–1884.

Naidu, A.S., W.R. Bidlack and R.A. Clemens. (1999). Probiotic spectra of lactic acid bacteria (LAB). Critical Review in Food Science and Nutrition 38: 26–34

Nielsen, B., G.C. Gürakan and G. Ünlü. (2014). Kefir: A multifaceted fermented dairy product. Probiotics & Antimicrobial Proteins 6: 123–135.

Ostadrahumi, A., A. Taghizadeh, M. Mobasseri, N. Farrin, L. Payahoo, Beyramalipoor Z. Gheshlaghi and M. Vahedjabbari. (2015). Effect of probiotic fermented milk (Kefir) on glycemic control and lipid profile in type 2 diabetic patients: A randomized double-blind placebo-controlled clinical trial. Iran Journal of Public Health 44(2): 228–237.

Otles, S. and O. Cagindi. (2003). Kefir: a probiotic dairy-composition, nutritional and therapeutic aspects. Pakistan Journal of Nutrition 2(2): 54–59.

Özer, B., and H. Kirmaci. (2014). Products of Eastern Europe and Asia. In: R. Robinson (ed.), Encyclopedia of Food Microbiology, Elsevier, London..

Perdigon, G., R. Fuller and R. Raya. (2001). Lactic acid bacteria and their effect on the immune system. Current Issues of Intestinal Microbiology 2: 27–42.

Pidoux, M. (1989). The microbial flora of sugary kefir grain (the gingerbeer plant): biosynthesis of the grain from *Lactobacillus hilgardii* producing a polysaccharide gel. MIRCEN Journal 5: 223–238.

Powell, J.E., R.C. Witthuhn, S.D. Todorov and L.M.T Dicks. (2007). Characterization of bacteriocin ST8KF produced by a kefir isolate *Lactobacillus plantarum* ST8KF. International Dairy Journal 17: 190–198.

Powell, J.E. (2006). Bacteriocins and bacteriocin producers present in kefir and kefir grains. M.S. Thesis, Stellenbosch University, Stellenbosch South Africa.

Powell, J.E., S.D. Todorov, C.A. van Reenen, L.N.T. Dicks and R.C. Witthuhn. (2006). Growth inhibition of *Enterococcus mundtii* in Kefir by *in situ* production of bacteriocin ST8KF. Le Lait 86(5): 401–405.

Punaro, G.R., F.R. Maciel, A.M. Rodrigues, M.M. Rogero, C.S.B. Bogsan, M.N. Oliveira, S.S.M. Ihara, S.R.R. Araujo, T.R.C. Sanches, L.C. Andrade, E.M.S. Higa. (2014). Kefir administration reduced progression of renal injury in STZ-diabetic rats by lowering oxidative stress. Nitric Oxide 37:53-60.

Rattray F.P., and M.J. O'Connell (2011). Fermented milks - Kefir. In: Fuquay J.W. (ed.), Encyclopedia of Dairy Sciences. Academic Press, San Diego.

Rocha, D.M.U.P., J.F.L. Martins, T.F.L. Santos and A.V.B. Moreira (2014). Labneh with probiotic properties produced from kefir: development and sensory evaluation. Food Science and Technology 34(4): 694–700.

Rodrigues, K.L., L.R.G. Caputo, J.C.T. Carvalho, J. Evangelista and J.M. Schneedorf. (2005). Antimicrobial and healing activity of kefir and kefiran extract. International Journal of Antimicrobial Agents 25: 404–408.

Rojas-Ronquillo, R., A. Cruz-Guerrero, A. Flores-Nájera, G. Rodríguez-Serrano, L. Gómez-Ruiz and J.P. Reyes-Grajeda, et al. (2012). Antithrombotic and angiotensin-converting enzyme inhibitory properties of peptides released from bovine casein by *Lactobacillus casei* Shirota. International Dairy Journal 26(2): 147–54.

Saavedra, J.M. (1995). Microbes to fight microbes: a not so novel approach to controlling diarrheal diseases. Journal of Pediatric Gastroenterology and Nutrition 21: 125–129.

Sablon, E., B. Contreras and E. Vandamme. (2000). Antimicrobial peptides of lactic acid bacteri: mode of action, genetics and biosyntesis. Advances in Biochemistry Engineering and Biotechnology 68: 21–60.

Saloff-Coste, C.J. (1996). Kefir. www.danonevitapol.com/extranet/vitapole/portail.nsf. ACCUEIL

Santos, A., M. san Mauro, A. Sanchez, J.M. Torres and D. Marquina. (2003). The antimicrobial properties of different strains of *Lactobacillus* spp. isolated from Kefir. Systematic and Applied Microbiology 26: 434–437.

Santos, J.P.V., T.F. Araújo, C.L.L.F. Ferreira and S.M. Goulard. (2013). Evaluation of antagonistic activity of milk fermented with kefir grains of different origins. Brazilian Archives of Biology and Technology 56(5): 823 –827.

Serpa, A.J. Kefir. Available at: http://chanasaude.no.sapo.pt/kefir.htm. Acessed on 05/09/2015.

Shahani, M., and R.C. Chandan. (1979). Nutritional and healthful aspects of cultured and culture-containing dairy foods. Journal of Dairy Science 62: 1685–1694.

Shiomi, M., K. Sasaki, M. Murofushi and K. Aibara. (1982). Antitumor activity in mice of orally administered polysaccharide from Kefir grain. Japan Journal of Medical Science and Biology 35: 75–80.

Simova, E., D. Beshkova, A. Angelov, Ts. Hristozova, G. Frengova and Z. Spasov. (2002). Lactic acid bacteria and yeasts in kefir grains and kefir made from them. Journal of Industrial Microbiology and Biotechnology 28(1): 1–6.

Tamime, A.Y. (2006). Production of kefir, koumiss and other related products. *In*: A.Y. Tamime (ed.), Fermented Milk. Blackwell Science Ltd, Oxford.

Todorov, S.D. and L.M.T. Dicks. (2008). Evaluation of lactic acid bacteria from kefir, molasses and olive brine as possible probiotics based on physiological properties. Annals of Microbiology 58(4): 661–670.

Van Wyk, J. (2001). The inhibitory activity and sensory properties of Kefir, targeting the low-income African consumers market. M.S. Thesis, Stellenbosch University, Stellenbosch, South Africa.

Vinderola, G., G. Perdigon, J. Duarte, D. Thaangavel, E. Farnworth and C. Mattar. (2006). Effects of kefir fractions on innate immunity. Immunobiology 211: 149–156.

Welman, A.D. and I.S. Maddox. (2003). Exopolyssacharides from lactic acid bacteria. Perspectives and challenges. Trends in Biotechnology 21(6): 269 –274.

[Yemoos]. Milk kefir history. Available at: http://www.yemoos.com/milkkefirhistory.html. Accessed on: 05/09/2015.

Yuksekdag, Z.N., Y. Beyatli and B. Aslim. (2004). Determination of some characteristics coccoid forms of lactic acid bacteria isolated from Turkish kefirs with natural probiotic. Lebensmittel Wissenschaft und Technologie 37: 663–667.

6

Brazilian Charqui

Vanessa Biscola[a]*, Bernadette Dora Gombossy de Melo Franco*[a] *and Tatiana Pacheco Nunes*[b]

Abstract

Brazilian charqui is a traditional fermented, salted and sun-dried meat product which is greatly appreciated in the whole country. This product represents an important animal protein source for the poor rural population from the northeast region of Brazil and occupies a prominent place among industrial meat products of this country. The preservation of Brazilian charqui is based on the hurdle technology, where salt, dehydration and fermentation are hurdles sequentially applied to prevent the growth of undesired microorganisms. However, if these steps are not applied in controlled ways, it may lead to variations in the sensory characteristics, chemical composition and shelf life of the products. The manufacturing process comprises two salting steps (wet salting with brine; and dry salting, with coarse marine salt), piling of the meat for juices draining and exposure to the sun for drying. The water activity (a_w) of the final product is around 0.7–0.8 and it is considered as an intermediate moisture product that is self-stable and can be stored without refrigeration up to six months. During piling, a fermentation step, carried out by lactic acid bacteria naturally present in the raw meat, is responsible for the unique flavor characteristics of charqui. Due to its a_w, the growth of microbial pathogens in charqui is very unlikely, but halophilic and halotolerant spoilage bacteria, coming from the salt used in the manufacture of the product, can grow, leading to the appearance of slime, red spots on the product surface and off-flavors. Therefore, the use of good-quality raw material and the application of good manufacturing practices (GMP) are essential to guarantee the quality of the final product. The great majority of Brazilian charqui production is destined for the internal market and the exportation remains at a small scale, but the product presents potential for international trade. However, to increase its acceptance in the market it is necessary to ensure that all charqui producers comply with the official regulations, applying GMP along the whole process and making sure that the final product is in accordance with the specified parameters. However, it is not an easy

---

[a] Department of Food and Experimental Nutrition, Faculty of Pharmaceutical Sciences from University of São Paulo, 580 Prof. Lineu Prestes Avenue, São Paulo – SP, Brazil – 05508-000. E-mail: vbiscola@usp.br; bfranco@usp.br; Phone: +55 11 30912493 and +55 11 30912199.
[b] Departament of Food Technology from Federal University of Sergipe. Marechal Rondon Highway, São Cristóvão – SE, Brazil - 49100-000. E-mail: tpnunes@uol.com.br; Phone: +55 79 21056903.
* Corresponding Author: vbiscola@usp.br

task, since the product is still manufactured in many small and clandestine processing plants that do not follow the recommended parameter of hygiene and standardization of the product.

Introduction

Salted and dried meat products have been produced all around the world, since time immemorial, as a simple and efficient way of preserving fresh meat. The first evidence of the use of salt and dehydration in the preservation of meat products dates from 3rd century BC. The characteristics and consumption habits of dried meat products are frequently associated with specific ethnic groups and geographic regions; this is what classifies them as traditional products. Their artisanal manufacture comprises of simple technological steps, passed along generations, resulting in the unique characteristics of each product. The lack of defined specifications for the manufacture of these artisanal products leads to a great variation in the sensory characteristics, chemical composition and shelf life among them, which may cause confusion in differentiation between different types of products (Ishihara and Madruga 2013).

In Brazil, the most popular salted and dried meat products are charqui, jerked beef and *carne de sol* (sun-meat) (Shimokomaki et al. 2006). However, only charqui and jerked beef are quoted in Brazilian regulations; even though jerked beef possesses its own regulated parameters, it is considered as a variation of charqui. As for *carne de sol*, there is no official parameter of quality, or any specific regulations for the product (Brasil 1952; Brasil 2000; Shimokomaki et al. 1987).

Brazil is not the only country where charqui is produced. It is also manufactured in many other regions of South America; however, the products present differences among them, mainly with regards to the raw meat used for its production. Other examples of greatly appreciated intermediate moisture (salted and dried) meat products can be found all around the world, as *bresaola* from Italy, *cecina* from Mexico and Spain, the *pastirma* from Turkey, Egypt and Russia; *jerky* from United States, *pemmican* from North America, *tasajo* from Cuba, *nikku* from Canada, *sou nan* and *rou gan* from China, *fenalår* from Norway, *biltong* from South Africa, *kilishi* from western Africa, and *kaddid* from Africa and South of Asia (Carvalho Junior et al. 2002).

Basically, charqui can be defined as large flat meat pieces (*mantas*), produced mainly from beef flank or plate cuts, preserved by salting and drying, which is characteristically fully dehydrated on its outer surfaces and has a typical strong odor and taste. Traditional charqui is made without the addition of nitrites or nitrates and salt is the only curing agent; nevertheless, microbial counts decrease during processing and storage, as a result of moisture and water activity (a_w) reductions (Pinto et al. 2002; Youssef et al. 2003, 2007).

This meat product is considered a very important source of animal protein, particularly for the poor rural population in the northeast region of Brazil. But, it is also greatly appreciated all around the country and occupies a prominent place among industrial meat products in Brazil, with great market and popular acceptance. It is used in the preparation of a great variety of typical dishes, among which, the most popular is *feijoada*, made with black beans, charqui and other desalted dried parts of beef and pork carcasses. Prior to consumption charqui must be desalted and cooked (Torres et al. 1994).

This chapter will be focused on the relevant aspects of Brazilian charqui meat. The chapter will cover the origins and historical aspects, production and marketing data, steps of its manufacture, principles of its preservation and microbiological aspects. It also includes some interesting information about other salted and dried meat products of relevance in Brazil.

Origin and Historical Aspects

The original name of this meat product (*xarqui*) comes from the dialect *quíchua*, spoken by the native Indian populations that inhabited the Andean region of South American continent (Gouvêa and Gouvêa 2007). Brazilian charqui is a fermented, salted and sun-dried meat product greatly appreciated in the country. It represents one of the most consumed industrial meat products in Brazil. Sometimes also referred to as *jabá*, charqui is mainly consumed in the northeast, center-east and south regions of the country, and, according to some authors, it has a great potential for expansion in the market. Since its preservation does not require low temperatures, shipping charqui to other regions and countries would not account for the high costs and other drawbacks that are associated with keeping the product cold (Shimokomaki et al. 2006; Abrantes et al. 2014).

It is believed that Brazilian charqui started to be produced in northeast of Brazil in an attempt to overcome the difficulties of preserving fresh meat in this part of the country. The low income of the population, coupled with the favorable climatic conditions and the great amounts of available marine salt, found in northeast coast of Brazil, made the processes of drying and salting the meat an easy choice to preserve the excess of fresh meat that could not be readily consumed (Souza 2007; Gouvêa and Gouvêa 2007).

Charqui represented an important source of good-quality protein for the poor population and also had a crucial role in Brazilian economy, not only in northeastern parts but also in central east and mainly in the south of the country. During the colonial period of Brazil, the tradition of making this product enabled the expansion of cattle breeding and the settlement of rural population in regions where it was produced (Vargas 2014). Historically, Brazilian charqui has already occupied a solid position in the international trade of dried meat products. However, by the end of 19th century a crisis, caused mainly by the end of slavery in the country, affected the Brazilian charqui industries and resulted in a great loss of market for the products manufactured by other countries from South America (Vargas 2014).

Due to its low cost, charqui meat represented one of the most important protein sources in the diet of the slaves brought to Brazil during to the colonization period to work at both the sugar-cane and coffee fields. This fact boosted its production during the late colonial period. In the 18th century, it was observed that there was a great increase in demand at sugar-cane plantations in northeast and south east of Brazil, as well as on the Caribbean coast, this was due to the arrival of hundreds of thousands of African slaves to these regions, causing an increase in the demand for food. By this time, the great majority of charqui farms were located in northeast of Brazil, however, this region was suffering from adverse environmental conditions (long dry periods), which led to a decrease in the production and failure to supply the demand

(Osório 2007). As a consequence, other regions saw an opportunity to expand their markets and, by the end of 18th century, the cities of Pelotas (south region of Brazil) and Montevideo in Uruguay became the main producers of salted and dried meat products in South America. During the second half of 19th century, the charqui produced in south of Brazil lost its market in the Atlantic trade to products manufactured in Uruguay and Argentina, which were tastier and more affordable. With time, the Argentine producers improved their factories, started to export to Europe and overtook the other producers from South America. In the 1880s, the producers from the south region of Brazil were unable to compete during this crisis that affected the industry and lost their place in the market (Vargas 2014).

Nowadays, Brazilian charqui is mainly produced and consumed in northeast of Brazil, but it is also appreciated in the whole country, due to its high nutritive value and exquisite taste. Since 1880s, the techniques of manufacturing charqui have greatly improved, as well as the quality and safety of the product, which led some authors to suggest that Brazilian charqui presents the potential to regain relevance in the international market (Lira and Shimokomaki 1998; Felicio 2002).

Data of Charqui Consumption

Despite the advances in the technologies for food preservation, the consumption of traditional dried and salted meat products has persisted over time. Even though the steps of salting and drying of meat contribute to its preservation, they are currently applied more as a means to develop the unique sensory characteristics of the final products, which are greatly appreciated. The cultural aspect of such traditional products is also relevant in influencing the consumption habits. However, when these products are manufactured under non-controlled conditions, the final products may present variations in taste and poor quality, which leads to its rejection by consumers (Ishihara 2012).

According to the National Association of Dried Meat Industries (ANICS), 95% of Brazilian dried meat production is destined for the internal market and the export accounts for small quantities of the produced charqui (to some countries in Africa and Central America). However, the product has the potential to be better explored on the international trade market (ANICS 2012).

Brazil is a world leading country in bovine meat production. It ranks second in the world for cattle herd and meat production and is the first world's source for these products. Its cattle herd in 2013, accounted for more than two million of livestock, according to the Brazilian Institute of Geography and Statistics (IBGE 2010a). In the same year, beef production rates accounted for more than nine million tons and the export levels reached more than 1.5 million tons (Data from Brazilian Ministry of Agriculture – MAPA). Therefore, bovine meat and meat products are greatly consumed and appreciated in Brazil. Bovine cured meat accounts for 25% of the total meat produced in Brazil and charqui represents the main cured product manufactured by the establishments under federal inspection (Ishihara et al. 2013)

The annual production of charqui in Brazil is estimated at 600 thousand tons (representing a market worth two billion dollars), a number that would certainly be higher if we take into account the many informal producers that also contribute to the market (Fayrdin 1998). The per capita charqui consumption varies according to the regions of the country. The highest consumption is observed in northeast of Brazil (approximately three kilograms per capita/year), mainly in the states of *Pernambuco* and *Paraíba*; this data represents the values reported for both charqui and jerked beef, since the latter is considered a variation of the former. As for *carne de sol*, the estimated annual consumption is of 0.890 kilograms per capita. Once more, the states from northeast region of Brazil (*Bahia, Rio Grande do Norte and Paraíba*) are responsible for most of consumption (IBGE 2010b).

This data highlights the relevance of dried meat products, especially in the northeast region of Brazil, and supports the need for more investments, which would allow the improvement of the manufacturing technologies, product safety and quality, as well as its commercialization.

Charqui Manufacture

With the advances in the technologies of food preservation, the reasons for the application of salting and drying steps in meat products became more related to sensory purposes then to preservation ones. However, it still plays an important role in avoiding the growth of both pathogenic and spoilage microorganisms in processed foods (Ishihara et al. 2013). Besides, if these steps are not applied in a controlled way, it might lead to variations in the sensory characteristics, chemical composition and shelf life of the products, which might lower its quality and, consequentially, the consumer's acceptance of the finished product (Ishihara and Madruga 2013).

Charqui manufacture is an artisanal process based on the application of the hurdle technology. Salt, dehydration and fermentation are hurdles sequentially applied to prevent the growth of undesired microorganisms. Large pieces of meat are deboned, opened into flat flaps (*mantas*) and submitted to two salting steps (wet salting, with brine; and dry-salting, with coarse marine salt). During this process, meat pieces are stacked to enable the draining of juices, and then dried on racks exposed to the sun. These piles are inverted every 24 hours, and the entire process is conducted over several days. All steps are carried out at a non-controlled temperature and the final product is self-stable and can be stored without refrigeration up to six months (Youssef et al. 2007; Biscola et al. 2013). During piling, a fermentation step, carried out by the lactic acid bacteria (LAB) naturally present in the raw meat, is responsible for the unique flavor characteristics of charqui (Pinto et al. 2002; Biscola et al. 2014). Due to its low water activity (a_w), the growth of microbial pathogens in charqui is very unlikely, but halophilic and halotolerant spoilage bacteria, coming from the salt used in the manufacture of the product, can grow (Biscola et al. 2014).

Here, we describe the main steps of Brazilian charqui manufacture, highlighting the critical points that must be observed at each step in order to guarantee the safety and quality of the final product. The following information is based on what was described in Canhos and Dias (1985), Pardi et al. (1996), Riedel (1996), Paiva et al. (1997), Shimokomaki et al. (1998), and Gouvêa and Gouvêa (2007).

Raw Material

Meat and salt are the main raw materials used in charqui manufacture. Both must present good quality and follow the parameters established by the regulatory agencies. Any meat cut can be used for charqui manufacture. However, the tenderest and well-appreciated cuts are usually destined for other preparations and are more profitable when sold as fresh meat. Usually, plate and flank cuts are chosen to be processed in charqui manufacture. Market leftovers of forequarters and hindquarters, such as chuck and knuckle, may also be used in charqui production.

Preparation of Meat

Manual deboning and cutting of meat pieces are the first steps of charqui manufacture. At this point of the process, large meat pieces are opened into flat flaps, popularly known as *mantas*, that are approximately two to three centimeters thick. This increases the surface to be exposed to the salt in further steps in order to standardize the pieces. To help the salt to diffuse through the tissues, small fissures may be made in the thickest parts of the muscle.

Wet-salting

In this step, only pure salt (NaCl) is used. The concentration of the brine may vary, but is usually about 23.5 °Baumé (33.5% of NaCl). This is performed in special tanks with constant shaking to ensure circulation of the brine, which must be kept at 15 °C to guarantee an optimal penetration of salt into the muscle. The addition of lactic acid in the brine is allowed, up to a maximum concentration of 2% over the weight of the salt. This process is carried out until the meat absorbs approximately 20% of brine (in weight), which takes from 30 to 50 min. In order to speed up the process, an automatic injection of brine and tumbling systems may be applied. It allows a quicker and more standardized diffusion of the salt, increasing the efficiency of the process and avoiding fermentation of the brine and spoilage of the meat during this step.

Dry-salting

After wet-salting, the meat pieces are transferred to pallets, and left for a few minutes, to allow the excess brine to drip off. Then, the meat flaps are laid out on a concrete floor, covered with coarse marine salt, and piled up. These piles reach from 1.20 to 1.80 meters high and the pieces are separated from each other by a 1 mm thick layer of coarse marine salt. The first meat flap is positioned with the fat portion upwards, while the second flap has the fat turned downwards, the pile is built successively in this arrangement. This step is carried out for 12-24 hrs. To standardize the salt diffusion, the pressure over the pieces of meat and the draining of the juices, a re-salting step takes place. The piles are periodically inverted and restacked, so that the uppermost pieces are repositioned at the bottom of the new piles. At each repositioning, the salt layers between the meat pieces are refreshed. This procedure is repeated every 24 hrs, for several days, according to the criteria established by each industry.

At this point of the process, certain precautions must be observed in order to guarantee the success of dry-salting. The quality of salt is of utmost importance. It must be renewed at every restacking, and should not be re-used. Hypochlorite solution (5 g/mL) may be sprinkled in the salt before its use. At the moment of restacking, the meat pieces must be completely strecthed, avoiding folds and creases that might hinder salt diffusion, and making sure that all surfaces are in contact with the salt.

Reversion of Meat Pieces Position

This step allows further draining of juices and standardizes the distribution of salt through the muscle. It consists of restacking of the meat piles, every 24–48 hrs, without renewing the salt layers between the pieces. The number of *revertions* may vary from two to four. At this point, it is possible to observe the appearance of red spots on the meat surface, indicating the beginning of the typical spoilage caused by the growth of halophilic microorganisms. When this happens, it is necessary to take the appropriate measures to avoid the spread of the contaminants. Removing the affected pieces and exposing the meat to aerobic conditions might help to inhibit further growth of the spoilage microorganisms.

At the end of the *revertion* step, the meat pieces should proceed to sun-drying. However, if the environmental conditions are not optimal (rainy days) or when there is an excess of the product in the market, the producers may choose to delay this step and keep the meat in the piles for longer periods, without restacking. At this point usually, a thick layer of coarse marine salt, treated with sodium hypochlorite, is added on top of the piles to seal the meat. This optional step may last up to four months and, depending on the duration of the delay before sun-drying, is called "waiting pile" (short time) or "winter pile" (long time). However, this step should be carried out with precautions because, as the moisture of the meat is still relatively high, the chances of spoilage with undesirable bacteria are increased.

Washing

After the *revertions*, the meat pieces are submitted to a washing step, to remove the excess of salt from the surfaces. This step can be carried out in tanks containing water that has an addition of active chlorine (500 ppm) or in current potable water. In both cases, after washing, the meat is piled up once more to allow the excess of water to drip off. By this point of the process, the meat should have lost 20% of its weight, in comparison to the weight of the deboned raw material.

Sun-drying

The drying of the charqui meat is usually carried out outdoors, by the direct exposure of the meat to the sun and the wind. The meat is placed on stainless steel rails that are parallel, at intervals of 1.50 to 1.80 meters, this facilitate the employees' s access during production. The rails should be oriented in the north-south direction, for a better exposure to the sun. The meat remains exposed for approximately 12 hours a day, for approximately three to five days (or more, depending on environmental conditions). This step is carried out until the meat reaches an approximate moisture content of 45% and a_w between 0.70–0.75.

At night, the meat pieces are removed from the rails, piled up on a concrete floor and covered with a canvas. This step is named "resting" and can last from 24 to 72 hours. This is a very important step for the development of the sensory characteristics of charqui, since the fermentation of the meat actually takes place at this point, stimulated by the heating accumulated during the exposure to the sun. Besides, during the first days of drying, the meat still presents a high moisture content, which allows the growth of its autochthonous microbiota. The main microorganisms responsible for the fermentation of charqui are the halophilic and halotolerant microorganisms from the salt, along with some species from the autochthonous microbiota of the meat. During the resting of the meat, some endogenous enzymes also participate in the ripening process.

The time of exposure to the sun, the arrangement of the meat pieces on the rails and the duration of the resting period may influence the final characteristics of charqui. The procedures vary according to each producer, but one of the most applied procedures is carried out as follows:

- First exposure to the sun, with the muscle portion upwards, followed by 2–3 days of resting;

- Second exposure to the sun, only for the morning period, with the fat portion upwards, followed by 2–3 days of resting;

- Third exposure to the sun, for the whole day, with the muscle portion upwards, followed by 2–3 days of resting.

In the past, rails used for sun-drying were made of wood. However, this is not an appropriate material for use in the food industry and, has over time been replaced by suitable materials, such as stainless steel. Wood has a porous nature and may present fissures, where meat remnants may collect, leading to poor hygiene conditions that increase the chances of spoilage and contamination of products over time. It is also recommended now that these rails should be placed in a space surrounded by metalic meshs, in order to avoid insects and other animals. However, these conditions are not always observed, mainly in small farms and by clandestine producers who are not being inspected.

These days, with the advances in the technologies of charqui production, the drying step may be conducted in artificial dryers where the conditions of ventilation, environment moisture and temperature (usually at 37 °C) are controlled. These conditions contribute significantly to improve the hygiene of the process, the standardization of the drying and, therefore, the overall quality of the final product.

Packing and Marketing

Once the drying step is finished, with the product reaching the appropriate parameters of moisture (45%), a_w (0.70–0.75) and NaCl content (10–20% in the intramuscular portion), it is ready to be prepared for commercialization. Since it is an intermediate moisture product, charqui can be preserved at room temperature, for up to 6 months.

It is not uncommon to find charqui meat that has been produced for the market with no packaging, especially in northeast region of Brazil. However, it is highly recommended to use plastic bags to protect the product against insects or the exposure to other impurities in the environment. As for presentation for retail the charqui can be commercialized in a great variety of shapes, from small packages of 500 g–1 kg, upto the big loads of 30–60 kg.

Jerked beef, which is a variation of charqui meat, is also stored at room temperature and is usually commercialized in small amounts (500 g–1 kg) in vacuum packs. *Carne de sol*, on the other hand, is not an intermediate moisture product and must be preserved at low temperatures.

Yield and Chemical Composition

The average yield of charqui, in relation to deboned raw meat, ranges from 45–50%. According to the Brazilian regulations, the final product must present a maximum moisture of 45%, a_w between 0.70–0.75 and no more than 15% of fixed mineral residue, with a tolerance of 5%. It's chemical composition varies according to the cut of meat used in the manufacture, the fat content and the techniques of drying applied. The average composition is 20–40% of protein, 1–19% of fat and 9–21% of NaCl.

In Fig. 6.1, we present a flowchart summarizing the main steps of charqui manufacture. It is important to highlight that some variations may be found in the procedures adopted in each step, by different producers. Important aspects of each step are discussed in more detail in the description of each step, above.

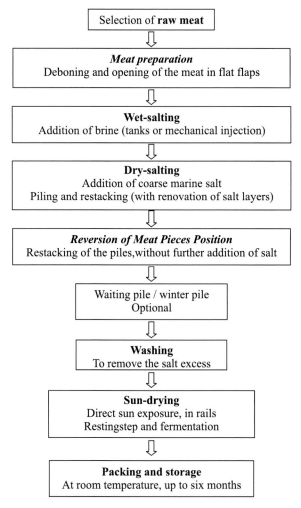

FIGURE 6.1 Flowchart summarizing the main steps of Brazilian charqui manufacture. Adapted from Shimokomaki et al. (1998).

Effect of the Salting Step on the Meat

The main effect of the addition of NaCl to the meat is the dehydration of the tissues and consequently reduction of a_w, which inhibits the growth of undesirable microorganisms and help in meat preservation. During charqui manufacture, the meat undergoes two salting steps: dry-salting, with coarse marine salt; and wet-salting, in brine. These processes remove 20 to 30% of the water from the product. Besides its preservation role, the dehydrating action of the curing steps decreases the volume of the meat, also reducing the costs of packaging, storage and transportation (Prata and Fukuda 2001; Gouvêa and Gouvêa 2007).

The size of the salt granules influences its diffusion into the muscle, mainly during the dry-salting step of charqui manufacture. Therefore, it has an effect in the preservation of the final product. Fine salt (one millimeter in diameter) penetrates quicker into the meat in the begining of the process; however, an increase in its concentration in the muscle leads to the coagulation of surface proteins, avoiding a deeper diffusion, which results in a shallow penetration and may cause inefficient and insufficient preservation of thick meat pieces (Pardi et al. 1996). Coarse marine salt is usually preferable for the dry-salting step. Its diffusion through the muscle is slower and does not cause the quick coagulation of surface proteins, thus, the penetration is deeper and more homogeneous. However, the long time required for the process might lead to the spoilage in the product, especially if this step is carried out under high temperatures (e.g. on hot summer days). The temperature of the meat also influences the diffusion of salt through the muscle. The use of refrigerated meat pieces is recommended, since the optimal penetration of NaCl occurs at 15 °C. For the wet-salting step, the concentration of the brine is a limiting factor for salt penetration in the muscle. The higher the concentration of salt, the quicker is its diffusion through the meat, up to the point when the osmotic equilibrium is reached (Gouvêa and Gouvêa 2007).

The purity of the salt is also an important parameter that influences the quality of the final product. It is recommended to use a salt with 99% of NaCl in its composition and levels no higher than 0.4% and 0.5% of calcium and magnesium salts, respectively. According to the regulations, salt destined for human consumption or used as an ingredient of any food product must be free from debris, pathogenic microorganisms or any other substances that might cause undesirable changes in the product or present a risk to human health (Brasil 1975).

It is important to highlight that the salt presents autochthonous microbiota, which is carried to the meat in the salting steps of charqui manufacture and, if growth conditions are favourable, those microorganisms may cause changes in the final product, influencing its quality (Biscola et al. 2014). Among the microbiota of salt, halotolerant bacteria, such as *Staphylococcus xylosus* and *Staphylococcus saprophyticus* can grow in NaCl concentrations higher than 10%; these microorganisms play an important role in the fermentation of dried and salted meat products and are related to the development of characteristic flavors. Additionally, the salt may contain spoilage microorganisms, such as the halotolerant *Micrococcus roseos* (able to grow in media containing 5 to 15%) and some halophilic bacteria as *Halobacterium cutirrubrum*, *Halobacterium salinarium*, *Sarcina litoralis* and *Serratia marcescens*, which require from 16% to 32% of NaCl for their growth. These microorganisms may multiply in charqui or other salted meat products and are frequently associated with a very characteristic spoilage, popularly known as *reddening*, when the products flavor goes off and it becomes slimy with the appearance of red spots on the product's surface (Pinto et al. 2002; Comi et al. 2005; Fontán et al. 2007; Talon et al. 2007)

The addition of salt to the meat causes a rapid acidification of the muscle, solubilizes the myofibrillar proteins, dehydrates the tissues, denatures proteins, causes loss of weight and changes in its color and texture (Gouvêa and Gouvêa 2007).

The salts ability to decrease water activity is thought to be due to the association of sodium and chloride ions with the water molecules that are available in the meat for chemical reactions. At 7–12% of NaCl, the water starts to exudate from the tissues, due to the osmotic pressure, and the meat shrinks. The exudate that comes out of the tissues is rich in soluble minerals and vitamins (especially those from B complex), non-protein nitrogen compounds, amino acids, albumin, globulin, myoglobin and hemoglobin (Gouvêa and Gouvêa 2007). It constitutes a very favorable environment for microbial fermentation, which takes place mainly during the dry-salting step of charqui manufacture and is responsible for the development of some of its sensory characteristics (Biscola et al. 2013).

The right concentration of salt in the brine, used for the wet-salting step of charqui manufacture, is very important to the success of the process. In order to maintain the appropriate concentration of NaCl, a deposit of salt might be settled in the pipes from where the brine is pumped to the tank, in constant circulation. During this step, the brine must be gradually replaced, since its volume is increased by the exudates that are carried from the meat. Since these exudates are rich in nutrients, if wet-salting is carried out under uncontrolled conditions, undesirable physicochemical and biological modifications may occur. Furthermore, the brine contains the microbiota brought in with the salt, which may grow and cause fermentation. Therefore, it is necessary to apply techniques to guarantee its preservation (Paiva et al. 1997).

According to the Brazilian Regulation of Industrial and Sanitary Inspaction of Products of Animal Origin - RIISPOA, the brine used in charqui manufacture must not be fermented, dirty, turbid, alkaline or have an ammonia odor (Brasil 1997). To guarantee its preservation and avoid fermentation, as well as the growth of halophilic and halotolerant microorganisms, the brine is maintained at 15°C, this also contributes to the penetration of salt into the meat. Acetic or lactic acids may be used to decrease the pH. The simple steps of heat treatment (boiling), followed by cooling and filtration (or sedimentation) are very efficient and the most used techniques to recover the brine (Gouvêa and Gouvêa 2007).

In the dry-salting step, the meat surface can be rubbed with salt or simply covered by it. In the manufacture of charqui only pure salt (NaCl) is used, as it is for jerked beef, nitrite and nitrate salts are also employed (Nobrega and Schneider 1983). Even though this step contributes to the preservation of the meat, by the reduction of product's a_w, some undesirable reactions may occur. At the high temperatures, reached during the exposure of the meat to the sun, the salt can act as an oxidative agent, activating some lipoxidases of the meat, leading to rancidity of the muscle's fat and to the appearance of off-flavors (Gouvêa and Gouvêa 2007).

The color of the meat is one of the first sensory characteristics observed by consumers during the moment of purchase, a basic criterion for its acceptance. The red color of the fresh meat comes from the muscular pigment myoglobine, which accounts for 80–90% of the total pigments, and is constituted by a protein portion (globyne) and a non-protein portion, the heme group that has an iron molecule (Sabadini et al. 2001).

The addition of salt to the meat causes the oxidation of the iron molecule from the heme group of myoglobin, originating from the metmyoglobin that presents brown coloration. This color is frequently associated with non-fresh meat (stored for long periods) and sometimes leads to rejection by the consumer. After cooking, the color of charqui meat is not distinguishable from that of freshly cooked meat. If the curing is carried out with nitrite and nitrate salts, other reactions take place. The nitrite is reduced to nitrous oxide and reacts with myoglobine to form nitrosyl myoglobin. With heat, this molecule is denatured into nitrosyl hemochrome, which is responsible for the characteristic pink coloration of cooked cured meat products (Sabadini et al. 2001).

The chemical composition of the meat after the dry-salting step suffers some modifications, such as a decrease in moisture, total proteins and collagen, followed by an increase in the levels of lipids, chlorides and ashes. The final composition of charqui will depend on the amount of salt used in its manufacture (Gouvêa and Gouvêa 2007).

The Role of Salt in Meat Preservation

The application of salting to preserve foods is a technique used since time immemorial. Since all metabolic reactions are water-dependent, the dehydrating characteristic of salt (NaCl) is the main ingredient responsible for the inhibition of both microbial growth and endogenous enzymes activity. The osmotic pressure exercised by saturated solutions is a physical action that causes the water to drain from the meat tissues. The water diffuses through the semipermeable membrane, while the solute (salt) enters into the cell until the osmotic equilibrium is reached (Fennema 1996; Potter and Hotchkiss 1995). Thus, the salting step of charqui manufacture is responsible for the reduction in the amount of water available for chemical reactions. Besides the a_w decrease in the meat, the salt also exerts osmotic pressure in the bacterial cells, leading to dehydration, with a bacteriostatic effect or even to plasmolysis and cell death (Gouvêa and Gouvêa 2007).

By the chemical effects of salting, NaCl forms a saline-proteic complex with the meat proteins and interferes in the activity of the endogenous proteolytic enzymes. In the microbial cells, it interferes both in the oxygen solubility and in the enzymes responsible for oxygen transport, delaying the growth of aerobic bacteria. At concentrations higher than 5%, it is also able to inhibit the growth of anaerobic microorganisms. The bacterial proteolytic enzymes are also affected. Besides, the increase in the concentrations of NaCl in the medium forces the cells to expend energy to exclude sodium ions from the cytoplasm, which can extend the duration of lag growing phase, reducing the growth rates (Shelef and Seiter 2005). Furthermore, being an ionic molecule, NaCl dissociates when in solution, releasing active chlorine, which possess bactericide action. However, some microorganisms are able to survive in high concentrations of NaCl. *Salmonella* spp. has already been reported to be able to do so for up to four weeks in brine; *Staphylococcus aureus* tolerates up to 10–20%

of NaCl, depending on other environmental conditions; *Mycobacterium tuberculosis* can remain viable in brine up to two months. Other microorganisms are able to grow in high salt concentrations. Halotolerant bacteria can multiply in the presence of 20% of NaCl, while halophilic microorganisms do not grow in concentrations under 15% of salt (Gouvêa and Gouvêa 2007).

Microbiological Aspects of Charqui

The radical conditions used during charqui processing are able to inhibit the development of pathogenic bacteria in the product; however, they allow the growth of halotolerant fermentative microorganisms. The selection of the autochthonous microbiota of charqui is an effect of the strict intrinsic factors that predominate in the product. These conditions are responsible for the stability of charqui at room temperature, since it inhibits the growth of undesired microorganisms and the activity of endogenous enzymes. Therefore, it is possible to affirm that charqui manufacture is based on hurdle technology (Lara et al. 2003; Pinto et al. 2002).

The hurdle technology was described by Leistner (1985, 1987) and consists of the simultaneous utilization of two or more barriers to control microbial growth in order to achieve the stability of the food. The ideal application of this principle occurs when the combined effect of the hurdles applied surpasses the isolate effect of each barrier alone, in a synergistic way. When this is achieved, the product obtained is stable at room temperature.

The most important hurdles that are often used in food technology are: temperature, water activity, pH, redox potential, competitive microorganisms and preservatives (Leistner and Gorris 1995). Note that salt and dehydration (water activity control) are the only hurdles sequentially applied to inhibit microorganisms that cause deterioration, while keeping the possibility of selecting desirable microbiota (Shimokomaki et al. 1998). In charqui manufacture, salting and drying steps are the main processes that ensure the stability of the product, all of them leading to a reduction in a_w. As many microorganisms present in raw meat can survive the dehydration process (not considering the use of high temperatures) it is important to ensure the quality of the raw materials just as it is important to maintain good hygiene conditions during the process (Sinigalia et al. 1998).

The microbiota of raw meat depends on the conditions in which the animals were raised and slaughtered, as well as on the way the meat is processed. The change of initial microbiota in the meat used for charqui manufacture occurs due to the salt addition, the decrease in water activity and the temperature used during the drying step (Silva 1997). An established a_w value between 0.70 and 0.75, maximum moisture content of 45% and salt concentration from 10–20% are recommended parameters by RIISPOA (Brasil 1952). The temperatures achieved during drying will depend on wheather this step is carried out indoors or outdoors and, in the last case, on the temperature reached by the sun heating (Silva 1997).

According to Bressan (2001), the drying step of charqui manufacture is crucial for a_w decrease; however it is also a critical point for contamination with undesirable microorganisms, especially if it is carried out on wooden rails exposed to the environment and without any barriers preventing contamination by insects and other animals. In order to reduce the chances of product contamination during this step, some precautions must be taken, such as the replacement of wooden rails with rails of stainless steel, protection of the meat from environment contamination (rain and standing waters, heavy weeds, insects, rodents, dust etc.).

Regulation RDC no. 12/2001 determines the microbiological criteria for foods in Brazil and establishes the parameters for processed meat products (raw hams, Salami, dehydrated sausages, charqui, jerked beef and similar products): *Salmonella* spp.–absence in 25 g; thermotolerant coliforms–10^3 MPN/g; and *Staphylococcus aureus* – 5×10^3 CFU/g (Brasil 2001).

The salt, is a unique ingredient, that acts to inhibit the growth of several microorganisms, but provides conditions for the development of a haloterant bacteria (which grow well in up to 15% salt) and halophilic microorganisms (need a minimum of 16% NaCl for growth). Besides, depending on the salt concentration, some pathogenic Gram-positive bacteria may also grow. As an example, *Staphylococcus aureus* that can grow in high salt content environments (tolerates 10–20% of NaCl) and produces heat stable enterotoxins, which are able to resist conventional thermal processing techniques and thus cause food poisoning. The presence of *S. aureus* in charqui indicates failure in good manufacturing practices and in the control of the hurdles applied for microbial inhibition in the product (Jay 2005; Pinto et al. 1998).

Another factor that influences the microbiota of charqui is the fermentation step that occurs at the end of *inversion of meat position* (during the resting of the meat pieces) and during the drying of the product. This is responsible for the development of the product's typical strong flavor and is carried out by the autochthonous microbiota, selected during the conditions of processing (Biscola et al. 2014), mainly *Micrococcus* spp., *Pediococcus halophyllus*, *Staphylococcus xylosus*, *Lactobacillus* spp. and *Streptococcus* spp. Besides its role in the development of sensory characteristics, the fermentation step may also inhibit the growth of undesirable microorganisms, such as *S. aureus* and other pathogens, which are usually weaker competitors (Sinigalia et al. 1998).

Despite the fact that charqui fermentation is carried out by the autochthonous microbiota that comes from the meat and from the salt used, some authors have already suggested the use of starter cultures of LAB in an attempt to better control this fermentation and improve charqui standardization, sensory characteristics, and overall quality. Among all the microorganisms that act in meat fermentation, LAB can improve safety, stability and product diversity (Pinto et al. 2002; Biscola et al. 2013).

Microbiota of Salted and Dried Meat Products

Carcass contamination occurs during the slaughter process with most contaminants coming from microorganisms in the skin, hide and intestinal content, including Enterobacteriaceae species, *Pseudomonass* spp., LAB and *Staphylococci*. During the processing of dried and salted meat products, some ingredients such as, sodium chloride, nitrate and nitrite are added to reduce water activity (a_w) and change the microbiota. Since many dried meats have low pH values (due to fermentation), the interacting effects of pH and a_w induce selective pressures that permit the survival and growth of LAB, yeasts, molds, *Micrococci* and non-pathogenic *Staphylococci*, while inhibiting Gram-negative bacteria (Blackburn et al. 1997).

Initial population of LAB in meat might range from 3 to 4 log CFU/g, but during fermentation can reach 7 to 8 log CFU/g. In addition, some strains of *Micrococcus* and *Staphylococcus* are lipolytic and proteolytic and contribute to the flavor development. For cured products, such as jerked beef, they reduce nitrate to nitrite, generating nitric oxide, which reacts with myoglobin to produce the characteristic pink color of cooked cured meats (Bacus 2005).

Staphylococcus aureus, *Escherichia coli* O157:H7, *Salmonella* spp., *Listeria monocytogenes*, *Campylobacter* spp. and the nematode *Trichinella spiralis* are the pathogens that cause great concern in meat. Some of them, such as *Staphylococcus aureus* can be controlled with the reduction of pH to less than 5.3; on the other hand, *Escherichia coli* O157:H7 is somewhat acid tolerant, but is inactivated by reducing the pH quickly followed by a heat treatment. Nonetheless, to reach the pH 5.3 as soon as possible it is necessary to use starter cultures, and neither starter cultures nor thermal treatment are used in charqui processing. In other words, food safety relies on reductions of a_w combined with salt addition (Ricke et al. 2007).

As a_w decreases due to the high salt concentration on the surface as well as the inside of products, along with the mechanical stress suffered during stacking, the only microorganisms able to grow are xerophilic molds, salt-tolerant yeasts and halophilic bacteria; only halophilic bacteria has been related to charqui deterioration (Blackburn et al. 1997).

Among the halophilic and halotolerant bacteria, *Halobacterium cutirrubrum* causes the appearance of red and pink spots on the product, and *Staphylococcus aureus* is the main representative; similarly *Debaryomyces hansenii* is an example of very salt-tolerant yeasts (Leistner and Bern 1970; Comi and Cantoni 1980; Biscola et al. 2014).

Another problem that can affect charqui is the occurrence of several myiasis, caused by fly larvae. In addition to the myiasis, insects and mites may also infest the product, therefore it is important to protect the large flat meat pieces (*mantas*) in order to avoid contamination (Pardi et al. 1996).

Other Salted and Dried Meat Products found in Brazil

Apart from charqui, two other salted and dried meat products are greatly appreciated in Brazil. One of these is jerked beef, which is a variation of charqui. The other is *carne de sol* (sun-meat), which presents a higher moisture and lower salt content than both charqui and jerked beef. Despite the fact that each one of these products presents its own chemical composition and unique sensory characteristics, the consumers are usually confused when trying to differentiate between them and, usually classify the three of them as "charqui" or just "dried meat".

In order to facilitate the differentiation of the three most popular salted and dried meat products from Brazil, Table 6.1 summarizes the distinct parameters regarding the chemical composition and the manufacture of charqui, jerked beef and *carne de sol*.

TABLE 6.1 Differences in the manufacture steps and chemical composition of the most consumed dried and salted meat products from Brazil

Parameters	Type of salted and dried meat product		
	Charqui	**Jerked beef**	**Carne de sol**
Raw material	Plate and Flank	Beef forequarters or hindquarters	hindquarters
Salting	Wet-salting, dry-salting and *inversion of meat pieces* / for several days	Wet-salting, dry-salting and *inversion of meat pieces* / for several days	Dry-salting / for 8–12 hours
Drying	Exposed to the sun	Exposed to the sun	Sheltered and well ventilated area
Shelf life at room temperature	6 months	6 months	3–4 days
Salt content	10–20%	18%	5–6%
Moisture	40–50%	Maximum 55%	Maximum 64–70%
Water activity	0.70–0.75	0.78	0.92
Additives	None	Sodium nitrite, maximum 10 ppm	None
Package	Plastic bags / with or without vacuum	Plastic bags with obligatory vacuum	None
Official regulations	* RIISPOA (Decree n° 30691 from 29 March 1952 – MAPA**)	(Normative ruling n° 22 from 31 July 2000 – MAPA**)	None

Adapted from Shimokomaki et al. (1998) and Salviano (2011); * RIISPOA: Regulation for Industrial and Sanitary Inspection of Products from Animal Origin; ** MAPA: Brazilian Ministry of Agriculture.

Even though the focus of this chapter is on charqui, the authors considered important to include some information about other similar products manufactured in Brazil. Due to its relevance in the national market, jerked beef and *carne de sol*, together with the traditional charqui, are considered the main dried and salted meat products produced in the country.

Jerked Beef

The history of Brazilian jerked beef began with the acceleration of migratory flows from the northeast to the southeast of the country that led to an increase in the demands for charqui meat in this region. Due to this fact, in the early 1970s, charqui's producers began to seek the approval of a new product to be regionally marketed with a higher moisture content than the maximum allowed for traditional charqui. Such approval would represent a reduction in the cost of manufacturing, increases in productivity and a higher yield in the manufacture process (Felicio 2002).

Due to its trade advantages, this modified charqui emerged on the market, without legal permission. The product presented similar characteristics, but a higher moisture content (about 55%, which was above the 45% required by the regulations). This product was known as "*fresh* charqui", which had a dark brown color and was easily spoiled. Since the high moisture accelerated the microbial spoilage of the product, some new steps (hurdles) were introduced in its manufacturing to increase shelf life. The new steps included the addition of nitrate and nitrite in the curing process, to improve the red color of the product, and the vacuum packaging to delay the deterioration process (Felicio 2002). However, the Agriculture Department refused to approve the use of curing agents, either because they wanted to preserve the identity of the traditional charqui, or because, at that time 1974/1975, and even today, there have been major concerns about the residual nitrite levels and the presence of nitrosamines in meat products (Felicio 2002). After several seizures of this adulterated product, the National Department for the Inspection of Products from Animal Origin (DIPOA) approved, in 1978, the use of nitrate/nitrite in the new product, but maintained the maximum value of moisture at 45% (same as the traditional charqui). This new regulation was clearly not respected. To differentiate this product from the original charqui it was necessary to give the product a new name. As it was also classified in the category of salted, cured and dried meat, the name jerked beef was chosen, since "jerky" was the name that British sailors used to refer to charqui, in the 18th century. Finally, in August 2000, the maximum moisture content was officially increased to 55% and jerked beef started to be commercialized according to the regulations (Felicio 2002). According to the current Brazilian regulations, jerked beef must have maximum moisture of 55.0%, 18.3% of ashes, 50 ppm of sodium nitrite and maximum value of a_w of 0.78. It must also be vacuum packed (Brasil 2000).

This is the history of the meat product jerked beef, that was developed to meet a contingency and became a success. In general, the vacuum packed jerked beef maintains a good quality and is compliant with the technical standards. It is important to notice that jerked beef has its identification on the label, so that consumers know they are not buying charqui, but a similar product that comes with its own sensory characteristics (Felicio 2002). Jerked beef processing steps are similar to those applied in charqui manufacture. It is also classified as an intermediate moisture meat product and can be stored at room temperature for several months. The raw material (bovine forequarter cuts) is stitched-pumped with brine containing salt and nitrite, followed by dry and wet curing in a refrigerated room, drying and vacuum packaging. Thus, the microbial stability and safety of jerked beef, as well as traditional charqui, is based on the hurdle technology. However, more hurdles are applied in the manufacture of jerked beef than in charqui production. These new technological barriers resulted in both safety and sensory quality improvements (Torres et al. 1994; Shimokomaki et al. 1998; Shimokomaki et al 2003).

Among the main functions of nitrite/nitrate, the antimicrobial activity is the most important, especially in relation to the inhibition of *Clostridium botulinum*. Besides, nitrite improves the appearance of jerked beef, giving it the characteristic red color of cured products, due to the reaction with myoglobin resulting in nitrosyl myoglobin. And also avoids lipid oxidation during storage, because of its antioxidant activity, since salt is considered a pro-oxidant agent. Moreover, during drying step, the meat is exposed to sunlight and afterwards covered with a canvas, for the night; these procedures can increase the temperature of the meat and catalyze the lipid oxidation reaction (Souza et al. 2013).

It is worth noting that lipid oxidation generates toxic products, such as malondialdehyde and cholesterol oxides, which are related to heart disease, cancer and premature aging. However nitrite not only has beneficial effects, but under certain conditions, this additive can react with secondary and tertiary amines resulting in carcinogenic compounds such as N-nitrosamines. Moreover, nitrates and nitrites can cause methemoglobinemia, especially in children. In order to reduce the prevalence of this problem, health authorities control nitrate concentrations in cured meat products (Souza et al. 2013).

Carne de Sol

Carne de sol is another popular salted semi-dried meat product produced in Brazil. It is also known by many names, such as: *carne de sertão, carne serenada, carne de viagem, carne de paçoca, carne-mole, cacina* or *carne acacinado*. This meat is a very popular and typical product of the Brazilian northeast region. Although this product is known and has been consumed since 17th century, the traditional manufacturing process still remains rudimentary and, unfortunately, little is known about the changes that occur during its processing. The manufacture is restricted to fast curing and exposure to the sun and the wind (Norman and Corte 1985).

This meat differs from jerked beef and charqui because it is lightly salted, partially dehydrated and cannot be characterized as an intermediate moisture product (Lira et al. 1998). Its shelf life is of about five days. *Carne de sol* is characterized by its distinctive shape (sold in flat flaps, *mantas*) and appearance (the dark brown color of its surface) (Norman and Corte 1985). As it has been observed for charqui, *carne de sol* has emerged as an alternative to preserve the excess from meat production, given the difficulties to refrigerate it, and associated with the high availability of sea salt. Despite the name *"carne do sol"* (sun-meat), this product is rarely exposed to the sun during the dehydration process; on the contrary, it is left in a covered and well ventilated area; due to this fact, the previous name *carne serenada* (meat of the dew) best expresses the process by which this meat is prepared (Gouvêa and Gouvêa 2007).

Carne de sol does not have technical regulations, so it's processing follows the same techniques applied since the beginning of its manufacture, which consists of curing and dehydrating the meat, by stacking of the flat pieces for a few hours. Despite the lack of official regulations, some parameters of the product can be found in the literature. According to Lira (1998), this product should have: moisture between 66.33 – 70.10 %, NaCl 4.69 to 8.45 % and a_w between 0.92 and 0.97. Note that these features make the *carne de sol* suitable for rapid microbial contamination and may serve as a source of foodborne diseases, since most pathogenic bacteria can grow until 0.94 a_w.

It is important to highlight that each Brazilian state developed its own technology for the production of *carne de sol*, which leads to variations among the products manufactured in different regions of the country. This variation comprises sensory characteristics (such as aspect, taste, color, amount of salt) and product shelf life. The states of *Rio Grande do Norte* and *Ceará* are the greatest producers of this product, mainly due to environmental conditions that are suitable for the dehydration step.

In spite of the poor standardization of the product and the lack of defined quality parameters, *carne de sol* overcame the status of a locally consumed product, which was used only in some typical regional recipes, has reached a wider market and has been appreciated by the whole country and included in the preparation of many dishes (Baroni et al. 2013).

In general, the production of *carne de sol* comprises four steps: preparing the raw material, salting, drying and marketing. In order to ensure the quality and the safety of this product, it is of utmost importance to standardize the steps of manufacture, mainly the drying period and sodium chloride content, which are the most important ones in inhibiting the development of undesirable microorganisms and the development of sensory characteristics (Costa and Silva 1999). *Carne de sol* is made with noble meats, tender and appreciated cuts, tender and appreciated cuts which is the opposite of charqui. The most used cuts are hindquarters, such as *coxão mole* - topside (*Pectineus, Adductor, Semimenbranosus, Gracilis*), *coxão duro* - flat (*Biceps femoris*), *patinho* - knuckle (*Quadriceps femoris, Rectus femoris, Vastus medialis, Vastus intermedius* and *Vastus lateralis*), *alcatra* - rump (*Gluteus* and *Tensor da fascia lata*) and *lombo* - sirloin (*Longissimus dorsi*) (Beefpoint 2010).

After the deboning of the raw material, the muscle pieces are cut in large flat pieces (*mantas*) with three to four centimeters of thickness. The flat pieces are submitted to salting, by rubbing coarse marine salt on its surface. This step is carried out manually, in order to ensure the complete and standardized coverage of the whole surface, as well as the even distribution of salt. After the addition of salt, the pieces are placed on a concrete floor also covered with a fine layer of coarse marine salt, with the fat layer upwards. The pieces are stacked into piles, separated from each other by a three millimeter layer of salt. The height and dimensions of the piles are likely to be oriented by operational scale, but in any case, it should not surpass 0.50 meters in height. After four to six hours, the piles are restacked and inverted, to permit an even pressure over the meat mass and to guarantee uniform salt penetration. The new pile is left standing, with a 10 cm layer of salt covering its surface, for a further period of four to six hours (Norman and Corte 1985). After salting, the meat pieces are rapidly washed to remove the excess of salt from its surface; this washing is followed by drainage and drying (Ishihara 2012). These steps are carried out over rails, in a sheltered and well-ventilated area. The meat may be also exposed to sun-drying, but this procedure is rarely adopted. As we can note, the steps of manufacture of *carne de sol* are very similar to those of charqui production. The main difference is the shorter duration of both salting and drying steps, which leads to a lower degree of dehydration. At the end of the process, *carne de sol* presents higher moisture and a_w, in comparison with charqui, and thus cannot be included in the intermediate moisture products category, needing preservation under low temperatures (refrigeration) (Gouvêa and Gouvêa 2007).

In order to reduce the chances of microbial growth in the product, an 1% solution of lactic or acetic acid may be incorporated in the washing water, used to rinse the meat pieces after salting. It decreases the pH of the meat to values lower than 5.5. However, this chemical control measure should be applied to increase the hurdles for microbial growth, but it does not substitute the need for the application of good manufacturing practices during the whole process and does not diminish the importance of the use of a good-quality raw material (Norman and Corte 1985).

The microbiota of *carne de sol* comes from the meat, the salt and from the environment in which it is manufactured. Therefore, it is influenced by the hygienic conditions of the whole process and the quality of the raw material employed (salt and meat). The conditions in which the animals were raised, slaughtered and processed have a decisive influence on the autochthonous microbiota of the meat (Silva 1997). Pathogenic microorganisms that may contaminate this product include *Clostridium perfringens, Staphylococus aureus, Salmonella* spp., verotoxin-producing *Escherichia coli, Campylobacter* spp., *Yersinia enterocolitica, Listeria monocytogenes, Aeromonas hydrofila*. The product can also support the growth of a great variety of spoilage bacteria, such as *Pseudomonas* spp. and *Brochothrix thermosphacta*, LAB, etc. (Gill 1998).

Data on the physical and chemical properties of *carne de sol*, sold in butcheries and supermarkets in *João Pessoa, Paraíba*–Brazil, showed that water activity in all samples was relatively high, between 0.898 and 0.967, and content of sodium chloride (NaCl) ranged between 3.73% and 9.79%. Consequently, NaCl employed in the process was insufficient to decrease water activity in the product and thus it did not have a significant inhibitory effect in the development of most microorganisms in the beef (Costa and Silva 1999).

The great majority of *carne de sol* available in the market comes from small butcher shops and is produced on small farms (or even homemade) without the observation of good manufacturing practices; this fact increases the incidence of foodborne illness associated with the product. Another fact that can be decisive for the high contamination of this product is its low salt content. The salting step decreases the water activity of meat to levels close to 0.96, which are insufficient to hinder the growth of many microorganisms. The main spoilage bacteria associated with this product is *Pseudomonas* spp. As for pathogens, *carne de sol* may offer favorable conditions for the development of Gram-positive bacteria, such as *Staphylococcus aureus* (Silva 1991).

In the literature, it is possible to find some data regarding the microbiological quality of *carne de sol*. Evangelista-Barreto et al. (2014) analyzed samples of *carne de sol* sold in *Cruz das Almas*, Bahia state (Brazil) and found *Escherichia coli* in 83.33% of the analyzed samples. On the other hand, coagulase positive *Staphylococci* was not detected, which is consistent with the legal requirements to this pathogen; however high counts of *Staphylococcus* spp. were observed (10^6–10^8 CFU/g), which suggested a risk for contamination with pathogenic species of this microorganism. Another alarming result reported by the authors of the studies was the presence of *Salmonellas* spp. in 25% of the analyzed samples. This data highlights the importance of a better regulation of the product, surveillance and inspection of the manufacturing plants and the adoption of good manufacturing practices.

Charqui from Other Countries in Latin America

It is possible to find charqui meat in other Latin American countries; however these products present variations in comparison to Brazilian charqui.

In Chile, charqui may be prepared with bovine or horse meat. The product undergoes a curing process that results in the reduction of water activity to 0.7 to 0.75 (values similar to those found in Brazilian charqui) and allows the occurrence of chemical and biochemical reactions, which are responsible for the development of its characteristic flavor (Martin et al. 2009). There is no clear regulation for its manufacture or for its quality parameters and final sensory characteristics. There is also a wide variety of ingredients, processing and proportions among the producers, which lowers the quality and acceptance of the final product. Besides, the non-controlled conditions for its manufacture may increase the risks of contamination of the meat with pathogenic bacteria, since the lack of standards may lead to an inadequate process, resulting in a_w values and NaCl contents insufficient to guarantee the stability of the products (Gianelli et al. 2012).

Another example of charqui meat produced in Latin America is the Andean charqui, also known as *Alpaca* charqui. This product is manufactured in some countries of the Andean region, such as Peru, Argentina, Bolivia and Chile. This is a salted and dried meat product, prepared by a process passed down by their ancestors, and is usually produced with the meat of llamas (*Lama glama*), alpacas (*Lama pacos*) or alpaca-llama hybrids. In Peru (the major producing country), the production of charqui ranges to 500 tons per year, and is concentrated in the rural zones of the Andean high lands (Pachao 2006; Salvá et al. 2012).

Its manufacturing process varies according to the different regions, but the typical steps are very similar to those employed for the production of Brazilian charqui. Firstly, the raw meat is cut in to flat flaps; following, an intensive salting step is carried out for several days and can be performed using ground salt or concentrated brine; then, the meat pieces are piled and pressed, also for several days, with periodical restacking of the flat flaps to guarantee homogeneity of draining; drying step is performed in trays, by exposing the meat to the air. At the very end, the final product is packed and commercialized at room temperature (Frenández-Baca 2005). Many different presentations and sizes may be found in the market. Andean charqui can either be prepared from deboned meat slices or from whole bone-in-carcass pieces (locally known as "*charqui completo*"). For the manufacture of *charqui completo*, part of the fat is removed and the muscle blocks are opened into a single flat flap (*manta*), by a series of deep incisions (Norman and Corte 1985). Similarly to Brazilian charqui, the Andean product also must be desalted, by soaking it in water for several hours, before cooking. After this step, the meat may be eventually cut into small pieces (Salvá et al. 2012).

This is different to what is observed for salted and dried meat products in Chile. Andean charqui manufactured in Peru follows national regulations that have established some compositional parameters. It is considered an intermediate moisture product and must present a minimum protein content of 45%, on a dry basis; the maximum values for fat and moisture contents are, respectively, 12% and 20%, on a dry basis (INDECOPI 2006; Fernández-Diez et al. 2012). As for its sensory parameters, Andean charqui should present a characteristic thatch coloration, no rancidity odor and a non-viscous texture. Besides this approximate composition and characteristics, further literature data on the chemical composition of Andean charqui is scarce (Salvá et al. 2012).

Conclusion

Salted and dried meat products have been produced since the ancient times. The simple techniques applied for its manufacture are based on the hurdle technology and ensure the stability of the products at room temperature. Therefore, they constitute an excellent alternative to preserving meat.

Brazilian charqui is greatly appreciated in the whole country, and presents an exquisite sensory characteristic and is an important source of good quality protein. Its commercialization is still restricted mainly to the national market, with incipient exports to a few countries in Latin America and Africa. However, this product has a potential to be better explored for international trading.

To increase its acceptance in the market it is necessary to ensure that all charqui producers comply with the official regulations, applying the good manufacture practices along all the processes and make sure that the final product is in accordance with the specified parameters. However, it is not an easy task, since the product is still manufactured in many small and clandestine processing plants that do not follow the recommended parameter of hygiene and standardization of the product. This practice leads to variations among the products available in the market and hinder both the quality and the health safety aspects of charqui.

As for the principles of charqui preservation, its good stability at room temperature is due to the effect of salting and drying steps on the meat. The addition of NaCl causes dehydration of the tissues, influencing the moisture and a_w of the product; induces biochemical changes in the muscle, which has an important effect on the development of the sensory characteristics; and inhibits the growth of undesirable microorganisms, selecting the autochthonous microbiota of charqui. The drying step is responsible for the reduction of moisture and a_w, and also has an effect over the microbiota in the final product.

Another important event that takes place during charqui manufacture is the natural fermentation of the meat. It makes a great contribution in the development of the unique sensory characteristics of charqui. The fermentation is carried out by the autochthonous microbiota composed mainly by lactic acid bacteria and non-spoilage halotolerant microorganisms, such as *Staphylococcus xylosus* and *Staphylococcus saprophyticus*. These bacteria are present in the raw materials and in the salt and are selected by the intrinsic factors predominating in charqui (low a_w and high NaCl content).

Even though the growth of pathogenic bacteria is very unlikely in this product, lack of hygiene and failures in the manufacture (insufficient moisture and a_w reductions) may lead to the contamination of the meat and growth of foodborne pathogens. Besides, under non-controlled conditions, halophilic microorganisms (such as *Halobacterium cutirrubrum*, *Halobacterium salinarium*, *Sarcina litoralis* and *Serratia marcescens*) can grow and spoil the product.

References

Abrantes, M.R., A.C.P Sousa, N.K.S. Araujo, E.S. Sousa, A.R.M. Oliveira and J.B.A. Silva. (2014). Microbiological evaluation of industrially produced charqui meat. Archives of Institute of Biology 81: 282–285.

Associassão Nacional das Industrias de Carne Seca - Anics. (2012). Carne seca no menu mundial. São Paulo, Brazil.

Bacus, J.N. (2005). Microbiology – Shelf-stable dried meats. Available at: http://www.fsis. usda.gov/PDF/FSRE_SS_5MicrobiologyDried.pdf. Acess: 05 April 2015.

Baroni, S., I.A. Soares, R.P. Barcelos, A.C. de Moura, F.G.S. Pinto and C.L.M.S.C. da Rocha. (2013). Microbiological contamination of homemade food. *In*: I. Muzzalupo (Ed.), Food Industry. Intech, Rijeka, pp. 241–260.

Beefpoint (2010). Entendendo as diferenças dos cortes de carne bovina nos EUA e Brasil. Available at: http://www.beefpoint.com.br/radares-tecnicos/qualidade-da-carne/entendendo-as-diferencas-dos-cortes-de-carne-bovina-nos-eua-e-brasil-59837/. Acess: 03 Jul 2015.

Biscola, V., S.D. Todorov, V.S.C. Capuano, H. Abriouel, A. Gálvez and B.D.G.M. Franco. (2013). Isolation and characterization of a nisin-like bacteriocin produced by a *Lactococcus lactis* strain isolated from charqui, a Brazilian fermented, salted and dried meat product. Meat Science 93: 607–613.

Biscola, V., H. Abriouel, S.D. Todorov, V.S.C. Capuano, A. Gálvez and B.D.G.M. Franco. (2014). Effect of autochthonous bacteriocin-producing *Lactococcus lactis* on bacterial population dynamics and growth of halotolerant bacteria in Brazilian charqui. Food Microbiology 44: 296–301.

Blackburn, C. de W., L.M. Curtis, L. Humpheson, C. Billon and P.J. McClure. (1997). Development of thermal inactivation models for *Salmonella enteritidis* and *Escherichia coli* O157:H7 with temperature, pH and NaCl as controlling factors. International Journal of Food Microbiology 38: 31–44.

Brasil. (1952). Ministério da Agricultura. Regulamento de Inspeção Industrial e Sanitária de Produtos de Origem Animal de 29 de Março de 1952, Art. 432. Rio de Janeiro, pp 154.

Brasil. (1975). Agência Nacional de Vigilância Sanitária. Decreto n° 75.697, de 06 de Maio de 1975 Padrões de identidade e qualidade para o sal destinado ao consumo humano. Diário oficial da República Federativa do Brasil. Brasília, DF.

Brasil. (1997). Ministério da Agricultura. Regulamento de Inspeção Industrial e Sanitária de Produtos de Origem Animal de 04 de Junho de 1997, Art. 375. Brasília, DF.

Brasil. (2000). Ministério da Agricultura. Instrução Normativa n. 22, de 31 de Julho de 2000. Diário Oficial República Federativa do Brasil, Brasília, DF. Seção 1, n. 16, p. 15–25.

Brasil. (2001). Agência Nacional de Vigilância Sanitária. Resolução da Diretoria Colegiada da Agência Nacional de Vigilância Sanitária. Regulamento Técnico sobre os Padrões Microbiológicos para Alimentos. RDC N° 12, de 2 de janeiro de 2001. Diário Oficial da República Federativa do Brasil, Poder Executivo, Brasília, DF.

Bressan, M.C. (2001). Tecnologia de carnes e pescados. Curso de Pós Graduação *"Lato Sensu"* a distância: Processamento e controle de qualidade em carnes, leite, ovos e pescados. Lavras, UFLA/FAEPE, 240 p.

Canhos, D.A.L. and E.L. Dias. (1985). Tecnologia de Carne Bovina e Produtos Derivados. Fundação Tropical de Pesquisa e Tecnologia, Campinas, São Paulo. pp. 239–255.

Carvalho Junior, B.C. (2002). Estudo da evolução das carnes bovinas salgadas no Brasil e desenvolvimento de um produto de conveniência similar à carne-de-sol. Ph.D. Thesis, State University of Campinas, Campinas, Brazil.

Comi, G. and C. Cantoni. (1980). I lieviti in insaccati crudi stagionati. Industrie Alimentari-Italy 19: 857–60.

Comi, G., R. Urso, L. Iacumin, K. Rantsiou, P. Cattaneo, C. Cantoni and L. Cocolin. (2005). Characterisation of naturally fermented sausages produced in the North East of Italy. Meat Science 69: 381–392.

Costa, E.L. and J.A. Silva. (1999) Qualidade sanitária da carne de sol comercializada em açougues e supermercados de João Pessoa – PB. Boletim do Centro de Pesquisa de Processamento de Alimentos 17: 137–144.

Evangelista-Barreto, N.M., P.C. Miranda, D.C. Barbosa, R.H. de Souza and M.S. Santos. (2014). Condições higiênicas sanitárias da carne de sol comercializada no município de Cruz das Almas, Bahia e detecção de cepas com resistência antimicrobiana. Semina: Ciências Agrárias 35: 1311–1322.

Fayrdin, A. (1998). O sucedâneo do charque ganha mais espaços no mercado. Revista Nacional da Carne 256: 8–12.

Felicio, P.E. (2002). Jerked beef - Um sucedâneo do charque criado a partir de uma fraude. Revista ABCZ 7: 98.

Fennema, O.R. (1996). Food Chemistry. Marcel Dekker, New York, New York.

Fernández-Diez, A., B.K. Salvá-Ruiz, D.D. Ramos-Delgado, I. Caro and J. Mateo. (2012). Technological quality traits and volatile compounds of Andean alpaca (*Vicugna pacos*) charqui freshly prepared and after five months of storage. Anales de Veterinaria (Murcia) 28: 97–109.

Fontán, M.C.G., J.M. Lorenzo, A. Prada, I. Franco and J. Carballo. (2007). Microbiological characteristics of "androlla", a Spanish traditional pork sausage. Food Microbiology 24: 52–58.

Frenández-Baca, S. (2005). Situación actual de los Camélidos Sudamericanos en Perú. Proyecto de Cooperación Técnica en apoyo a la crianza y aprovechamiento de los Camélidos Sudamericanos en la Región Andina TCP/RLA/2914. Rome: FAO.

Gianelli, M.P., V. Salazar, L. Mojica and M. Friz. (2012). Volatile compounds present in traditional meat products (charqui and longaniza sausage) in Chile. Brazilian Archives of Biology and Technology 55: 603-612.

Gill, C.O. (1998). Microbiological contamination of meat during slaughter and butchering of cattle, sheep and pigs. *In*: A. Davies and R. Board (Eds.), The Microbiology of Meat and Poultry. Blackie Academic & Professional, London.

Gouvêa, J.A.G. and A.A.L. Gouvêa. (2007). Tecnologia de fabricação do charque. Technical Dossier, Rede de Tecnologia da Bahia, Bahia, Brazil.

INDECOPI. (2006). Norma Técnica Peruana 201.059. Carne y Productos Cárnicos. Charqui. Requisitos. Lima: Instituto Nacional de Defensa de la Competencia y de la Protección de la Propiedad Intelectual.

Instituto Brasileiro de Geografia e Estatística - IBGE. (2010a). Produção da pecuária municipal. Rio de Janeiro, Brazil.

Instituto Brasileiro de Geografia e Estatística - IBGE. (2010b). Pesquisa de orçamentos familiares 2008–2009. Rio de Janeiro, Brazil.

Ishihara, Y.M. (2012). Estudo da maciez em carne de sol. Ph. D. Thesis, Federal University of Paraíba, João Pessoa, Brazil.

Ishihara, Y.M. and M.S. Madruga. (2013). Tenderness indicators in salted and dried meat: a review. Semina: Ciências Agrárias 34: 3721–3738.

Ishihara, Y.M., R. Moreira, G. Souza, A. Salviano and M. Madruga, (2013). Study of the Warner-Bratzler force, sensory analysis and sarcomere length as indicators of the tenderness of sun-dried beef. Molecules 18: 9432–9440.

Jay, J.M., M.J. Loessner and D.A. Golden. (2005). Modern food microbiology. Springer, New York, USA.

Lara, J.A.F., S.W.B. Senigalia, T.C.R.M. Oliveira, I.S. Dutra, M.F. Pinto and Shimokomaki, M. (2003). Evaluation of survival of *Staphylococcus aureus* and *Clostridium botulinum* in charqui meats. Meat Science 65: 609–613.

Leistner, L., and Z. Bern. (1970). Vorkommen und bedeutung von hefen bei pokelfleischwaren. Fleischwirtsch. 50: 350–35l.

Leistner, L. (1985). Hurdle technology applied to meat products of the shelf stable product and intermediate moisture food types. *In*: D. Simatos and J.L. Multon (Eds.), Properties of water in foods in relation to quality and stability. Springer, New York, pp. 309–329.

Leistner, L. (1987). Shelf-stable products and intermediate moisture foods based on meat. *In*: L.B. Rockland and L.R. Beuchat (Eds.), Water activity: Theory and applications to food. Marcel Dekker, New York, pp. 295–327.

Leistner, L., and L.G.M. Gorris. (1995). Food preservation by hurdle technology. Trends Food Science and Technology 6: 41–46.

Lira, G.M. (1998). Avaliação de parâmetros de qualidade da carne-de-sol. Ph.D. Thesis, University of São Paulo, São Paulo, Brazil.

Lira, G.M., and M. Shimokomaki. (1998). Parâmetros de qualidade da carne-de-sol e dos charques. Higiene Alimentar 12: 33–35.

Martín, D., T. Antequera, E. Muriel, T. Pérez-Palacios and J. Ruiz. (2009). Volatile compounds of fresh and dry-cured loin as affected by dietary conjugated linoleic acid and monounsaturated fatty acids. Meat Science 81: 549–556.

Nóbrega, D.M. and T.S. Schineider. (1983). Contribuição ao estudo da carne-de-sol visando melhorar sua conservação. Higiene Alimentar 2: 150–152.

Norman, G.A. and O.O. Corte. (1985). Dried and salted meats: Charque and Carne de Sol. FAO production and health paper no. 51. Rome: FAO.

Osório, H. (2007). O império português no sul da fronteira: estancieiros, lavradores e comerciantes. UFRGE, Porto Alegre, Brazil.

Pachao, N. (2006). Characteristics of the supply and demand of charqui. *In*: M. Gerken and C. Renieri (Eds.), South American Camelids research. Wageningen Academic Publishers, Wageningen, pp. 261–270.

Pardi, M.C., I.F. Santos, E.R. Souza and H.S. Pardi. (1996) Ciência, higiene e tecnologia da carne. volume II. Editora da UFG, Goiânia, Goiás.

Pavia, P.C., L.A.T. Oliveira and R.M. Franco. (1997). Recuperação de salmouras utilizadas no preparo do charque. Revista Nacional da Carne 248: 41–44.

Pinto, M.F., E.H.G. Ponsano, B.D.G.M. Franco and M. Shimokomaki. (1998). Controle de *Staphylococcus aureus* em charques (jerkedbeef) por culturas iniciadoras. Ciência eTecnologia de Alimentos 18: 200–204.

Pinto, M.F., E.H.G. Ponsano, B.D.G.M. Franco and M. Shimokomaki. (2002). Charqui meats as fermented meat products: Role of bacteria for some sensorial properties development. Meat Science 61: 187–191.

Potter, N.N., and J.H. Hotchkiss. (1995). Food science. Food science texts series. Chapman & Hall, New York, USA.

Prata, L.F., and R.T. Fukuda. (2001). Fundamentos de higiene e inspeção de carnes. Funep, Jaboticabal, Brasil, 326p.

Ricke, S.C., I.Z. Diaz and J.T. Keeton. (2007). Fermented meat, poultry and fish products. *In*: M.P. Doyle and L.R. Beuchat (Eds.), Food Microbiology: Fundamentals and Frontiers. ASM Press, Washington, pp. 795–815

Riedel,G. (1996). Controle sanitário dos alimentos. Atheneu, São Paulo, Brazil.

Sabadini, E., M.D. Hubinger, P.J.A. Sobral and B.C. Carvalho Junior. (2001). Alterações da atividade de água e da cor da carne no processo de elaboração da carne salgada desidratada. Ciência e Tecnologia de Alimentos 21: 14–19.

Salvá, B.K., A. Fernández-Diez, D.D. Ramos, I. Caro and J. Mateo. (2012). Chemical composition of alpaca (*Vicugna pacos*) charque. Food Chemistry 130: 329–334.

Salviano, A.T.M. (2011). Processamento da carne-de-sol com carne maturada: qualidade sensorial e textura. M.S. Thesis, Federal University of Paraiba, João Pessoa, Brazil.

Shelef, L.A., and J. Seiter. (2005). Indirect and miscellaneous antimicrobials. *In* : P.M. Davidson, J.N. Sofos and A.L. Branen (Eds.), Antimocrobials in food. CRC Press, Boca Raton, pp. 573–598.

Shimokomaki, M., B.D.G.M. Franco and B.C. Carvalho Junior. (1987). Charque e produtos afins: tecnologia e conservação - uma revisão. Boletim da Sociedade Brasileira de Ciência e Tecnologia de Alimentos 21: 25–35.

Shimokomaki, M., B.D.G.M. Franco, T.M. Biscontini, M.F. Pinto, N.N. Terra and T.M.T. Zorn. (1998). Charqui meats are hurdle technology meat products. Food Reviews International 14: 339–349.

Shimokomaki, M., E.Y. Youssef and N.N. Terra. (2003). Curing. *In*: B. Caballero, L. Trugo and P.M Finglas (Eds.), Encyclopedia of Food Sciences and Nutrition. Elsevier, London, pp. 1702–1708.

Shimokomaki, M., R. Olivo, N.N. Terra and B.D.G.M. Franco. (2006). Atualidades em ciência e tecnologia de carnes. Livraria Varela, São Paulo, Brazil.

Silva, M.C.D. (1991). Incidência de *Staphylococcus aureus* enterotoxigênicos e coliformes fecais em carne de sol comercializada na cidade do Recife- PE. M.S. Thesis, Federal University of Pernambuco, Recife, Brazil.

Silva, J.A. (1997). Microbiologia da carcaça bovina: Uma revisão. Revista Nacional da Carne, 24: 62–87.

Sinigalia, S.W.B., T.C.R.M. Oliveira, L.O.P. Popper and M. Shimokomaki. (1998). Implementação do HACCP no processamento do charque. Revista Nacional da Carne 251: 30–36

Souza, D.R.S. (2007). Aspectos industriais na produção de charque. M.S. Thesis, Castelo Branco University, São Paulo, Brazil.

Souza, M.A.A., J.V. Visentainer, R.H. Carvalho, F. Garcia, E.I. Ida and M. Shimokomaki. (2013). Lipid and protein oxidation in charqui meat and jerked beef. Brazilian Archives in Biololgy and Technology 56: 107–112.

Talon, R., S. Leroy and I. Lebert. (2007). Microbial ecosystems of traditional fermented meat products: The importance of indigenous starters. Meat Science 77: 55–62.

Torres, E.A.F.S., M. Shimokomaki, B.D.G.M. Franco and M. Landgraf. (1994). Parameters determining the quality of charqui, an intermediate moisture meat product. Meat Science 38: 229–234.

Vargas, J.M. (2014). Supplying plantations: The insertion of dried beef produced in Pelotas (RS) in the meat atlantic trade and its competition with producers from River Plate (nineteenth century). História (Sao Paulo) 33: 540–566.

Youssef, E.Y., C.E.R. Garcia, M. Shimokomaki, M. (2003). Effect of Salt on Color and Warmed over Flavor in Charqui Meat Processing. Brazilian Archives of Biology and Technology 46: 595–600.

7

Biopreservation of Salami– Brazilian Experience

Sávio Guimarães Britto[a]*, Elisabetta Tome*[b]
and Svetoslav Dimitrov Todorov[c]*

Abstract

Consumption of fermented meat products has always been a part of human food culture of human's through out the world. Besides preservation and safety of the final product, fermentation processes play an essential role in the organoleptic properties of the meat products. In this process, important roles are played by different starter cultures and autochthonous microbiota of animal origin products, different additives of organic and inorganic nature, essential oils, non-digestive supplements, fibers and by-products.

Application of different starter cultures, most of them belonging to the group of the lactic acid bacteria (LAB), have a long history of application in the production of fermented meat products. Based on their metabolic activity LAB are capable of producing lactic acid, resulting in reduction of pH and solubilisation of proteins. The drop in pH to values close to the isoelectric point of protein reduces the water activity, facilitating the drying process and weight loss of the fermented meat products. Such alterations confer a firmer texture (consistence) of the final product, improve the organoleptic characteristics of the fermented meat product and are important the in process of control of the different food-borne pathogens, including *Listeria monocytogenes*.

Production of fermented food products in Latin America and Brazil is related to the rich heritage of the pre-Colombian indigenous tribes including in also knowledge of the emigrants who colonised the New World. The demographic amalgam of the Latin American region was not only resulted in extremely rich in human genetic mixture, but humans had different habits, cultures and traditions which include preparation of different fermented food products. Specific climates, socio-cultural conditions and influence of other cultures resulted in formation of a new variety of Latin American cuisine and the preparation of New World Traditional Fermented

[a] FACISA, Departamento de Medicina Veterinária, 36570-000, Viçosa, Minas Gerais, Brazil.
[b] Instituto de Ciencia y Tecnología de Alimentos, Escuela de Biología, Universidad Central de Venezuela, Apartado 47.097, Caracas 1041 A, Venezuela
[c] Universidade Federal de Viçosa, Veterinary Department, Campus UFV, 36570-900, Viçosa, Minas Gerais, Brazil
[*] Corresponding author: slavi310570@abv.bg

Food Products. The majority of emigrants arriving in Latin America have introduced their specific cuisines, and frequently the names were related to the original foods from Europe, Asia or Africa. However, over time, during their production in Latin America, they received their new "face" and nowadays can be considered traditional Latin American products. Fermentation processes were used for meat preservation for long time", and these products have been an important part of the human diet for centuries. During the Greatest Geographical Discovery of the new world in 14–16 century by Europeans, fermented meat products were an essential part of the diet of the explorers. In the Middle Ages it was wrongly believed that proteins were the most important part of the diet and had been associated with prosperity. Fruits and vegetables were ignored and this was a cause of vitamin deficiency especially in the sailors. After the discovery of South America new settlements were focused primarily on the meat diets based on the rich and easy access to these food sources. Even today in explored, discovered and colonized regions meat plays an important role in social events and is still considered a sign of prosperity.

Production of Salami in Brazil

According to the Technical Regulation of Identity and Quality (Brasil, 2000a), salami is defined as a "meat product obtained from commercially pork and beef meat with the addition of bacon, and other ingredients, and it is embedded in a natural or artificial sausage casings, dried, fermented, and left to mature; sometimes it is smoked" (Brasil 2000a,b; Marangoli and Moura 2011). The salami is fermented by the action of LAB, which transform carbohydrates in lactic acid. During this process, a pH value considered safe should be reached in a certain period of time in order to prevent the growth of pathogenic microorganisms. However, some food-borne pathogens such as *Staphylococcus aureus*, *Salmonella enteritidis*, *Listeria monocytogenes* and some biotypes of *Escherichia coli* can survive the low levels of pH and are considerade potentially hazardous risk in the production of fermented meat products. These microorganisms have been responsible for food-borne disease outbreaks caused by the consumption of fermented sausages (Adams 1986; Bryan 1980; Maciel 2003). The safety of the salami fermentation can be improved by the action of LAB belonging to the natural microbial population of the meat, and especially by the addition of the selected starter LAB cultures, with good fermentation characteristics and proven bio-preservation properties. The use of such starter cultures offers advantages for reducing the risks of food-borne diseases, along with improving the quality of the final product. During fermentation the LAB increase the oxygen in the raw matter, decreasing the redox potential and turning nitrite into a more effective tool in preventing the growth of aerobic spoilage and pathogenic bacteria. In addition, a low pH causes a decrease in the proteins water holding capacity, accelerating sausage dehydration and leading to low water activity (a_w) and high NaCl concentration in the final product (Työppönen et al. 2003; Degenhardt and Sant'Anna 2007).

The Italian immigration in western Paraná (Brazil) contributed to the development of small industries of fermented meat products. According to the report of the Brazilian Producers and Exporters of Pork Association (ABIPECS 2006), Brazil has the third largest herd of pigs after China and the United States. In 2006, Brazilian pork production was 2.87 million tons, and the State of Paraná was the third largest Brazilian producer of pork, with almost 500,000 tons, representing almost 16% of national pork production. The western Paraná region is well known for large scale industrial production of pork fermented products. However, because of Italian immigration to this region, small production facilities have come up which are still operating and well known for the preparation of traditional fermented meat products such as colonial salami, bacon, smoked pork, among others (Silva et al. 2011).

The colonial Italian salami is considered an important meat product. It is normally produced by small and medium scale production facilities and marketed in western Paraná (Brazil). However, this salami deserves special attention because it is manufactured with raw meat and requires a high level of hygienic standards and inspection to ensure food safety.

According to the RDC Resolution No. 360 of the National Health Surveillance Agency (Brasil 2003), processed foods sold in Brazil must provide nutritional labelling in order to inform the consumer of the nutritional properties of the product. Labelling should provide mandatory information about the energy value of food, as well as the proportions of nutrients such as protein, fat (total, saturated and trans), carbohydrates, sodium and fiber. Nutrients such as vitamins, calcium, iron and other minerals and nutrients, may be claimed as optional information.

According to the Ministry of Agriculture, Livestock and Supply (Brasil 2000), the term "Italian type salami", is the industrialized meat product, made with pork or pork and beef, bacon, with added ingredients ground into an average particle size between 6 and 9 mm, embedded in natural or artificial sausage casings, cured, smoked or not, and fermented, matured and dehydrated by the time indicated by the manufacturing process. The presence of "molds" is characteristic and a natural consequence of the technology of its manufacturing process. Therefore, it is a cured product, fermented, matured and dried. The variations between the different types of salami are due to the composition, added flavouring, additives and the type of meat used. Small producers of colonial salami products typically do not have the chemical laboratories capable of monitoring the quality of the raw material and the finished product. Thus, occasionally, samples are sent to laboratories providing this service in order to evaluate these parameters. However, it might be possible to have discrepancies or disagreements between the product and the parameters declared on the label. Some small scale food companies have access to the information in the Brazilian Table of Food Composition, however, significant differences can occur between this table and regional products such as colonial salami. On the other hand, large producers have chemical and microbiological laboratories to constantly monitor the quality of raw materials, product and product processing (Silva et al. 2011).

In the study by Silva et al. (2011), the authors determined the moisture, ash, protein, lipids and cholesterol content in different salami brands and compared them with the values indicated on the labels of their products, and with the values recommended by the Brazilian legislation (Brasil 2003).

Four different brands of colonial salami produced in the Toledo region, Parana State, and an industrialized brand of salami were evaluated in the work, for their moisture, ash, protein, lipids and cholesterol content. The results showed no disparities between the amounts presented in the nutritional information on the label and the values determined in the work, with the exception of industrialized salami (Silva et al. 2011). The moisture content was higher than the maximum levels recommended by the legislation, and the protein levels were lower than the minimum values; only the fat content agreed with the values, required by Brazilian law. The cholesterol analysis showed that the values were lower than those described in the labels when this substance was present. According to the compositions of salamis and Brazilian law, such salamis could not be classified as Italian salami (Silva et al. 2011).

Animal blood is an important source of proteins with important applications in the food industry. These proteins may be used as emulsifying agents, stabilizing and clarifying agents, nutritional components to enhance the properties of food and as source of lysine. However, only a small proportion of blood from slaughtered animalss is used, since most of it is discarded into the environment, generating pollution problems.

LAB can play an important role during the process of utilizing the blood from slaughter houses. The most promising bacteria for use as starter cultures in transforming "waste products" are those LAB, isolated from the indigenous microbiota of artisanal meat products. These microorganisms are well suited for meat, and several of them are capable of controlling pathogenic and spoilage microorganisms, due to the production of antimicrobial compounds such as organic acids, diacetyl and bacteriocins.

LAB starter cultures already have a long history of application in the production of fermented meat products. Based on their metabolitic activity LAB are capable to produce lactic acid, resulting in reduction of pH and solubilisation of proteins. The drop in pH to values close to the isoelectric point of protein facilitates the drying process and weight loss of fermented meat products, reducing the water activity. Such alterations confer a firmer texture (consistence) and improve the organoleptic characteristics of the final product. In addition to these technological advantages, the resulting acidity hinders the development of many pathogenic and spoilage microorganisms.

In the study of Campagnol et al. (2007), the medium was prepared with porcine plasma and distilled water (1:1, pH 11.0), with addition of glucose and potassium diphosphate. A strain of *Lactobacillus plantarum* isolated from artisanal salami (Sawitzki 2000) was used for the preparation of the starter culture. *Lactobacillus plantarum* presented a maximum growth of 9.82 Log UFC/mL, after 30 hours of fermentation. Salami produced with the starter culture had a significantly higher pH drop, and lower water activity than the other treatments. According to Rantsiou and Coccolin (2006), the LAB are the predominant microorganisms in fermented sausages, due to the anaerobic conditions of the medium and presence of sodium chloride, nitrate and nitrite, and can achieve a high number (log 7-8 CFU/g) even after three days of fermentation, and can remaining stable during maturation of the salami. It is important to state that according to Campagnol et al. (2007), throughout the studied period of production of salami, no fecal coliforms were detected nor any coagulase positive *Staphylococcus*. In addition, the salami produced with a starter culture *Lactobacillus plantarum* previously mentioned, in medium based on porcine plasma presented better pH characteristics and lower water activity compared to the control preparations and ensured greater microbiological safety features. Furthermore, the use of *Lactobacillus plantarum* resulted in a significant improvement in the taste of the salami (Campagnol et al. 2007).

Milano Type Salami Elaborated with Fiber of Red Wine By-products

Currently, dietary fiber is a widely used ingredient in development of products with nutritional appeal, due to its significance in health promotion and technological impact on the product. The American Dietetic Association recommends an intake of 25 to 30 grams of dietary fiber per day for adults (Verma and Banerjee 2010), this is important because we believe that it aids in the prevention of various diseases such as diabetes, irritable bowel, colon cancer, diverticulitis, gastrointestinal and cardio-vascular diseases among others (Rodríguez et al. 2006; Verma and Banerjee 2010).

Many agro-industrial by-products can be incorporated into meat products for improving their health benefits, whereas the fibers from fruits have shown most desirable effects on the fermented food products comparable to quality of cereal fibers (Garcia et al. 2002). In this sense, the industrial production of grape and wine generates large amounts of by-products with a high content of dietary fibers and rich in bioactive compounds with antioxidant properties (Cataneo et al. 2008). This, coupled with its low cost, availability and utilization of wine industry waste alleviates the problem of waste disposal while adding a nutritional enrichment and meeting the technological objectives, such as the inhibition of lipid oxidation in meat products, makes it feasible for use as a food ingredient.

In the work of Mendes et al. (2014), the effects from the addition of wine by-products flour in different concentrations (1, 2 and 3%) to the production of Milano type salami were evaluated. Mendes et al. followed the changes during the fermentation and drying processes on the physical and chemical characteristics of salami (pH and water activity), chemical (acidity, peroxide value, TBARS index and residual nitrite) and physical parameters (weight loss). They also evaluated the chemical composition of final products (Mendes et al. 2014). Additions of fiber reduced (P<0.05) pH values and increased the lactic acid content of the meat mixture during maturation, with a consequent increase in the weight loss and reduction in water activity products. The presence of fibers did not affect (P > 0.05) residual nitrite levels nor the peroxide index values, but reduced the development of lipid oxidation, measured by TBARS values of the product during aging (Mendes et al. 2014).

It is important to note that regardless of the effects of the addition of wine by-products flour, all the salami produced met the criteria set by the Ministry of Agriculture, Livestock and Supply (Brasil 2000), establishing 0.900 as the maximum value of a_w for the Milano type salami. The a_w values found are associated with the pH and make the product stable and able to be storable without refrigeration.

Mendes et al. (2014) pointed out that despite promoting changes in ash content (P < 0.05), the presence of fibers did not affect (P > 0.05) humidity values, protein and lipids, and this can be considered as a positive point, since the the addition of fibers should not change the typical organoleptic characteristics of the final fermented food products.

Based on the weight loss of salami during the maturation process, the addition of wine by-products at levels of 1, 2 and 3% to the raw salami, resulted in 1.56, 3.11 and 4.92% of flour in the final ready to eat product, respectively. Thus, based on the dietary fiber content (40.26%) present in the flour, it is estimated that the salami prepared with 1, 2, and 3% flour resulted in 0.63, 1.25 and 1.98% total dietary fiber in the final prduct, respectively. Although these values can be considered low, it should be consider that it represented a significant increase in the nutritional properties of the product, especially compared to a fiber free product (Mendes et al. 2014).

The presence of dietary fiber and antioxidant compounds in the flour of the cultivar 'Syrah' allowed its application in the processing of salami with technological and nutritional benefits. Salami can be enriched by adding nutritents obtained from red wine by-products flour, and also contribute to the process of fermentation and maturation of the product and reducing the effects of lipid oxidation, characteristic of fermented meat products (Mendes et al. 2014).

Replacement of fat

In search of functional fermented food products, nowadays research is focused on the reduction of animal fat and its replacement with plant lipids with functional properties. Additionally, it is important to note that the final products satisfy consumer's requests and conform to the safety standards.

Several studies have reported on the development of meat products with less saturated fat, through the use of vegetable oils, mainly soybean oil, flaxseed, rapeseed, sunflower and olive oil as a substitute for animal fat (Muguerza et al. 2004; Luruena Martinez et al. 2004; Pelser et al. 2007; Santos et al. 2008; Valencia et al. 2008; Del-Nobile et al. 2009; Choi et al. 2013; Yunes et al. 2013). The intake of unsaturated fatty acids, especially omega-3 fats, has positive effects on consumer health, as they contribute to the decrease in low-density lipoproteins (LDL) and an increase in high-density lipoprotein (HDL) contributing to the reduction in the incidences of coronary heart disease. In this context, canola oil has gained great prominence, particularly by promoting benefits to consumer health (McDonald et al. 1989), based on its hypocholesterolemic effect (Nydahl et al. 1995), antithrombotic effect (Kwon et al. 1991) and presence of a higher proportion of unsaturated fatty acids compared to saturated fatty acids from animal fat (Liu et al. 1991).

In work of Backes et al. (2013), application of canola oil in the production of salami was evaluated. This study evaluated the effects of the partial substitution of pork fat emulsion containing canola oil in the preparation of Italian type salami. The salami is well known to have high fat content, which can be seen even in the sliced product. Generally, this type of product in Brazil contains about 30% fat, and 35% is the maximum allowed by Brazilian law (Brasil 2000b). This fat contributes to the texture, juiciness and flavor of the product, factors that determine the quality and acceptability of fermented sausages (Wirth 1988). The production of salami is well known to involve three phases: a mixing of ingredients, fermentation and drying (Demeyer et al. 1986). Physical, microbiological and biochemical changes (Franco et al. 2002; Garcia-Fontan et al. 2007), involving meat enzymes and microbial enzymes, occur during fermentation and drying. These changes are influenced by the characteristics of raw material (Bacus 1984) and the process conditions (Soyer et al. 2005) and reflect in organoleptic properties of the final product.

To improve the stability of canola oil in the fermented meat product, an emulsion of oil in water and isolated soy protein was performed (Backes et al. 2013). Italian type salami was prepared in accordance with the formulation and procedures described by Backes et al. (2013), containing 65.00% pork, 20.00% beef, 3.00% salt, 0.30% glucose, 0.20% sucrose, 0.30% curing salts, 0.20% ground pepper, 0.20% garlic powder, 0.20% nutmeg, 0.25% colour fixative, and a commercial starter culture containing *Pediococcus pentosaceus* and *Staphylococcus xylosus* (Christian Hansen®). After the grinding process of pork and beef, mixing of the ground meat with the other ingredients was carried out, except the animal fat (Backes et al. 2013).

Three treatments were developed (Backes et al. 2013), among them: control (100% pork fat, without fat replacement), Experimental Preparation 1, where 15% of pork fat was replaced by an emulsion containing canola oil and Experimental Preparation 2, where 30% of pork fat was replaced by an emulsion containing canola oil. Backes et al. (2013) evaluated the physicochemical characteristics of the salami (pH, water activity, weight loss, colour and lipid oxidation) during the manufacture and storage, as well as a sensory evaluation after processing was completed. The addition of emulsified canola oil in different concentrations did not affect the pH and colour values during the processing period, despite significant differences in these parameters during the storage time were observed. Water activity did not differ significantly between treatments. However, treatments with addition of canola oil had lower weight loss than the control. It was possible to observe the increase in lipid oxidation values in Experimental Preparation 2 during processing and storage of salami, while the Experimental Preparation 1 values were not different from the control at the end of processing and remained lower than control during the storage time. Futhermore, the partial replacement of pork fat for emulsified canola oil did not affect product acceptance for aroma, flavour, colour, texture and visual appearance. Thus, the replacement of 15% pork fat emulsion containing canola oil does not compromise the sensory acceptance of the products. Therefore, despite the differences with the control batch's treatment, it is concluded that the replacement of 15% pork fat emulsion containing canola oil allowed the production of salamis, since the sensory attributes were retained, as well as better oxidative stability of salami during the storage period (Backes et al. 2013).

Problems with contaminations in salami

In Brazil, *Listeria monocytogenes* is a frequent contaminant detected in salami (Sakate et al. 2003; Petruzzelli et al. 2010; Okada et al. 2012), so the application of bacteriocins produced by LAB can be a technological alternative which are considered to improve the hygeinic status of these products. The control of *Listeria monocytogenes* in meat products is a problem, taken very seriously by the food industry, as this pathogen causes outbreaks with high fatality rates (20–30%), especially among high-risk groups, such as pregnant women, neonates, elderly and immunocompromised persons (Camargo et al. 2015). *Listeria monocytogenes* is an ubiquitous pathogen and may persist in the food industry environment due to its capability to produce resistant biofilms on equipment surfaces and premises (Barbosa et al. 2014). The introduction or recontamination of *Listeria monocytogenes* in the processing plants can have multiple sources, mainly through raw ingredients, and Good Hygiene Practices and HACCP systems may be inefficient to avoid persistence in the processing environment and presence of *Listeria* spp. in the final product (Tompkin 2002; Barbosa et al. 2014). Therefore, application of antimicrobial compounds may be necessary to inhibit the growth of pathogens. In this context, bacteriocins and bacteriocinogenic LAB can be explored as technological alternatives or ingredients for increasing the safety of the products manufactured in such conditions. However, *Listeria monocytogenes* is often isolated from fermented meat products, due to its capability to survive the adverse conditions of this type of product (Bolton and Frank 1999; Bonnet and Montville 2005; Incze 1998; Varabioff 1992).

There are many important explainations how *Listeria monocytogenes* survive during the manufacturing process of salami and at the final product. One of the most accepted reason is *Listeria monocytogenes'* ability in becoming acid-resistant (Bonnet and Montville 2005) and the ability of *Listeria monocytogenes* to survive under stress conditions (Johnson et al. 1988; Koutsoumanis and Sofos 2005, Nissen and Holck 1998). This ability is related to *Listeria monocytogenes* pathogenicity and it is normally found in strains isolated from fermented food or from meat processing facilities (Vialette et al. 2003).

Around 70,000 to 75,000 illnesses are caused due to *Escherichia coli* O157:H7 annually in USA. In addition foodborne pathogens such as *Salmonella* spp. and *Listeria monocytogenes* have been detected in finished dry and semi dry fermented sausages, including Italian-style salami, rendering the consumption of these ready-to-eat (RTE) products a potential food safety risk (Levine et al. 2001; Moore 2004). According to Di Pinto et al. (2010), *Listeria monocytogenes* was detected in 23/112 (20.5%) vacuum-packaged sliced salami samples from supermarkets in Southern Italy. In 2003, there was a recall of three processed (chicken franks, spiced ham and turkey ham) RTE meat products from a large processing plant in Trinidad as a result of contamination by *Listeria monocytogenes*, *Salmonella* spp., *Escherichia coli* and *Campylobacter* spp.

Biopreservation and Antimicrobial Peptides

Biopreservation processes and application of antimicrobial compounds were part of the preparation of fermented meat products, including various types of salami over the past few centuries. However, only after the scientific discovery of antibiotics produced from various fungi by Fleming in beginning of 20th century, and the detection of nisin a few years later, various antimicrobial peptides have been the subject of intensive research and scientific interest in biopreservation of fermented food products. Nowadays, we have sufficient scientific knowledge to state that antimicrobial peptides can be detected in all life forms, including not only microorganisms, but also animals, insects, fishes, birds and plants. The most frequent antimicrobial peptides produced by microorganisms are referred to as "bacteriocins" and enormous research was focused on those produced by lactic acid bacteria (LAB) (Cotter et al. 2005). According to Cotter et al. (2005), bacteriocins can be produced by almost all bacterial species as a part of the defending molecules. Several research groups have been focused on the inhibitory effect of bacteriocins from LAB against food-borne pathogens and spoilage bacteria (De Vuyst and Vandamme 1994; García et al. 2010). Several studies on the application of bacteriocins have been involved with their potential application in the treatment of food-borne contaminations, improving food quality and safety and the fight against antibiotic resistance problems.

LAB can produce various antimicrobial compounds, including organic acids (lactic, acetic, formic, propionic acids), however, their principal antibacterial action is related to the reduction of the pH of the media. Other methabolites like fatty acids, acetoin, hydrogen peroxide, diacetyl, antifungal compounds (propionate, phenyl-lactate, hydroxyphenyl-lactate, cyclic dipeptides and 3-hydroxy fatty acids), bacteriocins (nisin, reuterin, reutericyclin, pediocin, lacticin, enterocin and others) and bacteriocin-like inhibitory substances (BLIS) (Reis et al. 2012) can be actively involved in the antimicrobial action. Other antimicrobials entourage of LAB may include different biopolymers, sugars, sweeteners, nutraceuticals, aromatic compounds and various enzymes, and in this manner indicate that LAB have higher flexibility and a wider application than just as starter cultures. Study of the new antimicrobial compounds is very intensive in order to provide an alternative to the chemical additives, and to offer the market more natural food products. In addition a specific spectrum of activity of bacteriocins against certain emerging food-borne pathogens and spoilage microorganisms, their resistance to thermal processing and low pH, combined with their sensitivity to human proteolytic enzymes are important positive characteristics in the application of these compounds in food preservation (Masuda et al. 2011).

By definition, bacteriocins are ribosomally synthesized antimicrobial proteins (polypeptide or small proteins), usually active against genetically related species (Cotter et al. 2005). Based on the intensive research in area of bacteriocins since the discovery of nisin, we have sufficient examples of bacteriocins that may have applications in controlling Gram-negative bacteria, some yeast, *Mycobacterium* spp. and even viruses (Todorov et al. 2010; Schirru et al. 2012). However, amino-acid sequences of only a few of these unusual bacteriocins are provided (Todorov et al. 2010; Todorov et al. 2012). These reports need to be carefully screened, since only a few pieces of work explain the mechanisms of modes of action for this "unusual" bacteriocinogenic activity.

Many classifications have already been proposed (Cotter et al. 2005), but according to the most recent (Heng et al. 2007), bacteriocins of Gram-positive bacteria are grouped into four classes, based on their structure and function. Class I: lantibiotic peptides, class II: small non-modified peptides with < 10 kDa, class III, large proteins with > 10 kDa and class IV: cyclic proteins.

Application of Bacteriocins in Biopreservation of Salami

LAB are considered to be normal microflora of fermented meat products and bacteriocins from LAB have been tested for application in biopreservation of salami in only a limited number of studies.

One of the first reports in Brazil on bacteriocinogenic LAB, was by Maciel et al. (2003). In this work, bacteriocinogenic LAB isolated during the processing of Italian salami, obtained from two different processing plants, in the State of Paraná were reported. A total of 484 isolates were tested for their antibacterial activity against *Listeria monocytogenes*, *Staphylococcus aureus*, *Salmonella enteritidis* and *Escherichia coli*. From these samples, 115 isolates inhibited at least two of the pathogens. However, the authors of this study have not gone for a deeper and systematic study of the produced antimicrobial agents, and therefore it is only possible to speculate on a potential production of bacteriocin/s or other antimicrobial compounds.

Barbosa et al. (2015) isolated *Lactobacillus curvatus* from salami produced in Brazil and tested its potential biopresevation properties with the control of *Listeria monocytogenes* contaminants of salami. Sudirman et al. (1993) isolated *Lactobacillus* spp. strains obtained from semi-dry sausages; Cintas et al. (1995) reported LAB isolates from Spanish dry-fermented sausages. Aymerich et al. (2000) failed in the isolation of LAB from fuet, chorizo and salchichon. Belgacem et al. (2008) reported on LAB isolated from gueddid, a Tunisian fermented meat. In the work of Carvalho et al. (2006), anti-listerial activity of LAB isolated from salami and sausages and the study of the development of anti-Listerial resistance was evaluated. Vermeiren et al. (2004) obtained LAB from meat products. Todorov et al. (2013) reported on *Lactobacillus sakei* isolated from Portuguese fermented meat products. Todorov et al. (2007) reported the application of *Lactobacillus plantarum* in control batch of *Listeria monocytogenes* in salami prepared from game meat in South Africa. However, even if LAB are natural inhabitants of salami and have an important role in the fermentation processes and development of the organoleptical properties of salami, the role of the expressed bacteriocin/s during the maturation stage is difficult based on the negative effect of the limited conditions for production and expression of the bacteriocin/s. Some of these factors (nutritional, temperature, pH, presence of lipids, etc.) have been reviewed in detail by Favaro et al. (2015).

Despite the potential difficulties in the expression of bacteriocin/s during the fermentation and maturation of salami, some works have been showing the potential application of bacteriocinogenic LAB in control of food-borne pathogens in the final fermented products. Barbosa et al. (2015a) evaluated potential of bacteriocin produced *Lactobacillus curvatus* strain isolated from Italian-type salami on control of *Listeria monocytogenes* during manufacturing of salami in a pilot scale. In this study two isolates (differentiated by RAPD-PCR) showed activity against high numbers of *Listeria monocytogenes* in addition to several other Gram-positive bacteria. In addition, on the basic features of the expressed bacteriocins, Barbosa et al. (2015) performed a three-step purification procedure and indicated that both strains produced the same two active peptides (4457.9 Da and 4360.1 Da), homlogous to sakacins P and X, respectively. Addition of the semi-purified bacteriocins produced by *Lactobacillus curvatus* MBSa2 to the batter for the production of salami, experimentally contaminated with *Listeria monocytogenes* (10^4–10^5 CFU/g), caused 2 log and 1.5 log reductions in the counts of the pathogen in the product after 10 and 20 days respectively, highlighting the interest in application of these bacteriocins to improve the safety of salami during its manufacture (Barbosa et al. 2015a).

As has been shown previously (Favaro et al. 2015), environmental conditions can interfere with the stability and survival of the bacteriocins producers when applied in the real food production systems. In order to protect the LAB and to have a better effect of the expressed bacteriocin/s, Barbosa et al. (2015b) encapsulated the bacteriocin producer *Lactobacillus curvatus*, previously isolated from salami. *Lactobacillus curvatus* was entrapped in calcium alginate and tested for functionality in MRS broth and in salami experimentally contaminated with *Listeria monocytogenes* AL602/08 (a meat isolate), during 30 days of simulating manufacture process conditions, including fermentation and maturation steps. The entrapment process did not affect bacteriocin production by *Lactobcillus curvatus* MBSa2 in MRS broth and in salami. Both, free and entrapped *Lactobacillus curvatus* MBSa2 caused reduction in a similar manner in the counts of *Listeria monocytogenes* AL602/08 in salami during the manufacture process (Barbosa et al. 2015b). The entrapment of *Lactobacillus curvatus* in calcium alginate did not effect bacteriocin production when strain was applied in salami. Consequently, no improvement in inhibition of *Listeria monocytogenes* in this meat product could be achieved, when compared to a free and encapsulated bacteriocin producer, however, entrapped cells showed better survival (Barbosa et al. 2015b).

In a different study Barbosa et al. (2014) evaluated bacteriocin potential of *Lactobacillus sakei* MBSa1, isolated from that produced in Brazil salami, including the genetic features of the producer strain. Expressed bacteriocin by *Lactobacillus sakei* MBSa1 exhibited heat and pH stability with remarkable activity against *Listeria monocytogenes*. However, the expressed bacteriocin MBSa1 did not inhibit the tested probiotic strains (e.g. *Lactobacillus acidophilus* La 5) nor starter cultures (e.g. *Lactobacillus acidophilus* La-14). This suggests an interesting potential for technological applications in fermented foods for control of Listeria, without affecting starter or probiotic cultures. Expressed bacteriocin was purified by cation-exchange reversed-phase HPLC, molecular mass (4303.3 Da) and amino acid sequence (SIIGGMISGWAASGLAG), similar to that recorded to sakacin A, and which determined maximal production of bacteriocin MBSa1 (1600 AU/ml) in MRS broth and occurred after 20 hours at 25°C. The strain contained the sakacin A and curvacin A genes but was negative for other tested sakacin genes (sakacins Tα, Tβ, X, P, G and Q).

According to Castro et al. (2011), bacteria belonging to *Lactobacillus* species are common in fermented meat products. Particularly, *Lactobacillus sakei* is specially adapted to the meat environment and has already been used as a starter culture for the production of different meat products (Carr et al. 2002). Chaillou et al. (2005) reported on the determination of the complete genome sequence of the French sausage isolate of *Lactobacillus sakei* 23K, demonstrating that this particular strain has a specialized metabolic repertoire that may contribute to its competitive ability in these foods.

Based on the fact that *Lactobacillus sakei* can be a producer of different antimicrobial compounds, including lactic and acetic acids, diacetyl, hydrogen peroxide and bacteriocins, some *Lactobacillus sakei* strains possess interesting biotechnological potential application for food biopreservation (Carr et al. 2002). Numerous bacteriocins produced by *Lactobacillus sakei* strains have been identified, such as sakacin A (Schillinger and Lucke 1989; Holck et al. 1992); sakacin M (Sobrino et al. 1992), bavaricin A (Larsen et al. 1993; Messens and de Vuyst 2002); sakacin P (Holck et al. 1994; Tichaczek et al. 1994; Vaughan et al. 2001; Urso et al. 2006; Carvalho et al. 2010), sakacin K (Hugas et al. 1995), bavaricin MN (Kaiser and Montville 1996), sakacins 5T and 5X (Vaughan et al. 2001), sakacin G (Simon et al. 2002), sakacin Q (Mathiesen et al. 2005), sakacin C2 (Gao et al. 2010) and sakacin LSJ618 (Jiang et al. 2012).

Use of *Lactobacillus Plantarum* as Starter Culture in Fermentation of Salami

Dry Italian-type salami can be considered as a ready-to-eat product with a low risk of causing listeriosis due to the hurdles created during the manufacturing process such as low pH and a_w high salt concentration and presence of LAB. However, several studies have detected survival of *Listeria monocytogenes* in these products and also shown that process parameters, LAB and *Listeria monocytogenes* strains directly influence the results. In a study by Degenhardt and Sant'Anna (2007) survival of *Listeria monocytogenes* was followed in salami prepared with and without the additional starter culture of *Lactobacillus plantarum* and 2% of sodium lactate, using the manufacturing process usually employed in Brazil. Naturally contaminated sausages presented a small increase in the counts of *Listeria monocytogenes* during the first days of the process, however, this was followed by a gradual decrease by the end of the process (Degenhardt and Sant'Anna 2007). In experimentally contaminated samples containing *Lactobacillus plantarum*, the reduction in the counts of *Listeria monocytogenes* during processing was considerable, but there were no significant differences between the treatments. Although no significant differences were detected among the three artificially contaminated batches ($p > 0.05$), the batch inoculated with *Lactobacillus plantarum* presented a slightly different reduction in the counts of *Listeria monocytogenes* when compared to the standard batch, which had received no additional starter culture. In the case of the added sodium lactate, the results displayed very similar behaviour to the standard curve, this was without the added *Lactobacillus plantarum*. Therefore, the use of bio-protective cultures such as *Lactobacillus plantarum* is highly recommended in the commercial production of Italian salami. However, the use of sodium lactate must be better evaluated, especially when used with other inhibitory substances (Degenhardt and Sant'Anna 2007).

Application of Essential Oils in Bio-preservation of Salami

Application of different spices was part of the preparation of various fermented food products related to the gastronomic characteristics of the final product. However, application of spices has an influence on the safety characteristics of the products as well. Different works have demonstrated the effect of essential oils on the extension of the shelf life of the different fermented food products, including salami (Marangoni and Moura 2009; Marangoni and Moura 2011; Melo and Guerra 2002; Bertol et al. 2012). In addition essential oils can have antioxidant properties. Considerable emphasis has been granted to the application of natural products with antioxidant activity, which may act alone or synergistically with the other additives of the fermented food products, as an alternative to prevent oxidative deterioration of food and limit the use of synthetic antioxidants (Melo and Guerra 2002; Marangoni and Moura 2011). The replacement of chemical additives such as antioxidants (2-t-butyl-4-methoxyphenol, BHA, and 2,6-di-t-butyl-4-methylphenol, BHT, etc.), color enhancers, and control agents of undesirable microorganisms/pathogens (containing nitrite and/or nitrate curing salt) with natural products is a promising alternative (Bertol et al. 2012).

In order to look for a more natural and safe food products, several plant extracts have already been evaluated for their properties as natural antioxidants (Yen and Chen 1995; Lopez-Bote et al. 1998; Lee and Shibamoto 2002; Basmacioglu et al. 2004; Rababah et al. 2004), which are mainly attributed to the flavonoids present in the plants. Some extracts were classified by Lee and Shibamoto (2002) according to their antioxidant power in the following order: thyme (*Thymus vulgaris*), basil (*Ocimum basilicum* L.), rosemary (*Rosmarinus officinalis*), chamomile (*Matrica riarecutita*), lavender (*Lavandula* spp.), cinnamon (*Cinnamomum zeylanicum*) (Bertol et al. 2012).

The descriptive terminology and sensory profile of four samples of Italian salami prepared with coriander (*Coriandrum sativum* L.) essential oil was determined using a methodology based on the Quantitative Descriptive Analysis (QDA) in a study of Marangoli and Moura (2011). The salami prepared with a coriander essential oil, was compared with those prepared with BHT (butyl hydroxytoluene, synthetic antioxidant) had a lower rancid taste and rancid odour, whereas the control salami preparation showed high values of these sensory attributes. However, for the other attributes, all preparations showed similar acceptance by the sensory analysis panellists. The salami with the coriander essential oil exhibited reduction in lipid oxidation by increasing the shelf life of the product. The salami with the coriander essential oil and BHT showed no synergy between the antioxidants (Marangoli and Moura 2009). Based on the results obtained using coriander essential oil in Italian salami, Marangoli and Moura (2011) concluded that, Italian salami that had the additives, i.e. BHT or coriander essential oil, resulted in salami with lower scores for the sensory factor of stale taste and smell and confirmed the delay on lipid oxidation in the control salami. The sensory profile attributes for rancid aroma and taste indicate that the use of coriander essential oil presented a higher synthetic antioxidant effect of BHT delaying lipid oxidation. The coriander essential oil improved the sensory attributes of taste, odour, texture, brightness, and the intensity of red in Italian salami (Marangoli and Moura 2009; Marangoli and Moura 2011).

The study of Bertol et al. (2012) reported on the evaluation of the use of rosemary (*Rosmarinus officinalis*) extract, celery (*Apiumgrave olis*), and low levels of NO_3 and NO_2 as natural agents to enhance the quality of colonial salami. Salami was produced according to three treatments, including the control batch (which received only curing salt), salami supplemented with curing salt and rosemary extract and salami supplemented with sea salt, celery and rosemary extract. According to Bertol et al. (2012), addition of supplements (rosemary extract, salts, celery) did not interfere ($P > 0.05$) with the water activity, Na content, and residual NO_3 and NO_2 of salami. However, fatty acids C18:2 and C20:4 were reduced ($P < 0.05$) during the ripening period in the control treatment indicating the possible oxidation involved as a result of the addition of supplements. The use of celery resulted in lower pH values ($P < 0.05$) in the salami. Reduced addition of NO_3 and NO_2 resulted in changes of salami colour and were recorded as being lighter compared to the controls (higher L* values, $P < 0.05$) on the 12th day of ripening. The authors pointed out that celery-based products proved to be an effective source of NO_2 and NO_3 for colour development, but the low pH of the product indicates the need for a better evaluation of its uses in fermented salami. The rosemary extract influenced the reduction of fat oxidation in salami. In addition, it was shown that addition of rosemary extract was not affecting the presence and development of the starter cultures (Bertol et al. 2012). The authors did not detect the presence of *Listeria* spp. and *Salmonella* spp. in all the tested products. However, there will be speculation if this can be said to be as the effect of rosemary extract, since similar results were recorded in the control samples (that did not receive the rosemary extract). Most probably as this was a more complex process involving the effect of starter culture, presence of salts, fat content, and the changes in the water activity of the product.

Probiotic Salami

LAB presents many important properties in the salami fermentation process and manufacturing. However, LAB have a very important application related to their health-benefit properties as probiotics. Probiotics are defined as live microorganisms which when administered in adequate amounts (10^6–10^7 CFU per gram of food) confer health benefits to the host (WHO 2002; Bertazonni-Minelli et al. 2004; Oelschaeger 2010;

Todorovet al. 2012). Probiotics may have a beneficial effect for the host, including the suppression of growth of pathogens, control of serum cholesterol level, modulation of the immune system, improvement of lactose digestion, synthesis of vitamins, increase in the bio-availability of minerals and possible anti-carcinogenic activity (Gomes and Xavier 1999; Kailasapathy and Chin 2000; Chan and Zhang 2005). These bacteria can confer health benefits to the host such as reduction of gastrointestinal infections and inflammatory bowel disease, modulation of the immune system, and defense against colonization by pathogenic microorganisms (WHO 2002; Galdeano et al. 2007; Oelschlaeger 2010). As probiotics, LAB have been investigated intensively in last few decades. Several food matrixes have been proposed as a vector for the delivery of probiotic LAB to the host, however, most of them are dairy based products. There is very limited research (compared to those performed on dairy food matrixes) related to use of salami as a vector for probiotic delivery.

Production of dry fermented salami and other raw cured sausages are the result of several factors that act in synergy to yield specific organoleptic characteristics during ripening. Based on the complex biochemical reactions, the result is the formation of metabolic end-products (Paramithios et al. 2010). Normally this type of product does not need thermal treatment during manufacturing or prior to consumption, as it is stable and safe for consumption (Vignolo et al. 2010). However, this fact is favourable for salami to be considered as a potentially good candidate for the delivery of probiotic LAB. However, to have a LAB culture that can be a good probiotic and at same time cover technological characteristics is not an easy task. Very frequently production of antimicrobial compounds, including bacteriocins are not in correlation with production of photolytic enzymes. However, specific proteases are required for the starter cultures applied in production of fermented dairy and meat products.

In the study of Ruiz et al. (2014), the viability of *Lactobacillus acidophilus* and *Bifidobacterium lactis,* both considered to be probiotics, and their effects on the technology and sensorial characteristics of fermented sausage was evaluated. The presence of both LAB probiotic cultures reduced water activity and promoted the faster pH reduction in salamis, which presented final pH values between 4.71 and 5.23 and water activity between 0.84 and 0.89. In addition, lactic acid content ranged between 0.19 and 0.29 g, and the samples lost up to 35% of their weight during ripening. With regards to colour, no differences were found between the probiotic salamis and the control, presenting an overall mean of 40.85 for L* (lightness), 14.48 for a* (redness) and 6.46 for b* (yellowness). High consumer acceptance was observed for the probiotic salamis, which showed an average acceptance of approximately 7.0 on a nine-point hedonic scale for all attributes evaluated, with no differences ($p \leq 0.05$) when compared with the control. However, the performance of *Lactobacillus acidophilus* was better, as the salamis treated with this microorganism presented less weight loss, better acceptance and greater purchase intention. Flavour and texture were the attributes that most influenced sensory acceptance of the salami. Salamis supplemented with probiotic cultures might be a viable option for the formulation of fermented sausages in the food industry. In addition, a good survival rate of both of the studied LAB was recorded during the manufacture and storage of the salami (Ruiz et al. 2014).

Moreover, the addition of probiotic cultures in the salami was advantageous because Italian salamis had a high acceptance among consumers and a similar performance to the control, prepared without adding *Lactobacillus acidophilus* or

Bifidobacterium lactis. These products have great sales potential in the consumer market. This theory is reinforced by the increasing global trend of consuming healthy foods (Ruiz et al. 2014).

Conclusions

Production of fermented meat products in Latin America was based on the knowledge of the emigrants and largely ignored the indigenous tribes of the region. This is why the rich heritage and knowledge of the pre-Colombian era was almost lost. However, based on the the local environmental conditions, new fermented meat food products have been born, combining European, Asian and African knowledge with a touch of the Latin American spirit.

References

ABIPECS-Associação Brasileira da Indústria Produtora e Exportadora de Carne Suína. Relatório Anual, 2006. available from: http://www.abipecs.org.br/.

Adams, M.R. (1986). Fermented flesh foods. In: Adams, M.R. (Ed.), Progress in Industrial Microbiology 23: 175–179.

Aymerich, M.T., M. Garriga, J.M. Monfort, I. Nes and M. Hugas. (2000). Bacteriocin producing lactobacilli in Spanish-style fermented sausages: characterization of bacteriocins. Food Microbiology 17: 33–45.

Backes, A.M., N.N. Terra, L.I.G. Milani, A.P.S. Rezer, F.L. Lüdtke, C.P. Cavalheiro and L.L.M. Fries. (2013). Physico-chemical characteristics and sensory acceptance of Italian-type salami with canola oil addition. Semina: Ciências Agrárias 34(6, suplement 2): 3709–3720.

Bacus, J. (1984). Update: meat fermentation 1984. Food Technology 38(6): 59–69.

Barbosa, M.S., S.D. Todorov, I. Ivanova, J.-M. Chobert, T. Haetle and B.D.G.M. Franco. (2015a). Improving safety of salami by application of bacteriocins produced by an autochthonos *Lactobacillus curvatus* isolate. Food Microbiology 46: 254–262.

Barbosa, M.S., S.D. Todorov, C.H. Jurkiewicz and B.D.G.M. Franco. (2015b). Bacteriocin production by *Lactobacillus curvatus* MBSa2 entrapped in calcium alginate during ripening of salami for control of *Listeria monocytogenes*. Food Control 47: 147–153.

Barbosa, M.S., S.D. Todorov, Y. Belguesmia, Y. Choiset, H. Rabesona, I.V. Ivanova, J.M. Chobert, T. Haertle and B.D.G.M. Franco. (2014). Purification and characterization of the bacteriocin produced by *Lactobacillus sakei* MBSa1 isolated from Brazilian salami. Journal of Applied Microbiology 116(05): 1195–1208.

Basmacioglu, H., O. Tokusoglu and M. Ergul. (2004). The effect of oregano and rosemary essential oils or alpha-tocopheryl acetate on performance and lipid oxidation of meat enriched with n-3 PUFA's in broilers. South African Journal of Animal Science 34(3): 197–210.

Belgacem, Z.B., M. Ferchichi, H. Prévost, X. Dousset and M. Manai. (2008). Screening for anti-*Listeria* bacteriocin-producing lactic acid bacteria from "Gueddid" a traditionally Tunisian fermented meat. Meat Science 78: 513–521.

Bertazzoni-Minelli, E., A. Benini, M. Marzotto, A. Sbarbati, O. Ruzzenente, R. Ferrario, H. Hendriks and F. Dellaglio. (2004). Assessment of novel probiotic *Lactobacillus casei* strains for the production of functional foods. International Dairy Journal 14: 723–736.

Bertol, T.M., A.M. Fiorentini, M.J.H. Santos, M.C. Sawitzki, V.L. Kawski, I.B.L. Agnes, C.D. Costa, A. Coldebella and L.S. Lopes. (2012). Rosemary extract and celery-based products used as natural quality enhancers for colonial type salami with different ripening times. Ciência e Tecnologia de Alimentos 32: 783–792.

Bolton, L.F., and J.F. Frank. (1999). Simple method to observe the adaptive response of *Listeria monocytogenes* in food. Letters in Applied Microbiology 29: 350–353.

Bonnet, M. and T.J. Montville. (2005). Acid-tolerant *Listeria monocytogenes* persist in a model food system fermented with nisin producing bacteria. Letters in Applied Microbiology 40: 237–242.

Brasil. (2000a). Instrução Normativa n. 22, de 31 de julho de 2000. Regulamento técnico de identidade e qualidade do salame tipo italiano. Diário Oficial [da] República Federativa do Brasil, Brasília, DF, 3 ago. 2000. Seção 1, n. 149, p. 24–25.

Brasil. (2000b). Ministério da Agricultura, Pecuária e Abastecimento (MAPA). Secretaria de Defesa Agropecuária (SDA). Instrução Normativa n. 22, 31 julho de 2000. Aprova os Regulamentos Técnicos de Identidade e Qualidade de Jerked Beef, de Presunto tipo Parma, de Presunto Cru, de Salame, de Salaminho, de Salame tipo Alemão, de Salame tipo Calabres, de Salame tipo Friolano, de Salame tipo Napolitano, de Salame tipo Hamburgues, de Salame tipo Italiano, de Salame tipo Milano, de Linguiça Colonial e Pepperoni. Diário Oficial da União, Brasília, seção 1, p. 15–28.

Brasil, 2003. Agência Nacional de Vigilância Sanitária (ANVISA), Resolução RDC no. 360, de 23 de Dezembro de 2003. Regulamento técnico sobre rotulagem nutricional de alimentos embalados, tornando obrigatória a rotulagem nutricional. available from: http://www.anvisa.gov.br/elegis/.

Bryan, F.L. (1980), Foodborne diseases in the United States associated with meat and poultry. Journal of Food Protection 43(2): 140–150.

Camargo, A.C., L.A. Nero and S.D. Todorov. (2015). Where the problem is with *Listeria monocytogenes*? Editorial. Journal of Nutritional Health and Food Engineering 1(6): 00035.

Campagnol, P.C.B., L.L.M. Fries, N.N. Terra, B.A. Santos and A.S. Furtado. (2007). Salami sausage prepared with *Lactobacillus plantarum* fermented in porcine plasma culture medium. Ciência e Tecnologia de Alimentos 27: 883–889.

Carr, F.J., D. Chill and N. Maida. (2002). The lactic acid bacteria: a literature survey. Critical Reviews in Microbiology 28: 281–370.

Carvalho, A.A.T., R.A. Paula, H.C. Mantovani and C.A. Moraes. (2006). Inhibition of *Listeria monocytogenes* by lactic acid bacterium isolated from Italian salami. Food Microbiology 23: 213–219.

Carvalho, K.G., F.H.S. Bambirra, M.F. Kruger, M.S. Barbosa, J.S. Oliveira, A.M.C. Santos, J.R. Nicoli, M.P. Bemquerer, A. de Miranda, E.J. Salvucci, F.J.M. Sesma, and B.D.G.M. Franco. (2010). Antimicrobial compounds produced by *Lactobacillus sakei* subsp. *sakei* 2a, a bacteriocinogenic strain isolated from a Brazilian meat product. Journal of Industrial Microbiology and Biotechnology 37: 381–390.

Castro, M.P., N.Z. Palavecino, C. Herman, O.A. Garro and C.A. Campos. (2011). Lactic acid bacteria isolated from artisanal dry sausages: characterization of antibacterial compounds and study of the factors affecting bacteriocin production. Meat Science 87:321–329.

Cataneo, C.B., V. Caliari, L.V. Gonzaga, E.M. Kuskoski and R. Fett. (2008). Atividade antioxidante e conteúdo fenólico do resíduo agroindustrial da produção de vinho. Semina: Ciências Agrárias 29(1): 93–102

Chaillou, S., Champomier-Vergaes, M.C., M. Cornet, A.M. Crutz-Le Coq, A.M. Dudez. V. Martin, S. Beaufils, E. Darbon-Rongnere, R. Bossy, V. Loux and M. Zagorec. (2005). The complete genome sequence of the meat-borne lactic acid bacterium *Lactobacillus sakei* 23K. Nature Biotechnology 23: 1527–1533.

Chan, E.S. and Z. Zhang. (2005). Bioencapsulation by compression coating of probiotic bacteria for their protection in an acidic medium. Process Biochemistry 40: 3346–3351.

Choi, Y.S., K.S. Park, H.W. Kim, K.E. Hwang, D.H. Song, M.S. Choi, S.Y. Lee, H.D. Paik and C.J. Kim. (2013). Quality characteristics of reduced-fat frankfurters with pork fat replaced by sunflower oils and dietary fiber extracted from *makgeolli* lees. Meat Science 93(3): 652–658.

Cintas, L.M., J.M. Rodriguez, M.F. Fernandez, K. Sletten, I.F. Nes, P.E. Hernandez and H. Holo. (1995). Isolation and characterization of pediocin L50, a new bacteriocin from *Pediococcus acidilactici* with a broad inhibitory spectrum. Applied and Environmental Microbiology 61: 2643–2648.

Cotter, P.D., C. Hill and R.P. Ross. (2005). Bacteriocins: developing innate immunity for food. Nature Reviews in Microbiology 3: 777–788.

De Vuyst, L.D., and E.J. Vandamme. (1994). Bacteriocins of lactic acid bacteria: microbiology, genetics and applications. Blackie Academic & Professional.

Degenhardt, R. and E.S. Sant'Anna. (2007). Survival of *Listeria monocytogenes* in low acid Italian sausage produced under Brazilian conditions. Brazilian Journal of Microbiology 38: 309–314.

Del Nobile, M.A., A. Conte, A.L. Incoranato, O. Panza, A. Sevi and R. Marino. (2009). New strategies for reducing the pork back-fat content in typical Italian salami. Meat Science 81(1): 263–269.

Demeyer, D.I., A. Verplaetse and M. Gistelink. (1986). Fermentation of meat: an integrated process. Food Chemistry Biotechnology 41(5):131–140.

Di Pinto, A., L. Novello, F. Montemurro, E. Bonerba and G. Tantillo. (2010). Occurrence of *Listeria monocytogenes* in ready-to-eat foods from supermarkets in Southern Italy.New Microbiologica 33: 249–252.

D., M.L. Delignette-Muller, S. Christieans and C. Vernozy-Rozand. (2005b). Prevalence of *Listeria monocytogenes* in 13 dried sausage processing plants and their products. International Journal of Food Microbiology 102: 85–94.

Favaro, L., A.L.B. Penna and S.D. Todorov. (2015). Bacteriocinogenic LAB from cheeses - application in biopreservation? Trends in Food Science and Technology 41(01): 37–48.

Franco, D., B. Prieto, J.M. Cruz, and J. Carballo. (2002). Study of the biochemical changes during the processing of Androlla, a Spanish dry-cured pork sausage. Food Chemistry 78(3): 339–345.

Galdeano, C., A. de Moreno, G. Vinderola, M.E. Bibas Bonet and G. Perdigón. (2007). A proposal model: mechanisms of immunomodulation induced by probiotic bacteria. Review. Clinical Vaccination and Immunology 14: 485–492.

Gao, Y., S. Jia, Q. Gao and Z. Tan. (2010). A novel bacteriocin with a broad inhibitory spectrum produced by *Lactobacillus sake* C2, isolated from traditional Chinese fermented cabbage. Food Control 21: 76–81.

Garcia, M.L., R. Domingez, M.D. Galvez, C. Casas and M.D. Selgas. (2002). Utilization of cereal and fruit fibres in low fat dry fermented sausages. Meat Science 60: 227–236.

García, P., L. Rodríguez, A. Rodríguez and B. Martínez. (2010). Food biopreservation: promising strategies using bacteriocins, bacteriophages and endolysins. Trends in Food Science & Technology 21(8): 373–382.

Garcia-Fontan, M.C., J.M. Lornzo, A. Parada, I. Franco and J. Carballo. (2007). Microbial characteristics of "androlla", a Spanish traditional pork sausage. Food Microbiology 24(1): 52–58.

Gomes, A.M.P. and M.F. Xavier. (1999). *Bifidobacterium* spp. and *Lactobacillus acidophilus*: biological, biochemical, technological and therapeutical properties relevant for use as probiotics. Trends in Food Science and Technology 10: 139–157.

Heng, N.C.K., P.A. Wescombe, J.P. Burton, R.W. Jack and J.R. Tagg. (2007). The diversity of bacteriocins in Gram-positive bacteria. In: Riley, M.A. and Chavan, M.A. (Eds.), Bacteriocins Ecology and Evolution. Springer, Berlin, pp. 45–83.

Holck, A., L. Axelsson, S. Birkeland, T. Aukrust and H. Blom. (1992). Purification and amino acid sequence of sakacin A, a bacteriocin from *Lactobacillus sake* Lb706. Journal of General Microbiology 138: 2715–2720.

Holck, A.L., L. Axelsson, K. Huhne and L. Kreockel. (1994) Purification and cloning of sakacin 674, a bacteriocin from *Lactobacillus sake* Lb674. FEMS Microbiology Letters 115: 143–149.

H. Thippareddi, R.K. Phebus, J.L. Marsden and A.L. Nutsch. (2006). Validation of traditional Italian-style salami manufacturing process for control of *Salmonella* and *Listeria monocytogenes*. Journal of Food Protection 69(4): 794–800.

Hugas, M., M. Garriga, M.T. Aymerich and J.M. Monfort. (1995). Inhibition of *Listeria* in dry fermented sausages by the bacteriocinogenic *Lactobacillus sake* CTC494. Journal of Applied Bacteriology 79: 322–330.

I., A. Adesiyun, N. Seepersadsingh and S. Rahaman. (2006). Investigation for possiblesource(s) of contamination of ready-to-eat meat products with *Listeria* spp. and other pathogensin a meat processing plant in Trinidad. Food Microbiology 23(4): 359–366.

Incze, K. (1998). Dry Fermented Sausages. Meat Science 49(1): 169–177.

Jiang, J., B. Shi, D. Zhu, Q. Cai, Y. Chen, J. Li, K. Qi and M. Zhang. (2012). Characterization of a novel bacteriocin produced by *Lactobacillus sakei* LSJ618 isolated from traditional Chinese fermented radish. Food Control 23: 338–344.

Johnson, J.L., M.P. Doyle, R.G. Cassens and J.L. Shoeni. (1988). Fate of *Listeria monocytogenes* in tissues of experimentally infected cattle and hard salami. Applied and Environmental Microbiology 54: 497–501.

Kailasapathy, K. and J. Chin. (2000). Survival and therapeutic potential of probiotic organisms with reference to *Lactobacillus acidophilus* and *Bifidobacteria* spp. Immunology and Cellular Biology 78: 80–88.

Kaiser, A.L. and T.J. Montville. (1996) Purification of the bacteriocin bavaricin MN and characterization of its mode of action against *Listeria monocytogenes* Scott A cells and lipid vesicles. Appled and Environmental Microbiology 62: 4529–4535.

Koutsoumanis, K.P., and J.N. Sofos. (2005). Effect of inoculum size on the combined temperature, pH and aw limits for growth of *Listeria monocytogenes*. International Journal of Food Microbiology 104: 83–91.

Kwon, J.S., J.T. Snook, G.H. Wardlaw and D.H. Hwang. (1991). Effects of diets high in saturated fatty acids, canola oil or safflower oil on platelet function, thromboxane B2 formation, and fatty acid composition of platelet phospholipids. The American Journal of Clinical Nutrition, 54(2): 351–358.

Larsen, A.G., F.K. Vogensen and J. Josephsen. (1993). Antimicrobial activity of lactic acid bacteria isolated from sourdoughs: purification and characterization of bavaricin A, a bacteriocin produced by *Lactobacillus bavaricus* MI401. Journal of Applied Bacteriology 75: 113–122.

Lee, K.G. and T. Shibamoto. (2002). Determination of antioxidant potential of volatile extracts isolated from various herbs and spices. Journal of Agricultural and Food Chemistry 50: 4947–4952.

Levine, P., B. Rose, S. Green, G. Ransom and W. Hill. (2001). Pathogen testing of ready-to-eat meat and poultry products collected at federally inspected establishments in the United States1990 to 1999. Journal of Food Protection 64: 1188–1193.

Liu, M.N., D.L. Huffman and W.R. Egbert. (1991). Replacement of beef fat with partially hydrogenated plant oil in lean ground beef patties. Journal of Food Science 56(3): 861–862.

Lopez-Bote, C.J., J.I. Gray, E.A. Gomaa and C.J. Flegal. (1998). Effect of dietary administration of oil extracts from rosemary and sage on lipid oxidation in broiler meat. British Poultry Science 39: 235–240.

Luruena-Martinez, M.A., A.M. Vivar-Quitana and I. Revilla. (2004). Effect of locust bean/xanthan gum addition and replacement of pork fat with olive oil on the quality characteristics of low-fat frankfurters. Meat Science 68(3): 383–389.

Maciel, J.F., M.A. Teixeira, C.A. Moraes and L.A.M. Gomide. (2003). Antibacterial acitivity of lactic cultures isolated of Italian salami. Brazilian Journal of Microbiology 34: 121–122.

Marangoli, C. and N.F. Moura. (2009). Antioxidant activity of essential oil from *Coriandrum sativum* L. in Italian salami. Ciência e Tecnologia de Alimentos 31: 124–128.

Marangoli, C. and N.F. Moura. (2011). Sensory profile of Italian salami with coriander (*Coriandrum sativum* L.) essential oil. Ciência e Tecnologia de Alimentos 31: 119–123.

Masuda, Y., H. Ono, H. Kitagawa, H. Ito, F. Mu, N. Sawa and K. Sonomoto. (2011). Identification and characterization of leucocyclicin Q, a novel cyclic bacteriocin produced by *Leuconostoc mesenteroides* TK41401. Applied and Environmental Microbiology 77(22): 8164–8170.

Mathiesen, G., K. Huehne, L. Kroeckel, L. Axelsson and V.G.H. Eijsink. (2005). Characterization of a new bacteriocin operon in sakacin P-producing *Lactobacillus sakei*, showing strong translational coupling between the bacteriocin and immunity genes. Applied and Environmental Microbiology 71: 3565–3574.

McDonald, B.E., J.M. Gerrard, V.M. Bruce and E.J. Corner. (1989). Comparison of the effect of canola oil and sunflower oil on plasma lipids and lipoproteins and on *in vivo* thromboxane A2 and prostacyclin production in health young men. The American Journal of Clinical Nutrition 50(6): 1382–1388.

Melo, E. and N.B. Guerra. (2002). Ação antioxidante de compostos fenólicos naturalmente presentes em alimentos. Boletim SBCTA 36(1): 1–11.

Mendes, A.C.G., D.M. Rettore, A.L.S. Ramos, S.F.V. Cunha, L.C. Oliveira and E.M. Ramos. (2014). Milano type salami elaborated with fibers of red wine byproducts. Ciência Rural 44: 1291–1296.

Messens, W., and L. de Vuyst. (2002). Inhibitory substances produced by Lactobacilli isolated from sourdoughs-a review. International Journal of Food Microbiology 72: 31–43.

Moore, J.E. (2004). Gastrointestinal outbreaks associated with fermented meats. Meat Science 67: 565–568.

Muguerza, E., D. Gimeno, D. Ansorena and I. Astiasaran. (2004). New formulations for healthier dry fermented sausages: a review. Trends in Food Science &Techonology (15)9: 452–457.

Nissen, H. and A. Holck. (1998). Survival of *Escherichia coli* O157:H7, *Listeria monocytogenes* and *Salmonella kentucky* in Norwegian fermented, dry sausage. Food Microbiology 15: 273–279.

Nydahl, M., I.B. Gustafsson, M. Ohrvall and B. Vessby. (1995). Similar effects of rapeseed oil (canola oil) and olive oil in a lipid-lowering diet for patients with hyperlipoproteinemia. Journal of The American College of Nutrition 14(6): 643–651.

Oelschlaeger, T.A. (2010). Mechanisms of probiotic actions – A review. International Journal of Medical Microbiology 300: 57–62.

Okada, Y., S. Monden, S. Igimi and S. Yamamoto. (2012). The occurrence of *Listeria monocytogenes* in imported ready-to-eat foods in Japan. Journal of Veterinary Medicine and Science 74: 373–375.

Paramithios, S., E.H. Drosinos, J.N. Sofos and G.J. Nychas. (2010). Fermentation: microbiology and biochemistry. p. 185–198. In: Toldrá, F., (Ed.), Handbook of meat processing. Wiley-Blackwell, Hoboken, NJ, USA.

Pelser, W.M., J.P. Linssen, A. Legger and J.H. Houben. (2007). Lipid oxidation in n-3 fatty acid enriched Dutch style fermented sausages. Meat Science 75(1): 1–11.

Petruzzelli, A., G. Blasi, L. Masini, L. Calza, A. Duranti, S. Santarelli, S. Fisichella, G. Pezzotti, L. Aquilanti, A. Osimani and F. Tonucci. (2010). Occurrence of *Listeria monocytogenes* in salami manufactured in the Marche Region (Central Italy). Journal of Veterinary Medicine and Science 72: 499–502.

Rabanah, T., N. Hettiarachchy, R. Horax, S. Eswaranandam, A. Mauromoustakos, J. Dickson and S. Neibuhr. (2004). Effect of electron bean irradiation and storage at 5°C on thiobarbituric acid reactive substances and carbonyl contents in chicken breast meat infused with antioxidants and selected plant extracts. Journal of Agricultural and Food Chemistry 52: 8236–8241.

Rantsiou, K. and L. Cocolin. (2006). New developments in the study of the microbiota of naturally fermented sausages as determined by molecular methods: A review. International Journal of Food Microbiology 108(2): 255–267.

Reis, J.A., A.T. Paula, S.N. Casarotti and A.L.B. Penna. (2012). Lactic acid bacteria antimicrobial compounds: characteristics and applications. Food Engineering Reviews 4(2): 124–140.

Rodriguez, R., A. Jimenez, J. Fernadez-Bolanos, R. Guillen and A. Heredia. (2006). Dietary fibre from vegetable products as source of functional ingredients. Trends in Food Science and Technology 17: 3–15.

Ruiz, J.N., N.D.M. Villanueva, C.S. Favaro-Trindade and C.J. Contreras-Castillo. (2014). Physicochemical, microbiological and sensory assessments of Italian salami sausages with probiotic potential. Scientia Agricola 71: 204–211.

Sakate, R.I., L.C. Aragon, F. Raghiante, M. Landgraf, B.D. Franco and M.T. Destro. (2003). Occurrence of *Listeria monocytogenes* in pre-sliced vacuum-packaged. Archives of Latinoamerican Nutrition 53: 184–187.

Santos, C., L. Hoz, M.I. Cambero, M.C. Cabeza and J.A. Ordonez. (2008). Enrichment of dry-cured ham with α-linonelic acid and α-tocopherol by the used of linseed oil and α-tocopheryl acetate in pig diets. Meat Science 80(3): 668–674.

Sawitzki, M.C. (2000). Caracterização de bactérias ácido láticas isoladas de salames artesanais e aplicadas como cultivos iniciadores em salame tipo Italiano. Santa Maria, Dissertação (Mestrado em Ciência e Tecnologia de Alimentos) – Universidade Federal de Santa Maria (UFSM).

Schillinger, U. and F.-K. Lucke. (1989). Antibacterial activity of *Lactobacillus sake* isolated from meat. Applied and Environmental Microbiology 55: 1901–1906.

Schirru, S., S.D. Todorov, L. Favaro, N.P. Mangia, M. Basaglia, S. Casella, R. Comunian, B.D.G.M. Franco and P. Deiana. (2012). Sardinian goat's milk as source of bacteriocinogenic potential protective cultures. Food Control 25(1): 309–320.

Silva, C., F.C. Savariz, H.M. Follmann, L. Nunez, V.M. Chapla and C.F. Silva. (2011). Análise físico-química de salames coloniais comercializados no minicipio de Toledo, Estado do Parana. Acta Scientiarum and Technology 33: 331–336.

Simon, L., C. Fremaux, Y. Cenatiempo and J.M. Berjeaud. (2002). Sakacin G, a new type of anti listerial bacteriocin. Applied and Environmental Microbiology 68:6416–6420.

Sobrino, J., J.M. Rodriguez, W.L. Moreira, L.M. Cintas, M.F. Fernandez, B. Sanz and P.E. Hernandez. (1992). Sakacin M, a bacteriocin-like substance from *Lactobacillus sake* 148. International Journal of Food Microbiology 16: 215–225.

Soyer, A., A.H. Ertas, Ü. Üzumcuoglu. (2005). Effect of processing conditions on the quality of naturally fermented Turkish sausages (sucuks). Meat Science 69(1): 135–141.

Sudirman, I., F. Mathier, M. Michel and G. Lefebvre. (1993). Detection and properties of curvaticin 13, a bacteriocin-like substance produced by *Lactobacillus curvatus* SB13. Current Microbiology 27: 35–40.

Tichaczek, P.S., R.F. Vogel and W.P. Hammes. (1994). Cloning and sequencing of sakP encoding sakacin P, the bacteriocin produced by *Lactobacillus sake* LTH673. Microbiology 140: 361–367.

Todorov, S.D., L. Favaro, P. Gibbs and M. Vaz-Velho. (2012). *Enterococcus faecium* isolated from Lombo, Portuguese traditional meat product: Characterization of antibacterial compound and study of the factors affecting bacteriocin production. Beneficial MMicrobes 3(4): 319–330.

Todorov, S.D., P. Ho, M. Vaz-Velho and L.M.T. Dicks. (2010). Characterization of bacteriocins produced by two strains of *Lactobacillus plantarum* isolated from Beloura and Chouriço, traditional pork products from Portugal. Meat Science 84: 334–343.

Todorov, S.D., K.S.C. Koep, C.A. Van Reenen, L.C. Hoffman, E. Slinde and L.M.T. Dicks. (2007). Production of salami from beef, horse, mutton, blesbok (*Damaliscus dorcas phillipsi*) and springbok (*Antidorcas marsupialis*) with bacteriocinogenic strains of *Lactobacillus plantarum* and *Lactobacillus curvatus*. Meat Science 77 (3): 405–412.

Todorov, S.D., M. Vaz-Velho, B.D.G.M. Franco and W.H. Holzapfel. (2013). Partial characterization of bacteriocins produced by three strains of *Lactobacillus sakei*, isolated from salpicao, a fermented meat product from North-West of Portugal. Food Control 30: 111–121.

Tompkin, R.B. (2002). Control of *Listeria monocytogenes* in the food-processing environment. Journal of Food Protection 65: 709–725.

Työppönen, S., A. Markkula, E. Petäjä, M.L. Suihko and T. Mattila-Sandholm. (2003). Survival of *Listeria monocytogenes* in North European type dry sausages fermented by bioprotective meat starter cultures. Food Control 14: 181–185.

Urso, R., K. Rantsiou, C. Cantoni, G. Comi and L. Cocolin. (2006). Sequencing and expression analysis of the sakacin P bacteriocin produced by a *Lactobacillus sakei* strain isolated from naturally fermented sausages. Applied Microbiology and Biotechnology 71: 480–485.

Valencia, I., M.N. O'Ggrady, D. Ansorena, I. Astiasaran and J.P. Kerry. (2008). Enhancement of the nutritional status and quality of fresh pork sausages following the addition of linseed oil, fish oil and natural antioxidants. Meat Science 80(4): 1046–1054.

Varabioff, Y. (1992). Incidence of *Listeria* in small goods. Letters in Applied Microbiology 14: 167–169.

Vaughan, A., V.G.H. Eijsink, D. van Sinderen, T.F. O'Sullivan, and K. O'Hanlon. (2001). An analysis of bacteriocins produced by lactic acid bacteria isolated from malted barley. Journal of Applied Microbiology 91: 131–138.

Verma, A.K., and R. Banerjee. (2010). Dietary fibre as functional ingredient in meat products: a novel approach for healthy living – a review. Journal of Food Science & Technology 47(3): 247–257.

Vermeiren, L., F. Devlieghere and J. Debevere. (2004). Evaluation of meat born lactic acid bacteria as protective cultures for the biopreservation of cooked meat products. International Journal of Food Microbiology 96: 149–164.

Vialette, M., A. Pinon, E. Chasseignaux and M. Lange. (2003). Growths kinetics of clinical and seafood *Listeria monocytogenes* isolates in acid and osmotic environment. International Journal of Food Microbiology 82: 121–131.

Vignolo, G., C. Fontana and S. Fadda. (2010). Semidry and dry fermented sausages. p. 379–398. In: Toldrá, F., (Ed.), Handbook of meat processing. Wiley-Blackwell, Hoboken, NJ, USA.

WHO (World Health Organization). (2002). Guidelines for the Evaluation of Probiotics in Food. Available at: www.who.int/foodsafety/fs_management/en/probiotic_guidelines.pdf.

Wirth, F. (1988). Technologies for making fat-reduced meat products. Fleischwirtsch 68(9): 1153–1156.

Yen, G.C. and H.Y. Chen. (1995). Antioxidant activity of various tea extracts in relation to their antimutagenicity. Journal of Agricultural and Food Chemistry 43: 27–32.

Yunes, J.F.F., N.N. Terra, C.P. Cavalheiro, L.L.M. Freis, H.T. Godoy and C.A. Ballus. (2013). Perfil de ácidos graxos e teor de colesterol de mortadela elaborada com óleos vegetais. Ciência Rural 43(5): 924–929.

8

Influence of New Trends in Wine Technology on the Chemical and Sensory Profiles

Maurício Bonatto Machado De Castilhos[a]*, Vanildo Luiz Del Bianchi[b] and Isidro Hermosín-Gutiérrez[c]

Abstract

The search for nutritional and nutraceutical foods and their health benefits beyond basic nutrition is constantly increasing. Among beverages, many of these beneficial characteristics are present in wine. Wines contain phenolic compounds in their composition, which are responsible for several health benefits due to their antioxidant capacity. However, as wine technology is complex and involves several steps, wineries encounter problems with the degradation of these compounds, which adversely affect wine quality. Basically, wine is composed of water and ethanol, but it is nevertheless considered to be one of the most complex beverages due to the presence of minor compounds such as phenolic and volatile substances, which interfere in both the chemical and sensory profiles. Wine technology demands intensive care since it involves great numbers of chemical reactions during the two fermentative steps: alcoholic fermentation which involves the yeast metabolism of sugars, and malolactic fermentation which involves malic acid decarboxylation into lactic acid by lactic acid bacteria. In this context, new trends in wine making are the object of several scientific studies. Among them, thermovinification, grape pre-dehydration, cold soaking, carbonic maceration, submerged cap and the application of selected yeasts and pectinases are examples of techniques applied by Latin American wineries in order to study the reactions involving changes in these minor compounds, assessing the antioxidant capacity as the main nutritional factor and providing results about the improvement of wine sensory quality. The wine making techniques reported in this chapter bring relevant results about the chemical behavior of all wine composition, mainly regarding phenolic compounds, such as anthocyanins and tannins. In

---

[a] Vinification and Bioprocess Laboratory, Engineering and Food Technology Department, São Paulo State University, Cristóvão Colombo Street, 2265, São José do Rio Preto, São Paulo, Brazil.
[b] Engineering and Food Technology Department, São Paulo State University, Cristóvão Colombo Street, 2265, São José do Rio Preto, São Paulo, Brazil.
[c] Instituto Regional de Investigación Científica Aplicada (IRICA), Universidad de Castilla-La Mancha, Avda. Camilo José Cela S/N, 13071, Ciudad Real, Spain.
* Corresponding Author: mbonattosp@yahoo.com.br

addition, all the chemical changes described were correlated with the wine sensory features, which allow for the assessment of sensory quality. The present chapter aims to present a discussion of the state of the art of alternative wine making technologies and their effects on the chemical, sensory and nutritional profiles of the wines, the latter through the analysis of the antioxidant capacity.

Introduction

Wine making is a complex process involving several steps which frequently overlap. The transformation of the grapes into wine presents variations according to the region, due to the climate, soil type, grape cultivar and vine management (Jackson 2008). The traditional wine making process (Fig. 8.1), basically, followed the steps of destemming and crushing the grapes, which allow the release of the juice (must). The must and pomace (solid part) are fermented together or not and the production of red or white wines depends on the choice of the wine maker. The mixture is placed in fermentative vessels and treated with sulfur dioxide in order to avoid opportunist contamination. The alcoholic fermentation is performed by yeasts inoculation (*Saccharomyces cerevisiae*) or takes place spontaneously.

At the time of the alcoholic fermentation, the must is separated from the pomace by the dejuicing step thus allowing the chaptalization, if necessary. The solid part can be pressed, which allows the release of 10 to 15% of the juice, which remain adhered in pomace. Usually, three racking are done during the wine making process aiming to separate all the compounds that can haze the wine. The first one is performed after the protein and the phenolic stabilization which is followed by filtration or centrifugation. After the first racking, the wine is submitted to the malolactic fermentation, spontaneously or induced by the inoculation of *Oenococcus oeni*, aiming to provide low acidity to the wine. After the second racking the wines are sulphitated and the blending step can be done in order to improve wine features, followed by the cold stabilization or tartrate crystallization. In order to separate the tartrate crystals from the wine, the third racking is carry out and the wines are submitted to the oak maturation followed by bottling.

The process described above was intensely mapped in order to obtain information about the changes in the chemical profile, which result in sensory changes, altering the final quality of the wines. The wines were generally composed of water and ethanol and other minor compounds such as acids, pectins, carbohydrates, minerals and phenolic compounds. The latter is responsible for the main changes in wine sensory features during the wine making process and wine aging (Jackson 2008; Coombe and Dry 2006). Table 8.1 summarizes the composition of wine and juice.

Amongst these chemical substances present in grapes, one can highlight the phenolic compounds, which comprise anthocyanins, flavonols and flavan-3-ols as well as other compounds such as hydroxycinammic acid derivatives and stilbenes. Other phenolic compounds can be incorporated into the wine from additional sources, such as the ellagitannins and volatile phenolic compounds released from the cooperage wood used for the aging process of several wines. Moreover, all the aforementioned phenolic compounds are subject to change, which can be intensified as a result of reactions occurred throughout the wine making process, and the main goal of current studies is to control them in order to obtain a wine with singular sensory features and high antioxidant capacity.

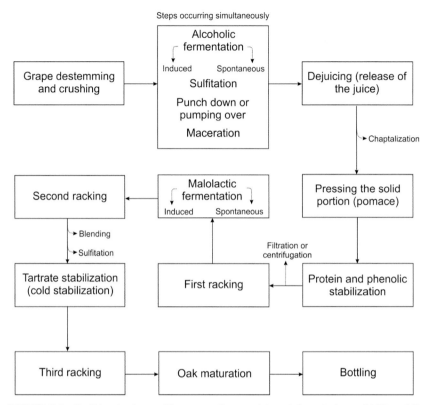

FIGURE 8.1 Traditional wine making steps. Source: Adapted from Jackson (2008) and De Castilhos et al. (2012).

The anthocyanins and other phenolic compounds present antioxidant capacity, and played an important role in the prevention of cardiovascular diseases, arteriosclerosis and thrombosis, controlling diabetes and reducing the risk of some types of cancer (Wang et al. 2006). In addition to the aforementioned antioxidant property, the phenolic compounds have been a focus of several other studies since they present intense correlation with astringency, mouth feel enhancement, color impact and also with the aging potential (Sacchi et al 2005; Gonzalo-Diago et al. 2014; Ma et al. 2014; Chira et al. 2011; Lago-Vanzela et al. 2014a). However, the wine antioxidant capacity, sensory attributes and aging potential of these compounds depend on the wine making procedure applied by the winery. In this context, several wine making techniques have been the focus of studies in order to enhance the concentration of these compounds and the sensory quality of the wines.

The largest wine producers in the world are located in Europe: France with 46.2 million hectoliters, Italy with 44.4 million hectoliters and Spain with 37.0 million hectoliters (OIV 2014). In these countries, the study of the various wine making procedures is carried out in a more intensive way, since they present a huge grape production and are known as the countries that produce the best quality wines in the world. However, Latin America has been associated with viticulture and wine making almost since its discovery and colonization by the Spanish and Portuguese.

Nevertheless, the emergence of Latin America as an important wine producer is recent when compared to the traditional aforesaid countries. While most Latin American countries present insignificant wine production, four South American countries stand out as great wine producers: Chile, Argentina, Uruguay and Brazil. In these countries, the climate and soil promote the production of *Vitis vinifera* cultivars, although, in Brazil, the cultivation of these species is restricted to the south of the country. As the greater part of Brazil has a tropical climate, the American grapes known as *Vitis labrusca* and their hybrids are gaining relevance on the wine panorama, because their wines present features truly appreciated by Brazilian consumers (De Castilhos et al. 2013; Jackson 2008).

TABLE 8.1 Typical concentration ranges of the major and minor chemical components of grape juice and dry wine. Source: Adapted from Coombe and Dry 2006.

Chemical component	Juice (g.L^{-1})	Wine (g.L^{-1})
Water	700–850	800–900
Carbohydrates	150–270	1–10
Pectins	0.1–0.8	Trace
Acids	3–15	4.5–11
Tartaric acid	3–12	1–6
Malic acid	1–8	0–8
Lactic acid	Trace	1–5
Acetic acid	Trace	0.2–1.5
Ethanol	Trace	80–150
Glycerol	Trace	3–14
Anthocyanins	0–0.5	0–0.5
Tannins	Trace–5	Trace–5
Nitrogen compounds	0.2–2	0.1–1
Inorganic constituents	3–5	1.5–4

Chile has a unique and extensive coastline that provides favorable conditions for the production of premium quality wines. Among the several wine regions in Chile, the most highly regarded section is called Regadio, which comprises the central region of Chile from the north of the Aconcagua River to the south of the Maule River. Most Chilean vineyards are associated with river valleys including the Maipo, Cachopoal, Tinguiririca, Lontué and Maule rivers. Most of the grape varieties cultivated in Chile is red type (75%), Cabernet Sauvignon being the most cultivated grape followed by Merlot and Carménère cultivars. Among the white varieties, Chardonnay and Sauvignon Blanc are of the most appreciated quality (Jackson 2008).

Viticulture and wine production are the third largest industry in Argentina. The vineyards are located mainly in the rain shadow of the Andes and the major wine producer region is the province of Mendoza. Vine management and wine making procedures follow the Spanish style and the grapes are harvested early in the Mendoza region to prevent malolactic fermentation. The grape varieties commonly cultivated in the region of Mendoza are Cabernet Sauvignon and Malbec, and in the central regions, Tempranillo and Semillon are the most representative cultivars. Nearly 50% of the grape varieties cultivated in Argentina produce red wines, 20% are used to produce white wine, and the remaining grape cultivars, such as Criolla and Cereza, are mainly cultivated in regions of the country. The slight oxidized character is appreciated by the Argentinean consumer. This feature can be mainly due to the prolonged aging of the white wines in oak cooperage (Jackson 2008).

Among the countries which produce grapes and wine in Latin America, Uruguay has the longest history, since it began its wine making history with the Spanish colonizers. Most of the vineyards are located near Montevideo in the southern part of the country. The most important cultivar is Tannat, accounting for approximately 32% of the total production (Jackson 2008).

In addition to these three Latin American countries, Brazil has been emerging as a wine producer due to its classic and emerging viticulture regions. Brazil is characterized by a considerable production of *Vitis labrusca* grapes and their hybrids, and has focused on the production of juices and wines. Among them, the Bordô, Isabel and Niágara grapes are the most important *Vitis labrusca* representative cultivars. However, these grapes present some disadvantages when compared with *Vitis vinifera* grapes, since they present weak color features and low soluble solids content in their optimal stage of ripening, and consequently, chaptalization is a necessary practice in Brazilian wines produced from these grape cultivars. In this context, the Brazilian Agro-farming Research Agency (Embrapa) has been developing new grape cultivars, known as BRS type grapes, produced by genetic improvement, in order to enhance the grape color features and the soluble solids content for avoiding the practice of chaptalization (De Castilhos et al. 2012, 2013; Lago-Vanzela et al. 2014a).

Among the Brazilian wine regions, the state of Rio Grande do Sul, located in the south of Brazil, is the most important and classical wine producer region. It is characterized by the production of wines from *Vitis vinifera* grape cultivars, since the climate allows for their cultivation. Many grape cultivars are harvested in this region, with a highlight on Cabernet Sauvignon and Merlot. Chardonnay and Sauvignon Blanc are the main white grape representative cultivars. However, other tropical regions of brazil are emerging as great wine producers, mainly for the vinification of *Vitis labrusca* wines, such as the Northwest of São Paulo state and the São Francisco river valley located in the Northeast of the country. In these regions, the BRS type cultivars have gained importance due to their intense color features and typical aromas related to red fruits such as strawberry and raspberry. Among them, BRS Cora (Camargo and Maia 2004), BRS Rúbea (Camargo and Dias 1999), BRS Violeta (Camargo et al. 2005) and BRS Carmem (Camargo et al. 2008) are the most important representative cultivars.

In this context, the present chapter summarizes and critically reviews the literature on the impact of wine making procedures on the chemical, mainly phenolic, and sensory profiles. Different procedures were considered in this chapter, describing their application at different moments in the wine making process. Many studies focused on the more important wine countries in the New World: Chile, Argentina, Brazil and Uruguay, but in addition, other studies from North America and Europe were included, in order to compare the results found in the different regions.

Wine making Procedures

Thermovinification and grape pre-dehydration

Two wine making procedures comprised the use of high temperatures: thermovinification and grape pre-dehydration. These thermal processes have the advantage of reducing the fruit microbial population, inhibiting undesirable enzymes, such as polyphenol oxidase (PPO) and laccase, and also reducing the wine making time (Andrade Neves et al. 2014).

Thermovinification is described as an alternative to the traditional maceration process for the extraction of phenolic compounds from the grape skins. Basically, it consists of submitting the grapes, after crushing, to high temperatures of around 60 to 70 °C for a short time, extracting the compounds from the grape skins with the juice, and cooling them before fermentation. The heat damages the cell wall membranes of the grape skins, releasing the phenolic compounds (mainly anthocyanins, but also flavonols, flavan-3-ols and hydroxycinnamic acid derivatives), and also avoiding the effect of enzimatic oxidation due to their thermal inactivation, as mentioned previously. Since this process is only applied for a short time, and there the absence of alcohol during heating, tannin extraction from the seeds is avoided and the extraction of the anthocyanins is improved, although the extraction of other phenolic compounds is reduced (Sacchi et al. 2005).

Thermovinification is also a wine making procedure indicated for red grapes that present a low color potential and bad sanitary conditions. The application of the usual range of temperatures allows for denaturation of the oxidative enzymes, but it preserves the proteolytic and pectinolytic enzymes, which present great technological importance in wine making and facilitates the fining process (Rizzon et al. 1999).

Grape dehydration before fermentation uses higher temperatures in order to enhance the soluble solids content of the grapes avoiding the step of chaptalization in wine produced from *Vitis labrusca* and its hybrids (De Castilhos 2012, 2013). This procedure is the focus of some studies which evaluated its influence on the extraction of phenolic compounds (De Castilhos et al. 2015a, b).

In addition, drying cause irreversible damage to the grape skins due to water evaporation from the grapes, and these changes decrease their resilience and allow for cell rupture, which in turn allow contacts between PPO and its substrates. Moreover, drying causes structural changes to the grape skins and allows for diffusion of anthocyanins and other phenolic compounds present in the skin to the pulp, thus transferring them during alcoholic fermentation. The temperature used in grape pre-dehydration leads to the formation of oxidative compounds due to the activity of PPO, and to other products resulting from the non-enzymatic Maillard reaction. Low temperatures ranging from 30 to 40 °C allow for PPO activity, causing the formation of oxidative compounds that lead to wine browning (Marquez et al. 2012, 2013); temperatures above 40 °C cause PPO denaturation, but the Maillard reaction occurs in an extensive way, allowing for the formation of melanoidins, which have antioxidant capacity (De Castilhos et al. 2015a).

Many studies have focused on these thermal processes. Rizzon et al. (1999) studied the effect of thermovinification on the chemical and sensory profiles of Cabernet Franc red wines. The grapes were harvested in their optimal ripening stage and under optimal sanitary conditions, de-stemmed, crushed and heated to 65 °C for

2 hours in stainless steel fermentation vessels, which contained a pumping device. The juice was then drained off and the pomace pressed. Alcoholic fermentation was carried out in the presence of the solid parts (pomace), which is necessary in red wine making processes. Fermentation was induced by the inoculation of 200 ppm of dry active yeast and the temperature controlled at 20 °C. Malolactic fermentation occurred spontaneously and was monitored by thin layer chromatography. At the end of the alcoholic fermentation the wines were treated with 50 mg.L^{-1} of sulfur dioxide and refrigerated at −4 °C for 10 days to allow for tartrate stabilization. They were then bottled.

In the aforementioned study, the thermovinification procedure provided wines with higher phenolic and anthocyanin contents, as well as greater color intensity in compared to the traditional treatment of wine. The sensory attributes were also assessed and the color intensity, aroma quality, body, acidity and flavor attributes obtained higher scores when compared to the other wine making procedures. The wines produced from thermovinification presented an intense color feature and high amount of dry extract, which was responsible for enhancement of the mouthfeel and body. As a result the thermovinification wines presented higher global acceptance.

Another study showed relevant results for the application of the thermovinification procedure as an alternative wine making for Cabernet Sauvignon and Pinot Noir grapes (Andrade Neves et al. 2014). The grapes were harvested in their optimal maturity stage and good sanitary conditions, and after crushing, the juice and pomace (skins and seeds) were separated into two different fermentation vessels. The solid part was then immersed in 10% juice and submitted to a heating process at 95 °C for 10 minutes. After cooling, the extract was added to the juice and completed the volume of the must. Alcoholic fermentation was induced by inoculating with dry active *Saccharomyces cerevisiae* yeasts at 25 ± 5 °C and monitoring at 12-hour intervals. After complete alcoholic fermentation, the must was transferred to PET bottles, 50 mg.L^{-1} of potassium metabisulfite added, and stored under refrigeration at 4 °C for 30 days for tartrate stabilization. The wines were then bottled and stabilized for 6 months at 22 °C.

In the aforementioned study the authors stated that the thermovinification process enhanced the extraction of the anthocyanins from the skins, which was responsible for the wine color. However, it was not effective in extracting the tannins from the seeds, and this low concentration of tannins negatively influenced wine aging, since there was a significant decrease in the anthocyanin and flavonol contents after six months of stabilizing. It has been also noted that the thermovinification significantly reduced the wine production time and the use of 95 °C for a short time provided good acceptability for the wines, and was considered an alternative for young wines with different features (Andrade Neves et al. 2014).

In addition to thermovinification, grape pre-dehydration is another alternative step in the wine making process that has been a focus of studies in order to avoid the chaptalization process and to enhance the extraction of the phenolic compounds responsible for color and mouthfeel. De Castilhos et al. (2013) reported that pre-drying significantly influenced the chemical profile and sensory acceptance attributes of Bordô (*Vitis labrusca*) and Isabel (hybrid *Vitis labrusca*) grapes. In the study, the grapes were dried to 22 °Brix of soluble solids content using a tray dryer at 60 °C and of air flow 1.1 m.s^{-1} in order to avoid chaptalization of the grape musts. As soon as the

22 °Brix was reached, the grapes were crushed and placed in fermentation vessels in order to begin the alcoholic fermentation. This process was induced by the inoculation of 200 ppm of dry active yeasts and the wines treated with 150 ppm of potassium metabisulfite. After 7 days of alcoholic fermentation they were racked and malolactic fermentation was induced by the inoculation of *Oenococcus oeni*. The final stage of the second fermentation process was monitored by thin layer chromatography. After this, the wines were stored in a refrigerator (3 °C) for 10 days to allow for tartrate stabilization, and then bottled in 750 mL glass bottles and stored at 18 °C until the chemical and sensory analysis. The grape pre-dehydration significantly increased almost all the chemical properties of the wines, highlighting the total acidity and the dry extract due to the evaporation of water during the drying process. In addition, the authors found that the drying process enhanced the total phenolic content of both Bordô and Isabel red wines. The dried grape wines also presented higher values for color intensity. The results showed that drying could improve the body of the wines, since the dry extract and acidity content were correlated with this sensory attribute. In addition, dried grape wines, regardless of the grape cultivar, presented higher sensory acceptance scores when compared with commercial wines. This result showed the great potential of the drying process as an alternative to produce red wines with good body and structure, despite their lower yields (De Castilhos et al. 2012, 2013, 2015a, 2015b).

A recent study presented data about the phenolic composition of red wines produced from the hybrid BRS Rúbea and BRS Cora grape cultivars submitted to pre-dehydration. The pre-dehydration process significantly degraded the anthocyanins, resulting in a decrease of their concentration; however, the grape pre-dehydration did not significantly affect the flavan-3-ols content (De Castilhos et al. 2015). The anthocyanin degradation was related to thermal degradation and the heat also affected the color of the red wines, changing the characteristic red-purplish color to a brown hue. The browning was produced by the Maillard reaction products, which took place in a more extensive way than the oxidative reactions caused by PPO, since at 60 °C the PPO loses its oxidative action (Patras et al. 2010).

According to Figueiredo-González et al. (2013), heating the grapes causes a loss of their physiological integrity, thus favoring the diffusion of anthocyanins and flavan-3-ols from the grape skin into the pulp. The same authors suggested that a prior contact between these compounds could cause well-known reactions between the anthocyanidin 3-glucosides and flavan-3-ols producing polymeric pigments, with a decrease in the flavan-3-ol contents. However, De Castilhos et al. (2015a) reported that the flavan-3-ols content was not significantly decreased by the grape pre-dehydration. The authors suggested that the copigmentation reaction between monoglucoside anthocyanins and flavan-3-ols was probably stopped because the wines produced using hybrid grape cultivars were composed mainly of 3,5-diglucoside anthocyanins.

De Castilhos et al. (2015) also reported that the antioxidant activity of dried wines was not significantly affected by the heat, suggesting that their antioxidant capacity could be explained by the balance between the loss of antioxidant phenolic compounds and reactions that produced compounds with antioxidant properties such as melanoidins, a Maillard reaction product (Delgado-Andrade and Morales 2005). In the same study, the authors reported on the potential of grape pre-dehydration as an alternative wine making procedure before alcoholic fermentation, in order to produce more structured wines with good body and intense astringency and bitterness, due to their high flavan-3-ol contents.

The results obtained in the above mentioned studies showed the potentials of grape pre-dehydration and thermovinification as alternative wine making procedures aiming at enhancing the phenolic compounds. In the latter case, the increase in the anthocyanin concentration resulted in the intense color of the red wines. Grape dehydration caused anthocyanin degradation and the flavan-3-ols content was apparently not affected by the heat. Sensory approaches showed that thermovinification could enhance the color features of the red wines and grape pre-dehydration could improve the body and structure of the wines, i.e., improving the mouthfeel sensations.

Submerged cap technique

The submerged cap technique aims to provide a constant contact between the pomace and must, avoiding the rise of solids due to the carbon dioxide produced by alcoholic fermentation. This procedure has been a focus of some studies aimed at assessing the enhancement of the extraction of the phenolic compounds as a result of the constant contact of the juice with the grape skins and seeds. Submerged cap technique limits contact with the existing oxygen inside the fermentation vessel and reduces the mechanical operations such as pumping down and pumping over (De Castilhos et al. 2012, 2013; Bosso et al. 2011; Suriano et al. 2012).

De Castilhos et al. (2013) described the influence of submerged cap on red wines produced from Bordô (*Vitis labrusca*) and Isabel (hybrid grape *Vitis labrusca* x *Vitis vinifera*) grapes. This procedure was carried out using stainless steel screens inside the fermentation vessels, aiming to hinder displacement of solids to the upper part of the vessel due to the carbon dioxide produced during alcoholic fermentation. The procedure was carried out on a microvinification scale and information about the enological parameters and sensory acceptance attributes was collected. The submerged cap wine making did not significantly change the results of the chemical analyses, since the acidity, dry extract and alcohol contents of the wines were lower than those of wines produced using the traditional treatment. In addition, the expected effect of enhanced extraction of the phenolic compounds due to the constant contact between the pomace and the must during alcoholic fermentation was not observed, since the total phenolic content and all the color indexes (absorbance at 420, 520 and 620 nm) presented no significant differences when compared with those of traditional wine making. The sensory acceptance of the submerged cap wines also presented no significant differences in comparison with those of the traditional treatment, although these wines were significantly better accepted when compared to the commercial wines. This result explained the modest potential of the submerged cap wine making when applied to Bordô and Isabel red wines, since there were no significant differences between them and the traditional treatment. A positive outcome of the study was the higher yield of the wine produced in relation to the initial amount of grapes when the submerged cap wine making procedure was used, i.e., about 1.5-fold that observed for the traditional treatment.

In another study, Bosso et al. (2011) applied the submerged cap technique in the production of wines from Barbera grapes, and compared it with the floating-cap wine making procedure. The method used to apply the submerged cap wine making technique was the use of a grate fixed at the top of the fermentation vessel to keep the cap submerged at approximately 10 cm depth in the must during alcoholic fermentation. They applied a short pump-over with 25 to 30% of the total volume of the must twice a day with no air, up to the end of maceration. The wine composition, and the polyphenolic and free volatile compounds were analyzed. The submerged cap technique increased the total extract and total acidity, and presented higher total anthocyanin and total flavonoid contents. All the color parameters were higher than those of the floating cap wines. The pressed wines and the pomace resulting from these two wine making procedures were also evaluated. The pressed wines and pomace from submerged cap presented significantly higher anthocyanin, total flavonoid and proanthocyanidin contents in comparison with the floating cap treatment.

Considering the obtained free volatile compounds, Bosso et al. (2011) noted lower C6 alcohol, acetate and higher alcohol concentrations in the submerged cap Barbera wines. The C6 alcohols are produced during the prefermentative phase by enzymatic lipoxidation of the unsaturated fatty acids, and the concentration of these compounds is closely related to the grape variety, mechanical operations and oxygen intake during the fermentative process. The lower concentrations of these compounds were due to the different oxygen intakes during alcoholic fermentation. In addition, the authors stated that the Barbera grape cultivar had poor amounts of terpenic and norisoprenoidic compounds such as linalool, α-terpineol, citronellol and β-damascenone; however, the submerged cap treatment increased the concentration of these compounds, and is considered an alternative wine making procedure to obtain wines with intense varietal features.

As another example, Suriano et al. (2012) provided relevant information about the chemical profile and sensory assessment of Nero di Troia wine produced using two alternative wine making procedures: submerged cap (horizontal winemaker-tank) and vertical winemaker-tank, and both procedures were compared to a traditional wine making technique. In their study, the submerged cap treatment was obtained using a horizontal winemaker-tank with a rotary steel vat composed of two horizontal and concentric cylindrical tanks. The tank worked with a 1 minute rotation every 3 hours at minimum speed, and after two turns, the direction of the rotation changed. The submerged cap treatment differed from those two aforementioned studies in which the behavior of the wine making procedure was assessed on an industrial scale using a rotary tank. The submerged cap technique significantly enhanced the dry extract of the wines as well as presented higher concentrations of total polyphenols, total flavonoids, flavans and proanthocyanidins. In addition, the wines presented higher concentrations of total anthocyanins and monomeric anthocyanins. The authors noted the potential of submerged cap wine making to produce high color wines, and to allow for a better dissolution of tannins and other substances responsible for the wine color intensity. The chemical results were confirmed by the sensory results, since the submerged cap wines presented relevant color intensity and more marked purple hues. In addition, they also presented intense cherry and dried prune olfactory features. With regard to the flavor descriptors, the wines were astringent, bitter and structured, due to the high extraction of catechins and tannins.

De Castilhos et al. (2015a) also evaluated the polyphenolic composition and sensory descriptors of wines produced from BRS Rúbea and BRS Cora hybrid grape cultivars, cultivated in Brazil using the submerged cap technique. They reported the potential of submerged cap as an alternative wine making procedure aiming at producing wines with intense color features, since they present higher anthocyanin concentrations. The findings showed the importance of the submerged cap to obtain highly colored wines, and this wine making procedure could be an alternative for grapes that present weak color features and varietal characteristics, since it could also enhance compounds that are intrinsic to the grape cultivar.

Cold soaking technique

The cold soaking procedure has received much attention worldwide and has been applied in most wine-growing regions, which are typically great producers of different grape cultivars. Prefermentative cold soaking consists of the contact of the fermentation solids (skins and seeds) with the must in a non-alcoholic and low-temperature environment (Casassa and Sari 2015). The absence of ethanol was assured by keeping the must at low temperatures, ranging from 5 to 10 ºC (Casassa et al. 2015) or from 10 to 15 ºC (Sacchi et al. 2005), for a period of time ranging from 3 to 5 hours per day for up to 10 days (Gil-Muñoz et al. 2009; Gordillo et al. 2010; Ortega-Heras et al. 2012). The chilling to which the pomace and the must were submitted ensured the lack of fermentative activity, avoiding the development of *Saccharomyces cerevisiae* (Zott et al. 2008) and, in contrast, favors the development of non-*Saccharomyces* yeasts, which would probably modify the volatile profile of the wine (Charpentier and Feuillat 1998). In addition, the absence of ethanol allows for a selective extraction of high water solubility compounds, including anthocyanins, free and glycosylated-bound aroma compounds and low molecular weight tannins (Apolinar-Valiente et al. 2013).

Some studies have reported that the extraction of anthocyanin and tannin due to the application of cold soak wine making was related to an intrinsic feature of the grape cultivar; however, several reports in the literature have presented controversial results. Some studies reported an increase in the extraction of phenolic compounds (Busse-Valverde et al. 2010; González-Neves et al. 2013; Favre et al. 2014), others reported a decrease (Budic-Leto et al. 2003; González-Neves et al. 2012) and no effect of cold soaking on the extraction of phenolic compounds (Ortega-Heras et al. 2012; Pérez-Lamela et al. 2007). The contradictory results found in the above mentioned studies led the researchers to search for an expansion of their knowledge in order to evaluate the real effects of cold soaking on the extraction of the phenolic compounds. Despite the results of the chemical assessment of the products from the cold soaking procedure, some authors were unable to agree about its effects on the sensory attributes. The authors reported that the negative impact of the non-*Saccharomyces* yeasts on wine flavor was due to the formation of ethyl acetate and acetaldehyde (González-Neves et al. 2013; Casassa and Sari 2015).

Casassa et al. (2015) provided information about the phenolic compound and chromatic compositions, Using the CIELab parameters, of six wines produced from Barbera, Cabernet Sauvignon, Malbec, Merlot, Pinot Noir and Syrah grape cultivars submitted to the cold soaking prefermentative treatment. The wine making procedure consisted of de-stemming the grapes and crushing them, allowing for the release of the must, which was treated with 80 mg.L^{-1} of sulfur dioxide. The cold soaking treatments consisted of 4 days at approximately 9 °C, achieved by the insertion of CO_2 pellets. The 4-day cold soaking period was followed by a 10-day maceration process carried out by two pumping-overs followed by two pumping downs. The alcoholic fermentation was initiated by the inoculation of active dry *Saccharomyces cerevisiae* yeasts, and after the maceration process, malolactic fermentation was also induced by inoculation with *Oenococcus oeni*. At the end of the malolactic fermentation, the wines were racked and stored at 1 °C for 45 days to allow for the tartrate stabilization. The free SO_2 content was then adjusted and the wines bottled and stored at 12 °C until analyzed. The cold soaking maceration technique significantly increased the anthocyanin concentration as compared to traditional wine making, as well as increasing the Chroma and redness of the wines. The total phenolic compounds, tannin concentration and the other CIELab parameters (luminosity, hue angle and yellowness) were not affected by the cold soaking procedure. The sensory approach only showed the influence of cold soaking on two attributes: color intensity and violet hue. Positive effects of the cold soaking treatment were observed for the Barbera and Cabernet Sauvignon wines; the Barbera wines obtained higher scores for color intensity and violet hue, and the Cabernet Sauvignon wines for color intensity. In contrast, cold soaking produced negative effect for the Pinot Noir wines, which showed lower scores for color intensity when compared with the traditional treatment. The cold soaking technique had also no effect on the aroma attributes, astringency, bitterness and body; these sensory attributes presented no significant differences when comparing the traditional and cold soaking treatments.

Casassa et al. (2015) have also applied the Principal Component Analysis to obtain relevant information about the relationship between the treatments and the wines produced from the different grape cultivars. The Principal Component Analysis explained the major data variance and allowed for differentiation of the grape cultivars, allocating them in two different clusters. The cold soaking was greatly influenced by the intrinsic features of the grape cultivars.

Other pertinent studies showed a lack of differences between the traditional and cold soaking wine making procedures, mainly concerning the sensory assessment. Using a triangle test, Gardner et al. (2011) described the failure of consumers to detect differences between Cabernet Sauvignon wines produced using the traditional and cold soaking treatments. In another study, Casassa and Sari (2015) reported lower color intensity and reduced fruity flavor in Malbec red wines submitted to cold soaking, as well as a noticeable acetaldehyde character.

Favre et al. (2014) described the application of a cold prefermentative procedure to Tannat wines and compared it to three other wine making procedures: traditional maceration, maceration with the addition of enzymes and maceration with the addition of seed tannins. The wine making procedure consisted of de-stemming the grapes followed by crushing, allowing for the release of the must. Alcoholic fermentation was induced by inoculation of yeasts, and the must was then drained and the pomace pressed. The free-run must was mixed with the pressed must and they were stored in stainless steel fermentation vessels until the end of the alcoholic fermentation. After this, sulfur dioxide was added to inhibit malolactic fermentation and the wine was bottled. The cold soaking procedure consisted of promoting the contact of the skins with the must at 10–15 °C for 5 days prior to alcoholic fermentation, using frozen water as the cooling agent. The total phenolic compound content, anthocyanidins, flavan-3-ols and proanthocyanidins were assessed, as well as the low molecular weight non-anthocyanin phenols. The color features were also evaluated according to the CIELab parameters. The authors demonstrated that cold soaking increased the total phenolic compound and anthocyanin contents of Tannat red wines. The five anthocyanidins (cyanidin, delphinidin, petunidin, peonidin and malvidin), detected and quantitated by HPLC in cold soaked Tannat wines, presented higher concentrations when compared with the traditional maceration process, with the exception of delphinidin. In addition, among the low molecular weight non-anthocyanin phenols, the cold soaked wines showed higher concentrations of stilbenes, flavan-3-ols and flavonols when compared with the traditional procedure. The color features of the cold soaked wines seemed to be less influenced by the treatment, since the CIELab parameters presented no significant differences when compared to the traditional wines. Another important result was the high concentrations of tyrosol and triptophol in the cold soaked wines, which are compounds derived from yeast metabolism, probably due to the activity of the native yeast strains in the prefermentative phase.

Based on the above mentioned studies, the application of cold soaking as a potential alternative wine making procedure needs to be further discussed and expanded by other studies. The grape cultivar used in cold soaking wine making technique seems to enhance the phenolic compounds extraction. In addition, important factors such as the application of low temperatures by the addition of CO_2 pellets or external refrigeration need to be intensely discussed (Casassa et al. 2015).

Carbonic maceration

Carbonic maceration aimed to produce lighter and fruitier wines that are indicated for young consumption. In this procedure, whole berries or grape clusters were submitted to a carbon dioxide saturated (CO_2) atmosphere, which allowed for the activity of the glycolytic enzymes proceed from the grapes (Sacchi et al. 2005). After one or two weeks, the grapes were then pressed and the must inoculated to complete the alcoholic fermentation. In fact, a natural enzymatic transformation occurs in the grape skins due to the anaerobic conditions promoted by the saturated CO_2 environment. The berries suffered an intracellular fermentation in which the malic acid is metabolized to ethanol and other substances (Rizzon et al. 1999).

According to several studies, it is difficult to draw secure conclusions about the effect of carbonic maceration procedure on the phenolic compounds of wines. Very different results have been found with different grape cultivars. Some studies showed that carbonic maceration caused a decrease in the phenolic compound concentrations (Sun et al. 2001; Timber lake and Bridle 1976), while other studies showed an increase in the phenolic composition (Lorincz et al. 1998). In addition, the carbonic maceration process could lead to the production of lighter wines with reduced body and structure and an intense fruity aroma, especially for strawberry and raspberry aromas (Etaio et al. 2008).

Rizzon et al. (1999) applied carbonic maceration to the production of Cabernet Franc wines, placing the grapes in a 4,000 liter fermentation vessel and injecting carbon dioxide to an internal pressure of 0.5 atm. This pressure was maintained for 10 days, and the temperature varied from 25 to 30 ºC. The whole grapes were then pressed to obtain the fermentative must and alcoholic fermentation was induced by the inoculation of dry active *Saccharomyces cerevisiae* yeasts. After the final stage of alcoholic fermentation, the wines were refrigerated at −4 °C to allow for tartrate stabilization, and then bottled.

According to the authors, the carbonic maceration wines presented lower anthocyanin concentrations and reduced color features. These wines showed an intense oxidized tonality when compared with the traditional and thermovinification procedures. In addition, they presented a higher pH and lower total acidity, since the malic acid was enzymatically transformed into ethanol inside the grape berries, and caused a decrease in the total acidity of the wines (Rizzon et al. 1999). The wines submitted to the carbonic maceration procedure presented weak aroma quality and a lower score for the flavor and olfactory balance. They were also lighter, less structured and with reduced body, due to the lower acidity and phenolic compound concentrations.

Bertagnolli et al. (2007) presented results concerning the influence of carbonic maceration and ultraviolet irradiation on the *trans*-resveratrol levels in Cabernet Sauvignon red wines. The grapes were submitted to different treatments and carbonic maceration was carried out at a temperature of 0.5 °C in an environment of 10% CO_2. After the treatment, the grapes were de-stemmed and crushed, allowing for the release of the must. The must was treated with SO_2, chaptalized, and alcoholic fermentation induced by inoculation with *Saccharomyces cerevisiae* yeasts. At the end of the alcoholic fermentation, the wines were racked, submitted to malolactic fermentation, racked again and bottled for 2 months. The carbonic maceration wines presented an increase in the *trans*-resveratrol concentration in the first days of alcoholic fermentation, but afterward it decreased. This behavior was probably due to the saturated CO_2 environment that changed the reaction for the synthesis of *trans*-resveratrol, since its synthesis needs the elimination of four molecules of CO_2. One positive outcome of the use of carbonic maceration was related to the reduction in processing time, i.e. the pre-fermentation occurred inside the grape berries due to the influence of the carbon dioxide, allowing for a decrease in the wine making time. In addition, Bertagnolli et al. (2007) also analyzed the *trans*-resveratrol concentration in the grapes right after the treatments, and reported that grapes submitted to carbonic maceration showed the lowest *trans*-resveratrol concentrations in comparison to the other treatments assessed.

Fuleki (1974) analyzed the effect of carbonic maceration on wines produced from the Concord grape cultivar (*Vitis labrusca*). The grapes were submitted to a saturated CO_2 environment at 15 °C for one or two weeks. After the treatment, the grapes were pressed and the must treated with sulfur dioxide. Alcoholic fermentation was then induced, the pomace was pressed and drained, the must was chaptalized and the alcohol content adjusted to 13% v/v. The wines were then maintained at 2 °C for two weeks to allow for tartrate stabilization, and then racked, bottled and stored at 15 °C. A chemical and sensory evaluation was carried out to evaluate the effect of carbonic maceration on the wines produced from the Concord grape cultivar.

The carbonic maceration has appeared as a suitable technique to produce Concord wines without the Concord character, i.e., the intrinsic aroma compounds of the Concord grape were degraded by carbonic maceration and a new flavor was developed. The taste of the wines was described as softer as a result of the decrease in acidity and lower tannin content. In addition, carbonic maceration probably degraded the methyl anthranilate, which is a volatile compound typical from *Vitis labrusca* grapes, including Concord, since the produced wine showed low concentrations of this compound when compared to the control treatment.

In summary, the studies concerning the application of carbonic maceration presented a slightly positive outcome, i.e. this technique reduced the production time because the pre-fermentation occurred in the grape berries before crushing and alcoholic fermentation. The wines presented low tannin concentrations, were light-structured and light-bodied, and the aroma profile was different from the one expected due to compounds which are naturally present in the grape cultivars.

Application of Different Yeast Strains for Alcoholic Fermentation

Important studies have provided information about the effect of the selected yeast on the chemical and sensory profiles of wines, mainly regarding the effect on the phenolic compounds and color features. It has been suggested that yeast lees absorb anthocyanins (Morata et al. 2003) and that macromolecules released on yeast autolysis can bind polymeric phenols, resulting in an effect similar to that provided by protein-fining agents (Sacchi et al. 2005). The mannoproteins released from different yeast species have been examined and there are hypotheses of the enhancement of the anthocyanin-tannin condensation reaction, resulting in a decrease in wine astringency (Escot et al. 2001). Mannoproteins are glycoproteins with a high glycosylation level, mainly composed of mannose (approximately 90%), glucose and proteins (approximately 10%) (Guadalupe et al. 2010; Vidal et al. 2003). The amount of mannoproteins released depends on the yeast strains and the wine making conditions (Giovani et al. 2010).

Studies have shown that the mannoproteins might improve certain negative sensory attributes such as bitterness and astringency, enhancing the persistence and mouthfeel of wines. In addition, they appear to improve color stabilization due to the colloid protector activity provided by these yeast metabolites (Del Barrio-Galán et al. 2012). Other studies have shown that these complex polysaccharides could avoid the interaction between the flavan-3-ols and the salivary proteins via different mechanisms, decreasing the astringency sensation (Carvalho et al. 2006).

Del Barrio-Galán et al. (2015) analyzed the chemical and sensory profiles of Syrah red wines fermented by different strains of *Saccharomyces cerevisiae*, which presented different capabilities for polysaccharide liberation. The wines were produced following a standard wine making procedure applied in the Caliterra winery, Chile. Briefly, the grapes were harvested at their optimal ripening stage and crushed allowing for liberation of the must, which was treated with sulfur dioxide. Alcoholic fermentation was induced by the inoculation of two *Saccharomyces cerevisiae* strains: a conventional one used in the Caliterra winery, and another one which produces high levels of complex polysaccharides during alcoholic fermentation. Fermentation was carried out at a controlled temperature ranging from 21 to 25 °C, and once fermentation was completed, the wines were racked, allowing for malolactic fermentation which took place spontaneously. After the malolactic fermentation, the sulfur dioxide concentration was corrected and the wines bottled. The different yeast strains resulted in no significant differences in the oenological parameters. The wine fermented by the yeast that produces high levels of complex polysaccharides presented lower contents of all phenolic compounds, except for the hydroxycinnamic acids, which presented similar amounts in both treatments. These results could be due to the interaction or adsorption phenomena that occurred between the mannoproteins released by the yeasts and the phenolic compounds, mainly anthocyanins, during alcoholic fermentation. In addition, the sensory results indicated lower values for alcohol, bitterness and astringency, and higher values for red fruit flavor, persistence and mouthfeel for the wine which was fermented with complex polysaccharides producing strains. These results were in agreement with other studies which reported the effect of mannoproteins in minimizing astringency by complexation with the tannins (Del Barrio-Galán et al. 2011; Del Barrio-Galán et al. 2012). In summary, the application of a large amount of mannoproteins producing yeast strain could improve the wine palate by decreasing astringency and bitterness. However, this alternative wine making procedure could present undesirable effects on the color intensity of red wines, since the phenolic compounds, mainly anthocyanins, were also complexed by the polysaccharides produced by these *Saccharomyces* yeast strains.

Saberi et al. (2012) reported on the volatile and sensory profiles of Chardonnay white wines fermented with individual commercial, individual Burgundian and mixed Burgundian yeast strains. The novel *Saccharomyces cerevisiae* strain was isolated from a vineyard in the Burgundy region, France, and the strains were mixed in different proportions–1:1, 1:2, 1:3 and 2:3. These strains were recommended for white wines, especially Chardonnay, aiming to increase the fruity aroma and complexity. The Chardonnay fermentation was carried out at 16 and 20 ºC aiming to optimize the retention of volatile compounds. The musts were inoculated using the ratios described above, but the yeast strains were not mixed before inoculation. The fermentation bottles were capped to provide an anaerobic environment and when fermentation was completed, the wines were treated with 100 mg.L^{-1} of potassium metabisulfite to avoid oxidation. The volatile compounds were analyzed via the headspace and the compounds identified and quantitated by gas chromatography-mass spectrometry. No significant differences were observed in volatile composition when the fermentation temperatures were compared. Eighteen compounds were identified and quantitated, including eight higher alcohols, five ethyl esters, three acetate esters, one aldehyde and one organic acid. The concentration of the volatile compounds for the individual and mixed Burgundian yeast strains was compared to industrial ones. The Burgundian strains produced more berry, fusel oil, candy and balsamic aromas, but the levels were below the human perception threshold. These strains also produced lower levels of nail polish and vinegar aromas. The sensory profile revealed that the individual and mixed Burgundian yeast strains produced wines with fruity aromas such as sweet fruit, strawberry, green apple, pear and banana. The Principal Component Analysis was successfully applied and it differented the Burgundian and commercial yeasts. The positive outcome of the mixed Burgundian strains may have been due to metabolic interactions between strains, allowing them to respond in a different form according to the different substrates. This study showed the high potential of applying an isolated yeast strain to obtain fruity, more balanced and complex wines with unique features (Saberi et al. 2012).

Pectinolytic Enzymes

The use of pectinolytic enzymes in the wine industry has been the focus of many studies since it maximizes the extraction of free-run juice during maceration and helps in the wine clarification and filtration (Ducasse et al. 2010). In addition, the pectinolytic enzymes, also known as pectinases, have been used in wine making trials to obtain an increase in wine color due to rupture of the skin cell walls, releasing the pigments. These enzymes are carbohydrases that catalyze the breakdown of pectic substances. Both the grape and the yeast showed pectinase activity, although their effect can be limited by the sugar and alcohol contents. The use of pectinases in wine making brings several benefits including the faster start of fermentation, higher yield of must flow from the dejuicing step, easier pressing of the pomace, enhancement of the clarification activity and an important extraction of phenolic and aroma compounds from the grape skin (Pardo et al. 1999).

Studies on the effect of enzymes on wine color and phenolic compounds have led to controversial results: some have indicated an increase in wine anthocyanin content and an improvement in wine color due to the application of pectinolytic enzymes (Kelebek et al. 2007; Bakker et al. 1999), some studies have shown a decrease in these compounds (Bautista-Ortin et al. 2005; Wightman et al. 1997), and others showed no significant effect (Ducasse et al. 2010). These discrepancies could be explained by several factors, such as different compositions of grape polyphenolics, the extraction rate in the maceration step and reactions between them during the wine making process and aging (Ducasse et al. 2010).

Many of the enzymatic preparations are enzyme complexes, which can promote some collateral effects in musts and wines, such as loss in the anthocyanins stability promoted by cinnamate-esterase activity (Günata et al. 1986), intense turbidity and loss of color promoted by β-glucosidase activity (Somers and Ziemelis 1985) and pectin hydrolysis in the grape skin by pectin methylesterase, promoting an increase in methanol in the juices and wines (Lago-Vanzela et al. 2014b). According to the latter authors, the toxicity of methanol is low; however, other compounds are produced in the metabolic process, such as formic aldehyde and formic acid. The methanol content and phenolic extraction depended on the type and amount of added enzymes, the grape cultivars and the time of contact between the grapes and the enzymes. Enzymatic preparations with low pectin methylesterase activity and high pectinlyase activity could minimize the formation of methanol in wines submitted to this treatment (Gómez-Plaza et al. 2001).

Echeverry et al. (2005) reported changes in the antioxidant activity of Tannat red wines during early maturation. The wines were submitted to different maceration times and treated with pectinolytic enzymes. The musts were treated with SO_2 and alcoholic fermentation was induced by the inoculation of *Saccharomyces cerevisiae* (15 g.hL^{-1}). The use of a pectinolytic enzyme with color extractor features enhanced the anthocyanin, flavonol, flavan-3-ol and proanthocyanidin contents, but the results provided no information about the presence of significant differences between the treatments. The authors also reported the higher antioxidant capacities of wines with higher polyphenol content, but stated that the wine antioxidant capacity was determined not only by the grape variety, vineyard methods and wine making techniques, but also by the reactions occurring during the wine making process.

Conclusions

Considering the alternative wine making techniques reviewed in this manuscript, thermovinification and submerged cap allowed for enhancement of the color features by increasing the anthocyanin contents. Cold soaking and carbonic maceration presented distinct results and seemed to be influenced by the intrinsic factors of the grape cultivars. Carbonic maceration could be considered as an alternative wine making procedure that degrades the varietal features of the grape, since these characteristics were not transferred to the wine. Further studies need to be developed aiming at evaluating the real effect of cold soaking and carbonic maceration on the chemical and sensory profiles as well as on the wine polyphenolic compound concentrations, since there seemed to be a strong effect of the grape on these two wine making procedures. The yeast selection procedure and the use of mixed yeast strains resulted in a decrease in the color features due to the prior stabilization of the phenolic compounds by the mannoproteins. In addition, the wines produced from isolated yeasts showed good structure and mouthfeel and enhancement of the fruity aromas. The application of pectinolytic enzymes could improve the color features of red wines, since they cause rupture of the skin cell walls releasing the anthocyanins, which are responsible for the attractive color of red wines. In addition, other phenolic compounds could be released by the use of pectinolytic enzymes, enhancing the antioxidant capacity of the wines. A negative outcome is related to the formation of considerable methanol by these enzymes, since methanol is highly toxic.

References

Andrade Neves, N., L.A. Pantoja and A.S. dos Santos. (2014). Thermovinification of grapes from the Cabernet Sauvignon and Pinot Noir varieties using immobilized yeasts. European Food Research and Technology 238: 79–84.

Apolinar-Valiente, R., P. Williams, I. Romero-Cascales, E. Gómez-Plaza, J.M. López-Roca, J.M. Ros-García and T. Doco. (2013). Polysaccharide composition of Monastrell red wines from four different Spanish terroirs: effect of wine-making techniques. Journal of Agricultural and Food Chemistry 61: 2538–2547.

Bakker, J., S.J. Bellworthy, H.P. Reader and S.J. Watkins. (1999). Effect of enzymes during vinification on color and sensory properties of port wines. American Journal of Enology and Viticulture 50: 271–276.

Bautista-Ortin, A.B., A. Martinez-Cutillas, J.M. Ros-Garcia, J.M. Lopez-Roca and E. Gomez-Plaza. (2005). Improving colour extraction and stability in red wines: the use of maceration enzymes and enological tannins. International Journal of Food Science and Technology 40: 867–878.

Bertagnolli, S.M.M., S.B. Rossato, V.L. Silva, T. Cervo, C.K. Sautter, L.H. Hecktheuer and N.G. Penna. (2007). Influence of the carbonic maceration on the levels of *trans*-resveratrol in Cabernet Sauvignon wine. Brazilian Journal of Pharmaceutical Sciences 43: 71–77.

Bosso, A., L. Panero, M. Petrozziello, R. Follis, S. Motta and M. Guaita. (2011). Influence of submerged-cap vinification on polyphenolic composition and volatile compounds of Barbera wines. American Journal of Enology and Viticulture 62: 503–511.

Budic-Leto, I., L. Tomislav and U. Vrhovsek. (2003). Influence of different maceration techniques and ageing on proanthocyanidins and anthocyanins of red wine cv. Babic (*Vitis vinifera L.*). Food Technology and Biotechnology 41: 203–299.

Busse-Valverde, N., E. Gómez-Plaza, J.M. López-Roca, R. Gil-Muñoz, J.I. Fernández-Fernández and A.B. Bautista-Ortín. (2010). Effect of different enological practices on skin and seed proanthocyanidins in three varietal wines. Journal of Agricultural and Food Chemistry 58: 11333–11339.

Camargo, U.A. and M.F. Dias. (1999). 'BRS Rúbea'. Comunicado Técnico, Embrapa Uva e Vinho, Bento Gonçalves, Rio Grande do Sul.

Camargo, U.A. and J.D.G. Maia. (2004). BRS Cora: Nova cultivar de uva para suco adaptada a climas tropicais. Comunicado Técnico, Embrapa Uva e Vinho, Bento Gonçalves, Rio Grande do Sul.

Camargo, U.A., J.D.G. Maia and J.C. Nachtigal. (2005). BRS Violeta: nova cultivar de uva para suco e vinho de mesa. Comunicado Técnico, Embrapa Uva e Vinho, Bento Gonçalves, Rio Grande do Sul.

Camargo, U.A., J.D.G. Maia and P.S. Ritschel. (2008). BRS Carmem: nova cultivar de uva tardia para suco. Comunicado Técnico, Embrapa Uva e Vinho, Bento Gonçalves, Rio Grande do Sul.

Carvalho, E., N. Mateus, B. Plet, I. Pianet, E. Dufourc and V. De Freitas. (2006). Influence of wine pectic polysaccharides on the interactions between condensed tannins and salivary proteins. Journal of Agricultural and Food Chemistry 54: 8936–8944.

Casassa, L.F. and S.E. Sari. (2015). Sensory and chemical effects of two alternatives of prefermentative cold soak in Malbec wines during wine making and bottle ageing. International Journal of Food Science and Technology 50: 1044–1055.

Casassa, L.F., E.A. Bolcato and S.E. Sari. (2015). Chemical, chromatic, and sensory attributes of 6 red wines produced with prefermentative cold soak. Food Chemistry 174: 110–118.

Charpentier, C. and M. Feuillat. (1998). Métabolisme des levures cryotolérantes: application à la macération préfermentaire du Pinot noir en Bourgogne. Revue Française d'Oenologie 170: 36–40.

Chira, K., N. Pacella, M. Jourdes and P.L. Teissedre. (2011). Chemical and sensory evaluation of Bordeaux wines (Cabernet-Sauvignon and Merlot) and correlation with wine age. Food Chemistry 126: 1971–1977.

Coombe, B. and P. Dry. (2006). Viticulture: Practices. Winetitles, Broadview, South Australia.

De Castilhos, M.B.M., M.G. Cattelan, A.C. Conti-Silva and V.L. Del Bianchi. (2013). Influence of two different vinification procedures on the physicochemical and sensory properties of Brazilian non-*Vitis vinifera* red wines. Lebensmittel-Wissenschaft & Technologie 54: 360–366.

De Castilhos, M.B.M., A.C. Conti-Silva and V.L. Del Bianchi. (2012). Effect of grape pre-drying and static pomace contact on physicochemical properties and sensory acceptance of Brazilian (Bordô and Isabel) red wines. European Food Research and Technology 235: 345–354.

De Castilhos, M.B.M., O.L.S. Corrêa, M.C. Zanus, J.D.G. Maia, S. Gómez-Alonso, E. García-Romero, V.L. Del Bianchi and I. Hermosín-Gutiérrez. (2015a). Pre-drying and submerged cap wine making: effects on poly phenolic compounds and sensory descriptors. Part I: BRS Rúbea and BRS Cora. Food Research International, 75: 374–384.

De Castilhos, M.B.M., O.L.S. Corrêa, M.C. Zanus, J.D.G. Maia, S. Gómez-Alonso, E. García-Romero, V.L. Del Bianchi and I. Hermosín-Gutiérrez. (2015b). Pre-drying and submerged cap wine making: effects on polyphenolic compounds and sensory descriptors. Part II: BRS Carmem and Bordô (*Vitis labrusca L.*). Food Research International 76: 697–708.

Del Barrio-Galán, R., M. Ortega-Heras, M. Sánchez-Iglesias and S. Pérez-Magariño. (2012). Interactions of phenolic and volatile compounds with yeast lees, commercial yeast derivatives and non toasted chips in model solutions and young red wines. European Food Research and Technology 234: 231–244.

Del-Barrio-Galán, R., M. Medel-Marabolí and A. Peña-Neira. (2015). Effect of different aging techniques on the polysaccharide and phenolic composition and sensory characteristics of Syrah red wines fermented using different yeast strains. Food Chemistry 179: 116–126.

Del-Barrio-Galán, R., S. Pérez-Magariño and M. Ortega-Heras. (2011). Techniques for improving or replacing ageing on lees of oak aged red wines: The effects on polysaccharides and the phenolic composition. Food Chemistry 127: 528–540.

Delgado-Andrade, C. and F.J. Morales. (2005). Unraveling the contribution of melanoidins to the antioxidant activity of coffee brews. Journal of Agricultural and Food Chemistry 53: 1403–1407.

Ducasse, M., R. Canal-Llauberes, M. de Lumley, P. Williams, J. Souquet, H. Fulcrand, T. Doco and V. Cheynier. (2010). Effect of macerating enzyme treatment on the polyphenol and polysaccharide composition of red wines. Food Chemistry 118: 369–376.

Echeverry, C., M. Ferreira, M. Reyes-Parada, J.A. Abin-Carriquiry, F. Blasina, G. González-Neves and F. Dajas. (2005). Changes in antioxidant capacity of Tannat red wines during early maturation. Journal of Food Engineering 69: 147–154.

Escot, S., M. Feuillat, L. Dulau and C. Charpentier. (2001). Release of polysaccharides by yeasts and the influence of released polysaccharides on colour stability and wine astringency. Australian Journal of Grape and Wine Research 7: 153–159.

Etaio, I., F.J.P. Elortondo, M. Albisu, E. Gaston, M. Ojeda and P. Schlich. (2008). Effect of wine making process and addition of white grapes on the sensory and physicochemical characteristics of young red wines. Australian Journal of Grape and Wine Research 14: 211–222.

Favre, G., A. Peña-Neira, C. Baldi, N. Hernández, S. Traverso, G. Gil and G. González-Neves. (2014). Low molecular-weight phenols in Tannat wines made by alternative wine making procedures. Food Chemistry 158: 504–512.

Figueiredo-González, M., B. Cancho-Grande and J. Simal-Gándara. (2013). Effects on colour and phenolic composition of sugar concentration processes in dried-on- and dried-off-vine grapes and their aged or not natural sweet wines. Trends in Food Science & Technology 31: 36–54.

Fuleki, T. (1974). Application of carbonic maceration to change the bouquet and flavor characteristics of red table wines made from Concord grapes. Canadian Institute of Food Science and Technology Journal 7: 269–273.

Gardner, D.M., B.W. Zoecklein and K. Mallikarjunan. (2011). Electronic nose analysis of Cabernet Sauvignon (*Vitis vinifera L.*) grape and wine volatile differences during cold soak and postfermentation. American Journal of Enology and Viticulture 62: 81–90.

Gil-Muñoz, R., A. Moreno-Pérez, R. Vila-López, J. Fernández-Fernández, A. Martínez-Cutillas and E. Gómez-Plaza. (2009). Influence of low temperature prefermentative techniques on chromatic and phenolic characteristics of Syrah and Cabernet Sauvignon wines. European Food Research and Technology 228: 777–788.

Giovani, G., V. Canuti and I. Rosi. (2010). Effect of yeast strain and fermentation conditions on the release of cell wall polysaccharides. International Journal of Food Microbiology 137: 303–307.

Gómez-Plaza, E., R. Gil-Muñoz, J.M. López-Roca, A. Martínez-Cutillas and J.I. Fernández-Fernández. (2001). Phenolic compounds and color stability of red wines: Effect of skin maceration time. American Journal of Enology and Viticulture 52: 266–270.

González-Neves, G., G. Gil, G. Favre and M. Ferrer. (2012). Influence of grape composition and wine making on the anthocyanin composition of red wines of Tannat. International Journal of Food Science & Technology 47: 900–909.

González-Neves, G., G. Gil, G. Favre, C. Baldi, N. Hernández and S. Traverso. (2013). Influence of wine making procedure and grape variety on the colour and composition of young red wines. South African Journal of Enology and Viticulture 34: 138–146.

Gonzalo-Diago, A., M. Dizy and P. Fernández-Zurbano. (2014). Contribution of low molecular weight phenols to bitter taste and mouthfeel properties in red wines. Food Chemistry 154: 187–198.

Gordillo, B., M.I. López-Infante, P. Ramírez-Pérez, M.L. González-Miret and F.J. Heredia. (2010). Influence of prefermentative cold maceration on the color and anthocyanic copigmentation of organic Tempranillo wines elaborated in a warm climate. Journal of Agricultural and Food Chemistry 58: 6797–6803.

Guadalupe, Z., L. Martínez and B. Ayestarán. (2010). Yeast mannoproteins in red wine making: Effect on polysaccharide, polyphenolic, and color composition. American Journal of Enology and Viticulture 61: 191–200.

Günata, Z.Y., J.C. Sapis and M. Moutounet. (1986). Substrates and aromatic carboxylic acid inhibitors of grape phenol oxidases. Phytochemistry 26: 1–3.

Jackson, R.S. (2008). Wine Science: Principles and Applications. Academic Press, San Diego, California.

Kelebek, H., A. Canbas, T. Cabaroglu and S. Selli. (2007). Improvement of anthocyanin content in the cv. *Okuzgozu* wines by using pectolytic enzymes. Food Chemistry, 105: 334–339.

Lago-Vanzela, E.S., M.A. Baffi, M.B.M. De Castilhos, M.R.M.R. Pinto, V.L. Del Bianchi, A.M. Ramos, P.C. Stringheta, I. Hermosín-Gutiérrez and R. Da-Silva. (2014b). Phenolic compounds in grapes and wines: chemical and biochemical characteristics and technological quality. In: J.S. Câmara (Ed.), Grapes: Production, Phenolic Composition and Potential Biomedical Effects. Nova Science Publishers, New York, pp. 47–105.

Lago-Vanzela, E.S., D.P. Procópio, E.A.F. Fontes, A.M. Ramos, P.C. Stringheta, R. Da-Silva, N. Castillo-Muñoz and I. Hermosín-Gutiérrez. (2014a). Aging of red wines made from hybrid grape cv. BRS Violeta: effects of accelerated aging conditions on phenolic composition, color and antioxidant capacity. Food Research International 56: 182–189.

Lorincz, G.Y., M. Kállay and G.Y. Pásti. (1998). Effect of carbonic maceration on phenolic composition of red wines. Acta Alimentaria 27: 341–355.

Ma, W., A. Guo, Y. Zhang, H. Wang, Y. Liu and H. Li. (2014). A review on astringency and bitterness perception of tannins in wine. Trends in Food Science & Technology 40: 6–19.

Marquez, A., M.P. Serratosa and J. Merida. (2013). Anthocyanin evolution and color changes in red grapes during their chamber drying. Journal of Agricultural and Food Chemistry 61: 9908–9914.

Marquez, A., M.P. Serratosa, A. Lopez-Toledano and J. Merida. (2012). Colour and phenolic compounds in sweet red wines from Merlot and Tempranillo grapes chamber-dried under controlled conditions. Food Chemistry 130: 111–120.

Morata, A., C. Gómez-Cordovés, J. Subervolia, B. Bartolomé, B. Colomo and J.A. Suarez. (2003). Adsorption of anthocyanins by yeast cell walls during fermentation of red wines. Journal of Agricultural and Food Chemistry 51: 4084–4088.

Organisation Internationale de la Vigne et du Vin (OIV). (2014). State of the vitiviniculture world market. 2 pp.

Ortega-Heras, M., S. Pérez-Magariño and M.L. González-Sanjosé. (2012). Comparative study of the use of maceration enzymes and cold pre-fermentative maceration on phenolic and anthocyanic composition and colour of a Mencía red wine. Lebensmittel-Wissenschaft & Technologie 48: 1–8.

Pardo, F., M.R. Salinas, G.L. Alonso, G. Navarro and M.D. Huerta. (1999). Effect of diverse enzyme preparations on the extraction and evolution of phenolic compounds in red wines. Food Chemistry 67: 135–142.

Patras, A., N.P. Brunton, C. O'Donnell, and B.K. Tiwari. (2010). Effect of thermal processing on anthocyanin stability in foods; mechanisms and kinetics of degradation. Trends in Food Science & Technology 21: 3-11.

Pérez-Lamela, C., M.S. García-Falcón, J. Simal-Gándara and I. Orriols-Fernández. (2007). Influence of grape variety, vine system and enological treatments on the colour stability of young red wines. Food Chemistry 101: 601–606.

Rizzon, L.A., A. Miele, J. Meneguzzo and M.C. Zanus. (1999). Effect of three processes of vinification on chemical composition and quality of Cabernet Franc wine. Pesquisa Agropecuária Brasileira 7: 1285–1293.

Saberi, S., M.A. Cliff and H.J.J. van Vuuren. (2012). Impact of mixed *S. cerevisiae* strains on the production of volatiles and estimated sensory profiles of Chardonnay wines. Food Research International 48: 725–735.

Sacchi, K.L., L.F. Bisson, and D.O. Adams. (2005). A review of the effect of wine making techniques on phenolic extraction in red wines. American Journal of Enology and Viticulture 56: 197–206.

Somers, T.C. and G. Ziemelis. (1985). Flavonol haze in white wines. Vitis 24: 43–50.

Sun, B., I. Spranger, F. Roque-do-Vale, C. Leandro and P. Belchior. (2001). Effect of different wine making technologies on phenolic composition in Tinta Miúda red wines. Journal of Agricultural and Food Chemistry 49: 5809–5816.

Suriano, S., G. Ceci and T. Tamborra. (2012). Impact of different wine making techniques on polyphenolic compounds of Nero Di Troia wine. Italian Food & Beverage Technology 70: 5–15.

Timberlake, C.F. and P. Bridle. (1976). The effect of processing and other factors on the colour characteristics of some red wines. Vitis 15: 37–49.

Vidal, S., P. Williams, T. Doco, M. Moutounet, and P. Pellerin. (2003). The polysaccharides of red wine: Total fractionation and characterization. Carbohydrate Polymers 54: 439–447.

Wang J., Ho L., Z. Zhao, I. Seror, N. Humala, D.L. Dickstein, M. Thiyagarajan, S.S. Percival, S.T. Talcott and G.M. Pasinetti. (2006). Moderate consumption of Cabernet Sauvignon attenuates Aβ neuropathology in a mouse model of Alzheimer's disease. The Faseb Journal 20: 2313–2320.

Wightman, J.D., S.F. Price, B.T. Watson and R.E. Wrolstad. (1997). Some effects of processing enzymes on anthocynains and phenolics in Pinot noir and Cabernet sauvignon wines. American Journal of Enology and Viticulture 48: 39–48.

Zott, K., C. Miot-Sertier, O. Claisse, A. Lonvaud-Funel and I. Masneuf-Pomarede. (2008). Dynamics and diversity of non-*Saccharomyces* yeasts during the early stages in wine making. International Journal of Food Microbiology 125: 197–203.

9

Amazon Fruits: Biodiversity, Regionalism and Artisanal and Industrial By-products Under Fermentation Processes

Carlos Victor Lamarão[a]*, Jane Maciel Leão*[b]*, Kirk Renato Moraes Soares*[b] *and Fábio Alessandro Pieri*[c]*

Abstract

The Brazilian Amazon is known for its wide biodiversity, especially those related to fruit and tubers. These raw materials are used by the local population in regional cuisine and also in processing of by-products, invariably using traditional knowledge and also with the use of rustic preservation methods. In this context, with the main aim of preserving foods, as well as adding nutrition value, fermentation techniques have been widely used over the years. In the Brazilian legislation, wines are beverages produced by the fermentation of grapes; however, in the Amazon, and in popular speech, wines can also be drinks derived from the fermentation of various fruits, including from their pulp, bark or seeds. The fermented wine produced with "buriti" (*Mauritia flexuosa* L.) and "caraná" (*Mauritiella armata*) shells, seeds of "cupuaçu" (*Theobroma grandiflorum*) and "bacuri" (*Plantonia insignis* Mart.), "corn aluá", "pineapple aluá", and "tarubá", an indigenous wine, can be highlighted as examples. Two particular fermented products are particularly used in the Amazon regional cuisine: "caxiri", derived from the fermentation of manioc, and "tucupi", derived from the fermentation of wild manioc. Caxiri is a very traditional drink for the indigenous population and is also known as "pajuarú", "aluá" and "mocororó". This drink, when slightly fermented, resembles a thick juice that is drunk by everyone due to its low alcohol content, however after ripening for many days, the alcohol content rises, becoming pure "caxiri", with a strong and bitter flavor. "Tucupi" is a

[a] Departamento de Engenharia Agrícola e Solos, Faculdade de Ciências Agrárias, Universidade Federal do Amazonas; Av Rodrigo Otávio, 3000, Bairro Coroado II, 69077-000, Manaus, Brazil. E-mail: victorlamarao@yahoo.com.br

[b] Curso de Agronomia, Faculdade de Ciências Agrárias, Universidade Federal do Amazonas; Av Rodrigo Otávio, 3000, Bairro Coroado II, 69077-000, Manaus, Brazil. E-mail: leao.ufam@hotmail.com; kirkrenato@ig.com.br

[c] Departamento de Ciências Básicas da Saúde, Universidade Federal de Juiz de For a, Campus Avançado Governador Valadares; Rua Israel Pinheiro, 2000, Bairro Universitário, 35032-200, Governador Valadares-MG, Brazil. Email: fabio.pieri@ufjf.edu.br.

* Corresponding Author: fabio.pieri@ufjf.edu.br.

fermented liquid extracted from wild manioc, and is a variety which also contains a poisonous substance, hydrogen cyanide. To obtain "tucupi", wild manioc tubers are scraped, and the juice is squeezed from it and allowed to stand. After extracting the starch, the content is boiled for a long time to remove hydrogen cyanide. The orange liquid obtained develops flavors that vary depending on the varieties of wild manioc used. The manioc tuber is also the substrate used for the preparation of both sweet and fermented manioc starch, manioc flour and tapioca. As well as the elimination of toxicity and preservation, the processing of this tuber refers to modifications of the rheological and sensory characteristics. There are studies that show the use of manioc starch and the Amazonian purple-yam for the production of sweet and sour starch on the development of various bakery products in food industry. In the beverage industry, especially in the soft drinks and beer industry, Amazonian fruits, through fermentation techniques, have also been used to produce soft drinks flavored with guarana and açaí which exist in the market, and alcoholic drinks flavored with bacuri, açaí, cupuaçu, taperebá (*Spondias mombin* L.), pripioca (*Cyperus articulatus*) and cumaru (*Dipteryx odorata*) in the categories "lager", "river", "stout", "witbier", among others, these have achieved recognition and awards in several beer festivals worldwide.

Amazonian Biodiversity: Food Potential of Native Fruits within the Regional Context

Native fruits and derived products from the Amazon region have become increasingly popular in Brazil and have aroused interest internationally. Nevertheless, scientific literature on some of them is scarce and there is a demand for more studies on their chemical characteristics, qualities, nutritional and commercial potential. The distinctive flavor, high energy and vitamin content of Amazonian fruits can be considered attractive to consumers, that has been revealed through the increase in demand by the food industry and increased consumption *in natura* of these fruits (Santana 2007).

In the dairy market, there are milk products with added pulp or pieces of fruit, especially fermented milk and fermented dairy drinks. The addition of regional fruits with different flavors and a high nutritional potential is an effective way to add nutritional value and increase product acceptance. The diversity and potential of Amazonian fruit, according to Herculano (2005), highlight optimistic prospects for the industrial use of tropical fruits, mainly as pulp. The *in natura* fruits are quite perishable, so the production of fruit pulps has become favorable for the full commercialization of fruit (Becker 2008). The main native fruits from the Amazon are "bacuri", "cupuaçu", "uxi", "camu-camu", "pupunha", "cibiu", "araça-boi", "açaí", and "buriti".

Bacuri (*Plantonia insignis* Mart.)

The name "bacuri" comes from the Tupi-Guarani language, in which "ba" means to fall and "curi" means soon, as "bacuri" is a fruit that falls just as it is matured. The "bacurizeiro" is a native tree of Pará state (Brazil), and the area with the highest concentration of these trees is the Amazon River estuary, with a stronger occurrence in the Salgado region and on the island of Marajó (Cavalcante 2010; Moraes et al. 1994).

Originally from Pará, "bacuri" spread to Maranhão, Piauí and other areas, but is rarely found in the Western Amazon. It occurs naturally in "capoeira" (pioneer vegetation that was established after the destruction of native forest) and other degraded and sandy areas, it is indifferent to soil types, such as poor in nutrients or clay. Occasionally, it is found in high forests. The fruit has a good aroma and tropical taste, qualities that allow for its use in the production of ice creams, juices, candies and yoghurt. It is rich in amino acids and minerals, but shows only traces of vitamin C, with average values of pH 3.3 and soluble solids 16.4 °Brix (Shanley 2005).

"Bacuri" has a great nutritional and economic potential, being an exotic alternative fruit, besides the use of the pulp, it is also possible to use the bark in the preparation of food products (Cavalcante 1991; Moraes et al. 1994; Brasil 2000; Rogez 2000; Herculano (2005).

The "bacuri" pulp is an important source of minerals and should be consumed by children during their growth phase, with effect to strengthening their bones and teeth. By consuming 100 grams of "bacuri" pulp, they would be taking in 105 calories, more than what they would get from "cupuaçu" and similar to "uxí" and "açaí". The pulp is also rich in carbohydrates, but contains few vitamins. "Bacuri" flavor has been extracted and used widely in yogurts in the Amazon region (Shanley 2005).

Cupuaçu (*Theobroma grandiflorum*)

"Cupuaçu" originates from the Amazon rain forest, and has a strong aroma and taste, along with a good consumer acceptance. The proteins present in "cupuaçu" show considerable nutritional potential because they have high biological value and an aminoacid composition that is superior to cocoa. The pulp of this fruit is used widely in Amazonian cuisine. "Cupuaçu" is an important agricultural export product with broad market prospects due to its high acceptability and to the intrinsic characteristics of the fruit. The pulp has a pH value equal to 2.60 and soluble solids of 9.0 °Brix (Moraes et al. 1994; Lopes et al. 2008).

Uxi (*Endopleura uchi* Cuatrec.)

"Uxi" is an Amazonian fruit that has gained great attention internationally; the use of its pulp is now greatly valued, and it has achieved a good price in the market. "Uxi" can be eaten raw, eaten as ice cream or a popsicle or drunk as a soft drink. In the city of Belem (Pará state, Brazil), the "uxi popsicle" is one of the most popular flavors. Besides the fruit, other parts of the plant are also used: the tree bark is used as medicine and the seed of the fruit as an amulet. The "uxi" tree is large, reaching about 25 to 30 meters high, one meter in trunk diameter, and 3 meters in the treetops. The "uxi" tree originated from the Brazilian Amazonian region. It is a typically wild species of high forest land and often occurs in the estuary of Pará, in Bragantina, Guamá and Capim regions, in western Marajó Island and in Furos region of Brazil (Shanley 2005).

"Uxi" is an excellent source of calories; 100 grams of pulp contains 284 calories, 6 times more than oranges. "Uxi" also strengthens the body with important vitamins; it has more vitamin B than many fruits: 0.13 milligrams of vitamin B_1 and 0.10 milligrams of vitamin B_2 per 100 grams of pulp. A 100 grams of "uxi" pulp contains 7.8 milligrams of iron and 33 milligrams of vitamin C. In addition, "uxi" pulp has between 10 and 21 grams of fiber per 100 grams, fiber is important for proper bowel function. "Uxi" also has many minerals: a 100 grams of pulp has 460 milligrams of potassium, 64–96 milligrams of calcium, 53–70 milligrams of magnesium, 39–46 milligrams of phosphorous and 22 milligrams of sodium. Finally, "uxi" oil is rich in phytosterols (1.378 milligrams per 100 grams of oil). The presence of phytosterols in food have a positive effect in the reduction in the levels of cholesterol in the blood (Gaia 2004).

Camu-camu (*Myrciaria dubia*)

"Camu-camu", a native of the Amazon floodplain, has a high nutritional potential due to the levels of vitamin C and anthocyanins. The importance of this wild fruit and food is due to its high vitamin C content, ranging from 1,600 mg to 2,994 mg per 100 g pulp. The fruit can reach twice the level of vitamin C compared to the acerola (Barbados cherry *Malpighia emarginata*). In relation to oranges, which are traditionally known as a source of vitamin C, the amount of ascorbic acid in "camu-camu" is about 60 times higher. Another component of interest in "camu-camu" is the anthocyanins, which are natural pigments with anti-inflammatory, antimicrobial and vasodilator properties, with concentrations of 0.54 to 0.74 mg per 100 g of fruit (Moraes et al. 1994; Cdib Taxi 2001; Maeda et al. 2007; Inoue et al. 2008).

Pupunha (*Bactris gasipaes* Kunth)

"Pupunha", a starchy fruit of the Amazonian region of Brazil, has fruits that have a great economic potential due to their chemical composition, agricultural productivity (13,500 tons/year) and broad regional consumption (Clement 2000). The "pupunha" fruit is rich in minerals and pro-vitamin A and, depending on the breed, carbohydrates or lipids (Arkoll and Aguiar 1984). From the "pupunha", "caiçuma", a drink consumed by the Indigenous people, is prepared. It is artisanal and prepared in different ways, according to ethnicity of the people producing it and their heritage. Some tribes chew "pupunha" and leave it to its natural fermentation process, usually for up to seven days, and thus the ptyalin contained in saliva is incorporated into the dough, and hydrolyzes starch as a result. Despite the enormous potential of "pupunha" in the region for the production and marketing of beverages, this prospect has been poorly explored, since most of the products from the fruit are processed in small scale and hand crafted (Clement 2000).

Cubiu (*Solanum sessiliflorum* Dunal)

"Cubiu" is one of the native unique genetic resources of the Amazon region, which has been completely domesticated by the indigenous peoples in the region from the Pre-Colombian period. From an economic view, "cubiu" has been an important raw material for modern agribusiness, because the plant is rustic, easy to grow, and very productive depending on the genotype, reaching 100 tons per hectare of fruit (Silva-Filho et al. 2005; Silva-Filho et al. 1996). It can be used in multiple ways (juices, jams, jellies, preserves, sauces, cosmetics and industrial or homemade drugs with hypoglycemic and hypocholesterolemic action) (Silva-Filho 2005; Silva-Filho 2002).

Thus, the existence of knowledge of their chemical and technological characteristics allows for a larger scale industrialization. As it has an annual production of fruits and is well suited for the soils of the floodplains of the Amazon, it is possible to produce fruits with low cost, allowing for it be marketed at very reasonable prices (Silva-Filho 2005; Silva-Filho 1998).

Araçá-boi (*Eugenia stipitata* McVaugh)

"Araçá-boi" is a native fruit from western Amazonia and the Guianas, and has adapted to tropical and subtropical climates. In Brazil, it is found in the Amazon region, Mato Grosso and Bahia states, but is still without commercial exploitation. The fruit is a globose berry, with a thin shell, and a strong yellow color when ripe and velvety in texture, weighing 30–800 g, with a rounded or flattened shape, with longitudinal and cross diameters of 5 to 10 cm and 5 and 12 cm, respectively. Its pulp is juicy, acidic, light yellow-colored, slightly fibrous, has 4 to 10 oblong seeds, measuring 0.5 to 1.0 cm in length (Sacramento et al. 2008). The fruits of "araçá-boi" present soluble solids content of 5.54° Brix, below the majority of tropical fruits, and an acidity of 2.38 g of citric acid per 100 g, higher than the acidity of fruits used for the agro-industrialization. Thus, the ratio Brix/acidity is extremely low (2.33), which impairs the acceptance of the fruits by Brazilian consumers. However, the pulp of "araçá-boi" added with sugar has been successfully used in ice cream, and has potential for use in juice formulations, jams and nectars, especially when associated with low-acid fruits such as mango, papaya and apple (Sacramento et al. 2008).

Açaí (*Euterpe oleracea* Mart. and *Euterpe precatoria* Mart.)

The *Euterpe oleracea* Mart species is native to Pará state, with most occurring in the Amazon River estuary where it covers an area of 10,000 km^2. It also occurs in Amapá, Amazonas and Maranhão states, Guyana and Venezuela. Dense "açaí" palm populations naturally occur in areas of floodplain and flooded forest. The responsibility for the spread of açaí seeds can be tracked to birds, monkeys, people and rivers in the region. The "açaí" palm grows better in open areas with plenty of sun for the develop ment of the fruit and well-drained soil. The palm reaches more than 25 meters in height, with trunks measuring between 9 and 16 cm in diameter, and has an average of 4 to 9 progenies (Weistein 2000; Shanley 2005).

"Açaí" "wine" is rich in calcium, iron, phosphorus and vitamin B$_1$. The level of vitamin A in "açaí" is much higher than in other tropical fruits. One hundred grams of "açaí" pulp contains 2 grams of protein, 12.2 g of lipid, 11.8 milligrams of iron, 0.36 milligrams of vitamin B$_1$ and nine milligrams of vitamin C. The level of protein in "açaí" is similar to that in dairy cattle. The "açaí" palm has fewer calories, but it is a good source of minerals, containing sodium, potassium, manganese, iron, phosphorus, copper and silicon (Weistein 2000).

The "açaí" species *Euterpe precatoria* Mart occurring in the Acre state, is different from the species of Acai in Pará state. The single Açaí (popular name of *Euterpe precatoria*), as the name implies, has only one stipe (trunk) and is generally larger than the Pará's "açaí" palm (can reach more than 23 meters high). It is native to the western Brazilian Amazon, and is typical of mature forests, and occurs both in flooded areas and on the mainl and. It rarely occurs in deforested areas. The fruits used to prepare the "açaí" "wine" using a similar process to that for the "açaí" wine from Pará. The exploitation of the single "açaí" palm heart was very intense and has caused a great reduction in the native population (Shanley 2005). The fruit can be used in the preparation of wine, ice cream, popsicles and "chicha" (fermented drink appreciated by local indigenous), the palm can be eaten fresh or in salads (Nogueira 1995).

Buriti (*Mauritia flexuosa* L.f.)

"Buriti" is one of the species of palm trees in the Amazon, with a diameter of 30 to 50 centimeters and a height of 20–35 meters. It offers an important nutritional fruit for the people and animals of the region. The geographical distribution of "buriti" covers the entire Amazonian region, northern South America, and extends to the Northeast and Mid-South of Brazil. This palm prefers the wetlands, flooded forest, creeks and river borders, where it is found in high concentrations. Water aids in seed dispersal, forming large populations of "buriti" palm forests. The fruits, leaves, oil, petiole and trunks are used for many purposes. Buriti is also known in Brazil as miriti, muriti and buriti do brejo; in Guyana, as "awuara" and "boche"; in Venezuela, as "moriche"; in Colombia, as "carangucha, moriche and nain"; in Peru, as "aguaje" and "iñéjhe"; and in Bolivia, as "kikyura" and "royal palm" (Lima 1987).

"Buriti" has one of the largest amounts of carotene or vitamin A of all plants in the world. There is 30 milligrams of vitamin A per 100 grams of pulp, which is 20 times the quantity found in carrots. vitamin A deficiency is a common problem in the popu- lation. Due to this, people develop diseases such as mouth infections, toothache, eye infection and night blindness. In the Northeast, sweets made with "buriti" have being used to address this deficiency. Furthermore, "buriti" can provide a decent amount of protein in the human diet. The pulp has 11% protein, a quantity similar to corn. The fruit is also used in the prevention of malnutrition and in aiding the recovery of under nourished children. Moreover, "buriti" oil has purifying and detoxifying actions. "Buriti" is an important component in the diet of Apinayé tribes. It is common to see the indigenous peoples walking with panniers full of "buriti"; they remove the peel of the fruit and suck the pulp (Moreira and Moraes 1998).

In certain regions of Pará state, people bore the trunk of male palm trees and collect 8-10 liters of sap to produce a light yellow sugar. This sap is thickened by evaporation, turning it into a honey-like product. From inside the trunk it is also possible to get starchy flour, which is used to prepare porridge, identical to "sago". The Indigenous people from the Amazon call this meal "ipurana" (Paula-Fernandes 2001).

The feeding habits of the people of the Amazon

The identity of a nation is mostly defined by its language and by its food culture. Eating habits determined over time by a society, begins to identify it and often when rooted, it becomes cultural heritage. The act of feeding, is more than biological, it involves forms and cultivation technologies, the management and collection of food, the selection, storage and preparation methods and presentation, and thus constitutes a social and cultural process (Carneiro 2003; Belluzo 2006).

The hot and humid climate of the Amazon region seems to contribute to the specificity of cooking habits. In turn, this has very strong indigenous features in its design, practice and products. Generally, this cuisine features products from the exuberant Amazonian nature, which is biologically diverse and multicolored. Precisely, resources mostly come from the rivers and the rainforest, and this forms the basis of their dishes and drinks (Belluzo 2006).

The simplicity and proximity to nature of Amazon cuisine is achieving new horizons. It is being increasingly refined and with higher quality because the consumer is also more demanding, wanting a quality product that surprises him. This consumer wants to eat well, and feel that the taste of the food reflects directly on the Amazon (Carneiro 2003).

The Amazon rainforest is a major source of many species of useful plants such as cocoa, Brazilian-nut, guarana, cupuaçu, pupunha, etc. Each of the products we know nowadays comes from a particular plant, but in the rainforest there are other usable species. Cocoa, for example, is the fruit of the species *Theobroma cacao*, and Cupuaçu is the *Theobroma grandiflorum* species, but there are other plants of the same genus (*Theobroma*) spread out inside the forest (Atroch et al. 2002; Arruda 2005).

Currently there are 150 known species of *Paullinia*, the genus of guaraná. The guaraná (*Paullinia cupana* var. Sorbilis) is an important and traditional crop in the state of Amazonas. It is a genuine Brazilian plant of great economic and social importance, especially in the Amazon region. This importance is evidenced in the increasing demand in its seeds by the beverage industry, in order to meet the needs of the soft drink and energy drink markets, both national and international (Atroch et al. 2002). Guaraná is used in the form of a powder, cane, in extracts and syrups. In soft drinks, the minimum content of guaraná seeds is 0.2 g and the maximum is 2 g/liter or its extract equivalent (Brasil 1994). Guaraná can also be used in the production of energy drinks, ice creams, creams, as well as pharmaceuticals, cosmetics, craft making, etc. (Arruda 2005).

The existence in the Amazon of wild relatives of many cultivated plants, and creole varieties (continuously selected products made by farmers for several generations), makes the region an important center of origin and diversity of a large amounts of vegetables appreciated for gastronomic reasons. This group of plants is strategic for the future development of food (Raud 2008).

In the context of fermented foods, the native indigenous population of the Northern region (Brazilian Amazon) have manioc (*Manihot esculenta*) as a staple food. This root is, to date, the typical local dish, and is used to prepare "tucupi", a sauce made from the broth of grated and squeezed manioc, which is then decanted and boiled. Thereafter, the broth obtained is added of basil (*Ocimum campechianum* P. Mill) and chicory (*Cichorium intybus*), which creates the famous dish known as "tucupi". Duck with "tucupi" sauce is the most famous dish of the Brazilian Amazon (Pereira 1974).

Fish also represent a significant part of the diet of the Amazon population, with "Tambaqui" (*Colossoma macropomum*), Traíra (*Hoplias* spp.), Piranha, Pescada (Teleosts), river sardines, "Tucunaré" (*Cichla* spp.), "Pacu" and the "Pirarucu" (*Arapaima gigas*) fishes being the most consumed in the Amazon region (Atroch et al. 2002; Arruda 2005). The latter is also known as Amazon codfish, which is kept edible by the salting process introduced by the Jesuits in the mid-seventeenth century (Murrieta 1998; Smith 2007).

The meat and eggs of aquatic turtles (hull animals) are considered a delicacy in the Amazon cuisine (Murrieta 1998; Rebêlo and Pezzuti 2000; Murrieta 2001). The hull animals that are preferred for consumption include "irapuca" (*Podocnemis erythrocephala*) and the "Cabeçudo" (*Peltocephalus dumeriliana*) (Smith 2007). Larger animals, like the turtle (*Podocnemis expansa*) and the "tracajá" (*Podocnemis unifilis*), are targeted to urban markets like Manaus as they are commercially valued, while smaller animals are consumed by the inland populations of the Rio Negro, as also observed by Pezzutti (2004) in the Jau National Park, Lower Rio Negro (Silva 2007). Nevertheless, it is important to note that the extraction of turtles and wild game animals occurs in the context of strict legal prohibition and in conflict with government environmental agencies (Silva 2007).

The capture of hull animals and collecting their eggs occurs in sandy places such as salt marshes, meadows, damiçás (land division at the time of the flooding becomes temporarily on the island) and forest fires, especially during the summer Amazon (dry season), between the months from September to December (Silva 2007). The eggs of the "cabeçudo" are more difficult to locate because their spawning occurs between trunks and branches of the flooded forest and in the holes of precipices on the edges of the lakes at the time of ebb (July), that is why this species is more abundant in nature (Rebelo and Pezzutti 2000 *apud* Silva 2007). According Silva (2007), the elders say that the oil extracted from the eggs was used in the local cuisine, but this habit was abandoned due to substitution from industrial oils. Additionally, the extraction of chameleon eggs for consumption on the islands of Rio Negro was reported by a family in Barcelos city, although its consumption is not usual.

Other Amazonian typical dishes and their ingredients are (Cascudo 2011):

- "Tacacá": broth Tucupi with leaves of Jambu (*Spilanthes oleracea* L.), a type of herb found in the region.
- All kinds of peppers with dried shrimps, served in bowls.
- "Maniçoba": stew of new cassava leaves crushed on a pylon or grinding machine and boiled for a full day. Then, corned beef, pig's head, "mocotó" (calf's foot jelly), bacon, salt, garlic, bay leaf and peppermint are added.
- Wild fruits like "açaí", "Murici", "Graviola", "Cupuaçu", "Mangaba" and "Pupunha" are found in ice creams, juices and creams. Pupunha is also cooked with salt to replace bread.
- Brazilian-nut, guaraná, green mango and avocado with flour and sugar.
- Turtle meat, a feeding habit that worries environmentalists.
- Alligator roasted or boiled with pepper.

Amazon Fermented food: Raw Materials and Major By-products.

Manioc

Manioc, also known as "aipim" and "macaxeira", in the North and Northeast regions of Brazil, is the most important source of energy-rich food. It is widely used in local cuisine, being consumed cooked, as shaped cakes, puddings and stuffed with meat and chicken; however, it should be noted that a popular variety called "wild manioc" exists which cannot be used to prepare the dishes mentioned above, as is contains two cyanogenic glycosides, lotaustralin and linamarin, which are both capable of generating hydrogen cyanide also known as hydrocyanic acid - HCN in the presence of enzymes or digestive juices this may cause death to the consumer (Albuquerque and Cardoso 1980).

Due to the presence of hydrocyanic acid in bitter manioc, this should not be used in the preparation of dishes before undergoing processing by heat. The manioc is used to make the by-products that eliminate hydrocyanic acid by the action of heat, in the form of products such as manioc flour, "tapioca" flour, gum, "chikki" (ground manioc cake), starch gum and "tucupi" (Chisté and Cohen 2006).

The cultivation process meets the temporary nature of Amazonian peoples. Thus, the cultural practices used for the cultivation of manioc is realized by the following methodology, overthrow of "capoeira" which can last 3–10 years. Soon after the overthrow comes the burning of this vegetation, and then proceeds to the planting that is made by vegetative propagation, using seedlings of 20 cm. After planting, the job is to control invasive plants to the crop establishment. The harvest varies from 8 to 24 months, depending on the cultivar, and may be achieved by manual or mechanical harvest (EMBRATER/CIAT 1982).

Tacacá

"Tacacá" (Fig. 9.1) is a food typical of the cuisine of Northern Brazil is characterized by its composition of yellowish broth accompanied with dried shrimp, gum starch, jambu (*Spilanthes oleracea* L.) and sometimes hot pepper. The commercialization of "tacacá" happens in the late afternoon in streets and squares, when it is served in bowls (Fig. 9.2), the bowl and the dish are traditional from Amazon people. (Santos and Pascoal 2013).

FIGURE 9.1   Brazilian hot broth known as "tacacá".

The main ingredient for the preparation of "tacacá" is "tucupi" (Fig. 9.3), a yellowish liquid extracted during the meal preparation process. In this process, the root of manioc is crushed and pressed to remove a fluid called manipueira. Thus, the formed mass is toasted, leaving the manioc meal, one of the main foods from Amazon people. The liquid is often discarded into the environment or through the decantation process for the removal of starch, known by the people of the Amazon as gum (goma in portuguese) (Fig. 9.2). The decantation process takes on average 1 to 2 days, which causes a mild fermentation (Borges et al. 2012).

FIGURE 9.2   The bowl: sort of regional plate in which is possible to observe the stickiness of "tacacá" broth.

FIGURE 9.3   The "tucupi", one of broth ingredients of "tacacá" is being adding to the bowl.

It is important to highlight that "tucupi" contains two glycoside precursors of hydrocyanic acid, making it not appropriate for consumption by animals and humans in its natural form. Therefore, Amazonians empirically observed that with application of the decanting and subsequent boiling process, "tucupi" ceases to be "poisonous" because the hydrocyanic acid is volatilized. After this step, the "tucupi" is seasoned with different spices and thus can be consumed in many different ways (Borges et al. 2012).

The "tacacá" is not a dish in which all of the ingredients are mixed and then served; in fact, ingredients are arranged in separate containers. The assembly of the dish is performed in front of the client at the time of consumption, and it must be consumed hot.

The preparation follows a specific order: the "tucupi" is placed in the bowl followed by the gum, which is extracted from the manioc starch decanting process. This extract passes through cooking to form a viscous and transparent mucilage. Then the "jambu" is added, which is also known as Pará's watercress (*Spilanthes oleracea* L.). Jambu is an unconventional vegetable with peculiar taste that causes a slight numbness of the mouth. Finally, the shrimp is added, followed by the pepper sauce. A spoon is not used to take the broth; it is consumed directly from the bowl, using only a small skewer to remove the shrimp and "jambu" (Fig. 9.4) (Borges et al. 2012; Santos and Pascoal 2013).

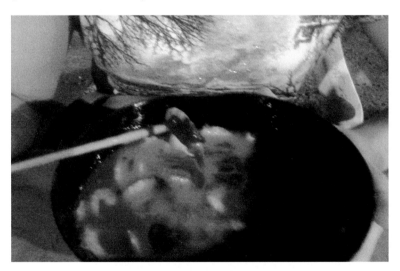

FIGURE 9.4 A way to consume "tacacá".

"Caxiri"

This is a beverage produced from the manioc root. Consumption of this beverage is a collective activity, and in some groups, events around the consumption of the drink can last for days, until the stock of drink is exhausted; this is often used in specific rituals. Other groups use this beverage to brew for therapeutic purposes and in shamanic rituals. Others drink it as food or for fun, often on the occasion of hunting, fishing, harvesting, war, birth rituals, tribe initiation, and funeral ceremonies and magical-religious celebrations (Santos 2010).

In Brazil, there are several indigenous groups that still consume "caxiri", which is a fermented alcoholic beverage. The preparation of the beverage is an exclusive activity of women; in order to enable them to produce a quality drink, in the context of their indigenous culture, it is important to practice sexual abstinence during the preparation period (Vidal 1999; Santos 2010).

The preparation of the drink differs according to the indigenous group, and the level of involvement that these groups have with non-indigenous people. In the traditional process, the preparation is by the use of microorganisms contained in the mouth; thus, the manioc is baked after being processed and passed through the chewing process by the women who return this to the pan to pass through the fermentation process. For non-indigenous people this process causes disgust, so some indigenous groups use beijú (manioc bread) to perform fermentation for consumption by non-indigenous people (Vidal 1999; Santos 2010).

The drafting process discussed here is the traditional one, highlighting there are other recipes for the preparation of the drink. After the wild cassava is collected, it is washed, peeled, and grated, subsequently becoming a pasty mass that is mixed with water. In the traditional preparation, the mixture should be cooked in a clay pot with fire coming from the burning of wood. The mixture should be stirred constantly with a wood palette, as it does not affect the taste of the beverage (Fernandes 2004; Santos 2010). The cooking can last for two to three days following the belief by women of sexual abstinence; otherwise, in their view, the beverage would be of low quality and too thick. Following in the ritualistic preparation of "caxiri", women get together and surround the pan to perform the mass chewing to provide a fermentation product (Assis 2001; Santos 2010). The fermentation time will determine the alcohol content of the drink. With this knowledge, the indigenous people leave the "caxiri" in the fermentation process for three days, since the alcohol content is low and can be consumed even by children, as it has high energy and can be considered an energy food (Assis 2001; Fernandes 2004; Vidal 1999).

Amazonian wines

The Amazon forest has wide flora diversity. For traditional people, the forest is considered as a large warehouse in which they have access to practice extractive activities, removing food, supplies for building, making kitchen utensils and crafts. Among the floristic families included in the Amazon flora, the *arecaceaes* stand out, since they have great social and cultural importance (Shanley 2005).

We will give increased focus to this family of flora because the wines produced by them are consumed by all of the people lives in the Amazon, and it is essential to highlight that some of these wines exceed the borders of the region and even the country. However, it is important to note that traditional people not only consume the drink of this botanical family, but also enjoy many wines produced with the use of other plants from the region (Santos 2010).

The leaves are used to cover houses, to make walls and prepare household items. Note, therefore, how indispensable this botanical family is to Amazonian people. However, it is necessary to highlight that these people live integrated with the environment surrounding them, being subjected to the seasonality of the region. Because of this, in these regional wines have a great importance in the absence of any food source, when the drink has the potential to supply the lack of energy and other nutrients. Amazonian wines are not fermented beverages when made from pulp, they are only fermented when made from the barks and seeds of some fruits. Among the palm trees that are used to produce wines, we highlight "açaí", "patauá", "bacaba" and "buriti" (Shanley 2005). The fruits are collected in the field, shooting a stick at bunches in order to select those bunches with ripe fruit; after harvest, it is necessary to wash the fruits to remove all dirt that may contaminate the drink. The fruits are put in hot water to soften; the time at which the fruit must be immersed varies with the species used (Santos 2010).

It is advised to monitor the process in order to determine the right time to move to the next step, which is separation of the seeds from the pulp; this step is performed with the help of a little bit of water and a pestle. Finally, the pulp is passed through a sieve to separate the seeds that remained and the "wine" is ready. In urban area nowadays, the "wine" is made using the electric removing device. For the alcoholic beverages, other by products are obtained from the barks and seeds of some of these fruits though fermentation techniques (Shanley 2005).

High Tech Fermented products: Amazonian Fruits and Roots Dictate new Soft Drinks, Beer and Yogurt Flavors.

The Amazon has a very modest food technology park, especially in the fermentation technology sector; however, their produce is quite well recognized on the national scene, since the region offers a range of plant varieties, including many fruits with exotic flavors. These products are differentiated by the development of by-products of fermentation, with great acceptability in the Brazilian and international markets, such as beer, soft drinks and yogurts (Shanley 2005).

In Brazil, "guarana" soda (*Paullinia kupana*) is already an established and highly popular product, generating foreign exchange for many large Brazilian and multinational companies which produce it, such as the "American Company of Beverages (Ambev®) and The Coca Cola Company®. "Guaraná" flavor is already among the 15 most consumed flavors in the world, with about 800 million liters being produced a year; half of this amount is consumed in Brazil (DeMelo 2005).

From the "guaraná" extraction in the forest, until its sale to the consumer in the form of soda, the fruit passes through many transformations. "Guaraná" soda produced by Ambev®/Antartica® started the production in 1921 and its formula is made in the company's safe-rooms in the cities of Manaus (AM) and Guarulhos (SP) (UOL 2013; DeMello 2005).

A large part of "guaraná" fruit production comes from the city of Maués, State of Amazonas, about 250 km from Manaus. Its inhabitants have their personal income based mainly in the production chain of this fruit, which is not highly consumed "in natura" due to its sour taste (SUFRAMA 2003).

Ambev® has a farm of 1007 hectares in the city of Maués (size equivalent to a thousand football fields) with monoculture plantations of "guaraná" (Ambev 2014). At this location, experiences are carried out for improving the agronomic treatment of the fruit, as well as the genetic bank of the species from plant breeding, obtaining "enhanced seedlings". It is noteworthy that each year about 60,000 seedlings are donated to local farmers for planting (SUFRAMA 2003).

In "Guaraná" plantations, the fruit harvest is performed manually between the months of October to February; after harvest, farmers separate the grain, which is the black portion of the fruit. In some cases, this separation can also be performed by the feet, using a technique similar to that used for wines. Then, these grains are washed in water tanks, with the help of baskets, so the thorough cleaning of this material occurs, separating them from other parts of the fruits that are still stuck in these grains (SUFRAMA 2003; UOL 2013). After this step, the grains are roasted in clay pots present on the farms, obtaining a color and odor similar to that of roasted coffee. After this step, the products are bagged in bags of 40 kg and taken to Manaus by the river (SUFRAMA 2003; UOL 2013). The transformation of "guaraná" is also performed in Maués (Amazonas state), where the toasted grains are used for the manufacture of a strong, concentrated extract, which is the basis of the soda. This extract is transferred to an aroma factory in Manus, where it is submited to a lot of evaluations (SUFRAMA 2003; UOL 2013).

After the evaluation, if approved, the extract is mixed with other ingredients of the company protected recipe, such as essences, dyes, fragrances and essential oils; this mix is the liquid concentrated of "guaraná", which is sent to the 30 Ambev® factories in Brazil (SUFRAMA 2003; UOL 2013). Only in the factories that the concentrated "guaraná" is added to the syrup, obtaining a sweeter flavor that is closer to the taste of the final product. The process ends with the addition of carbon dioxide, which confers a bubbly effect to the final product that is packaged in bottles or cans and labeled with the "Guaraná Antarctica" brand label, making it ready to dispense (SUFRAMA 2003; UOL 2013).

Also in the "guaraná soft drinks segment, the Amazon has small industries that produce this type of product and are well recognized in the local market as the brands "Baré" (Ambev®), "Real" (Real Bebidas®), "Boy" and "Magistral" (J. Cruz Industria e comércio®). This local soft drink segment recently launched an "açaí" soda, looking to the market sector that searches for this fruit in the improving of health, and an establishment of a lifestyle with energy and good nutrition (Author notes).

In the beer segment, fruits like "açaí", "bacuri", "tapereba" and roots as "cumaru" and "priprioca", have brought success to the Amazon beers brewed in the state of Pará. The success of Amazon beers is so large that can be seen not only by the R$ 17.5 million (almost US$ 6.6 million, in currency exchange quotation of December 31, 2014) profit obtained in 2014, but also by the numerous awards of excellence that it has gained in recent years, with the latter choosing "açaí" beer as the best craft beer in Brazil. The mark "Amazon Beer®" is responsible for the production of this product in the country, specifically in the city of Belém, Pará State capital (Author notes).

Finally, another prominent segment with fermented products involving Amazonian fruits refers to the yoghurt market and milk drinks, especially those involving the Amazon company "Flamboyant", in the city of Castanhal – Pará state. Among the products marketed in several product lines targeted at different consumers according to their profile, the company has a specific line for outstanding exotic products, with milk drinks prepared with the addition of Amazonian fruits like "guaraná", "cupuaçu" and "açaí"; the company uses the marketing phrase: "Iogurte com sabor da Amazônia" (yoghurt with Amazon flavor) (Author notes).

References

Albuquerque, M. and E.M.R.Cardoso. (1980). A mandioca no Trópico Úmido. Editerra, Brasília.

Ambev. (2014). Relatório anual. Companhia de bebidas das Américas, São Paulo.

Atroch, A.L., F.J. Nascimento-Filho, J. Ribeiro, L. Lima and J.O. Pereira. (2002). Agricultura familiar na Amazônia Brasileira: clones de guaraná: tecnologia sustentável para a Amazônia. Embrapa Amazônia Ocidental, Manaus.

Arkcoll, D.B. and J.P.L. Aguiar. (1984). Peach palm (*Bactris gasipaes*) a new source of vegetable oil from the tropics. Journal of Science and Food Agriculture 35: 520–526.

Arruda, M.R. (2005). Adubação do guaranazeiro: fontes, doses, época de adubação e localização dos fertilizantes. Embrapa Amazônia Ocidental. Comunicado Técnico, 31 Manaus.

Assis, L.P.S. (2007) Da cachaça à libertação: mudanças nos hábitos de beber do povo Dâw no Alto Rio Negro. Revista Antropos, 1(1): 101-173.

Becker, B.K. (2008). Serviços Ambientais e Possibilidades de Inserção da Amazônia no século XXI. T&C Amazônia, 4 (14): 3-10.

Belluzo, R., W.M.C. Araújo, C.M.R. Tenser. (2006). Gastronomia: cortes & recortes. SENAC-DF 1: 181–188.

Borges, L.S., R. Goto and G.P.P. Lima. (2012). Comparação de cultivares de jambu influenciada pela adubação orgânica. Horticultura Brasileira 30: 2261–2267.

Carneiro, H. (2003). Comida e sociedade: uma história da alimentação. Elsevier, Rio de Janeiro.

Cascudo, L.C. (2011). A História da alimentação no Brasil. Global Editora, São Paulo.

Cavalcante, P.B. (2010). Frutas comestíveis da Amazônia. 7 ed. CNPq/Museu Paraense Emílio Goeldi, Belém.

Chisté, R.C. and K.O. Cohen. (2006) Estudo do processo de fabricação da farinha de mandioca. Embrapa Amazônica Oriental. Documentos, Manaus.

Clement, C.R. (2000). Comunicação pessoal. Instituto Nacional de Pesquisas da Amazônia, Manaus.

Clement, C.R., E. Lleras-Pérez and J. Vanleeuwen. (2005). O potencial das palmeiras tropicais no Brasil: acertos e fracas s os das últimas décadas. Agrociências 9: 67-71.

Cdib-Taxi, C.M.A. (2001). Suco de camu-camu (*Myrciaria dubia*) micro encapsulado obtido através de secagem por atomização. Tese, Universidade Estadual de Campinas, Campinas, São Paulo, Brazil.

DeMello, Y.E. (2005). Guaraná - Prospects and Geographic Indicator Status. TED Case Studies n.780. Available in: http://www1.american.edu/ted/guarana.htm#r1. Accessed in: 2016 February 9th.

[EMBRATER/CIAT] (1982). Seleção e preparo de estacas de mandioca para o plantio. Brasília.

Fernandes, J.A. (2004). Selvagens bebedeiras: álcool, embriaguez e contatos culturais no Brasil colonial. 302 p. PhD Thesis (Doctor degree in History). Universidade Federal Fluminense, Rio de Janeiro.

Gaia, G.A., M. Alexiades and P. Shanley. (2004). Productos forestales, medios de subsistencia y conservación: estudios de caso sobre sistemas de manejo de productos forestales no maderables. Cifor 3: 219–240.

Gama, M.M., G.D.Ribeiro, C.F. Fernades and I.M. Medeiros. (2005). Açaí: características, formação de mudas e plantio para a produção de frutos. EMBRAPA-Circular Técnica 80: 1-6.

Herculano, F.E.B. (2005). A SUFRAMA e a dinâmica do desenvolvimento regional nortista. Prêmio Professor Samuel Benchimol 2005 - Relatório, Brasília/DF, 1: 121.

Inoue, T., H. Komoda, T. Uchida and K. Node. (2008). Tropical fruit camu-camu (*Myrciaria dubia*) has anti-oxidante and anti-inflammatory properties. Journal of Cardiology 52: 127–132.

Lima, M.C.C. (1987). Atividade de Vitamina A do Doce de Buriti (*Mauritia vinifera* Mart.) e Seu efeito no tratamento e prevenção da hipovitaminose A em crianças. M.S. Thesis, Universidade Federal da Paraíba, João Pessoa, Paraíba, Brazil.

Lopes, A.S., N.H. Pessoa-Garcia and J. Amaya-Farfán (2008). Qualidade nutricional das proteínas de cupuaçu e de cacau. Ciência e Tecnologia de Alimentos 28: 263–268.

Maeda, R.N., L. Pantoja, L.K.O. Yuyama and J.M. Chaar. (2007). Estabilidade de ácido ascórbico e antocianinas em néctar de camu-camu (*Myrciaria dubia* (H.B.K.) McVaugh). Ciência e Tecnologia de Alimentos 27: 313–316.

Moraes, V.H.F., C.H. Muller, A.G.C. Souza and I.C. Antônio. (1994). Native fruit species of economic potential from the Brazilian Amazon. Angewn date Botanik 68: 47–52.

Moreira, S.G.C., A.V. Moraes. (1998). Light absorption coeficiente using He-Ne laser (630 nm and 543 nm) to buriti, copaíba, babaçu and andiroba oils. Acta Amazonica, 28(4):101-104.

Murrieta, R.S.S. (1998). O dilema do papa-chibé: consumo alimentar, nutrição e práticas de intervenção na Ilha de Ituqui, Baixo Amazonas, Pará. Revista de Antropologia 41: 97–145.

Murrieta, R.S.S., D.L. Dufour and A.D. Siqueira. (2001). Dialética do sabor: alimentação, ecologia e vida cotidiana em comunidades ribeirinhas da Ilha de Ituqui, Baixo Amazonas, Pará. Revista de Antropologia 44: 39–88.

Nogueira, O.L.A. (1995). Cultura do açaí. Coleção Plantar. Embrapa Amazônia Oriental, Brasília.

Paula-Fernandes, N.M. (2001). Estratégias de produção de sementes e estabelecimento de plântulas de *mauritia flexuosa* L. f. (Arecaceae) no Vale do Acre/Brasil. Ph. D. Thesis, Instituto Nacional de Pesquisa da Amazônia, Manaus, Amazonas, Brazil.

Pereira, N. (1974). Panorama da alimentação indígena: comidas, bebidase tóxicos na Amazônia brasileira. São José, Rio de Janeiro, Rio de Janeiro, Brazil.

Pezzuti, J. (2004). Tabus alimentares. Hucitec, São Paulo.

Raud. C. (2008). Os alimentos funcionais: a nova fronteira da indústria alimentar: Análise das estratégias da Danone e da Nestlé no mercado brasileiro de iogurtes. Revista de Sociologia e Politica 16: 85–100.

Rebêlo, G. and Pezzuti, J. (2000). Percepções sobre o consumo de quelônios na Amazônia. Sustentabilidade e alternativas ao manejo ambiental, Ambiente e Sociedade 6: 85–104.

Rogez, H. (2000). Açaí: preparo, composição e melhoramento da conservação. Editora UFPA, Belém.

Sacramento, C.K., W.S. Barreto and J.C. Faria. (2008). Araçá-boi: uma alternativa para agroindústria. Bahia Agrícola 8: 22–24.

Santana, A.C. (2007). Índice de desempenho competitivo das empresas de polpa de frutas do Estado do Pará. Rer 45: 749–775.

Santos, C.C.A.A. (2010). Identificação e caracterização físico-químico da bebida fermentada cair produzidas pelo povo Juruna (Yadjá). M.S. Thesis, Universidade Federal de Lavras, Lavras, Minas Gerais, Brazil.

Santos, V.F.N. and G.B. Pascoal. (2013) Aspectos gerais da cultura alimentar paraense. Revista da Associação Brasileira de Nutrição 1: 73–80.

Silva, A.L. (2007). Comida de gente: preferências e tabus alimentares entre os ribeirinhos do Médio Rio Negro (Amazonas, Brasil). Revista de Antropologia 50: 125–179.

Silva-Filho, D.F. (1998). Cocona (*Solanum sessiliflorum* Dunal): Cultivo y utilizacion. Tratado de Cooperacion Amazonica. Caracas, Venezuela.

Silva-Filho, D.F. (2002). Discriminação de etnovariedades de cubiu (*Solanum sessiliflorum* Dunal) da Amazônia, com base em suas características morfológicas e químicas. Ph. D. Thesis, Instituto Nacional de Pesquisa da Amazônia, Manaus, Amazonas, Brazil.

Silva-Filho, D.F., L.K.O. Yuyama, J.P.L. Aguiar, M.C. Oliveira and L.H.P. Martins. (2005). Characterization and evaluation of the agronomic and nutritional potential of ethn-ovarieties of cubiu (Solanum sessiliflorum Dunal) in Amazonia. Acta Amazônica, 35 (4): 399-405.

Silva Filho, D.F., C.J. Anunciação Filho; H. Noda and O.V. Reis, O.V. (1996). Variabilidade genética em populações naturais de cubiu da Amazônia. Horticultura Brasileira, 14 (1): 9-15.

Shanley, P. and G. Medina. (2005). Frutíferas e plantas Úteis na vida amazônica. Cifor, Imazon, Belém, Pará.

SUFRAMA. (2003). Projeto potencialidades regionais estudo de viabilidade econômica: Guaraná. Superintendência da Zona Franca de Manaus - Suframa, Manaus. 34p.

UOL. (2013). Veja como é produzido o guaraná. São Paulo, Universo Online. Available in: <http://economia.uol.com.br/album/2013/01/14/veja-como-e-produzido-o-guarana. htm>. Accessed in 2016 february 9th.

Vidal, L.B. (1999). O modelo e a marca, ou o estilo dos "misturados". Cosmologia, História e Estética entre os povos indígenas do Uaçá. Revista de Antropologia 42 (1-2): 29-45.

Weinstein, S. (2000). Causes and consequences of açaí palm management in the amazon estuary. Thesis, University of Florida, Florida.

10

Fermented Foods and Beverages from Cassava (*Manihot esculenta* Crantz) in South America

Marney Pascoli Cereda* and
Vitor Hugo dos Santo Brito

Abstract

Fermentations are considered pre-scientific technologies that were used to prepare traditional foods and drinks in order to preserve them, improve palatability, flavor or nutritional enrichment and reduce its toxicity. It is therefore natural that the fermentation of food started with the most abundantly available and more often consumed raw materials in each place. The cassava (*Manihot esculenta* Crantz) was domesticated in the tropical zone of the planet and used as food and as a substrate for fermentation processes. The increased acidity, developed by microbial activity, inhibits the growth of pathogenic microorganisms for humans and favors the growth of lactic acid bacteria (LAB). Despite the numerous products that have been developed and were in people's memory, only a few remained. Specifically, in South America many types of cassava fermented products have been developed and many disappeared. This chapter selected tiquira and fermented cassava starch as representatives of South America as they have remained in the market because they have been popular with consumers. To remain as commercial products they need to be competitive. For this it is necessary not only to research but also to understand the basis of their production methods. It is also necessary to develop processes and products that are compatible with modern life.

Introduction

Cassava (*Manihot esculenta* Crantz) originated from the American continent, from where it spread to Asia and Africa. Amerindian people used it as food because it was easy to cultivate and because predatory insects did not attack its starchy roots, as was the case with cereals such as maize (*Zea mays* L.).

Catholic University of Campo Grande (UCDB), Centre of Technology and Agribusiness Analysis. Avenida Tamandaré, 8000, Zip Code 79117-900, Phone 55 (67) 331-3913, cereda@ucdb.br; britovhs@gmail.com
* Corresponding Author: cereda@ucdb.br

Many food products were developed in the pre-scientific phase as a result of microbial metabolism. Increased acidity which is developed by microbial activity as a consequence of the growth of lactic acid bacteria (LAB), inhibits the growth of pathogenic microorganisms harmful to humans and animals. In addition, fermentation processes improves the appearance, British English colour, and flavour and also soften the rigid texture of cassava roots. This colonisation by microorganisms was a natural process as there were no methods of conservation of cooked products once it had cooled (Chuzel and Cereda 1995).

Lancaster et al. (1982) published a complete overview of the literature bringing together information regarding the diversity of processing techniques and the wide variety of cassava-based foods and beverages that are available. The aforementioned authors point out that in terms of fermented foods most technologies are related to the removing of cyanide present in cassava. They also note that despite the diversity of products made in various countries and regions of Africa, South America and Asia, many of these technologies originated among the Amerindian peoples of South America. Although it is over 30 years since this review was published, these technologies still remain in use in many countries, and most of them are still at virtually the same level of advancement as reported by Lancaster et al. (1982).

In Brazil also the technology for processing cassava in fermented foods originates from the Amerindian people, as was pointed in an extensive review by Gonçalves de Lima (1974). The complex techniques that are used around the world today were introduced to other regions with the cassava plant or, in some cases, at a later date (Lancaster et al. 1982).

The products discussed in this chapter are either not mentioned in the review by Lancaster et al. (1982) or only merit one line of description. However, in Latin America and the Caribbean in the twenty-first century it is still possible to find some of these food products that are still marketed for sale and widely consumed.

In popular markets in Bolivia and Paraguay it is possible to find various types of beverages derived from cassava flour. When they are lightly fermented, they are gaseous and similar to beer. Another product made from fermented cassava starch can be found with different local names in many different countries. Some of them are known as *polvilho azedo* in Brazil, and as *almidon agrio* in Colombia, Bolivia, Argentina and Paraguay. Distilled drinks made from cassava are rare and only one is still found in Brazil, which goes under the name of *tiquira*. Last two mentioned products (*tiquira* and *polvilho azedo*) demonstrate that it is possible for artisanal-type products to remain commercially viable.

Composition of Cassava Roots

In order to better understand how food products made from fermented cassava are produced it is important to be aware of the composition of cassava roots, which are the raw material for these foods. Cassava is primarily a source of carbohydrates and contains little protein (Lancaster et al. 1982). The chemical composition of the roots varies depending on the variety, conditions of cultivation and climatic specifics of the production region, but the edible portion typically comprises 62% water and 35% carbohydrates. According to Jones (1959), the food energy of cassava roots is around 5,800 Mjoules kg^{-1}. Cereda et al. (2003) commented that the potential acidity varies from 1.29 to 2.91 (ml of NaOH 100 g^{-1} of sample) and that the effective acidity is close to 5.9 (pH). The roots contain the sugars sucrose, glucose and fructose and these do not exceed 2%. Because there is limited opportunity for the sugars to be easily metabolised by microorganisms, starch is the main available substrate with a carbon/nitrogen (C/N) ratio close to 32, which is a very high value. Dried cassava roots are almost wholly made up of starch. The cyanide present in cassava appears in analyses as non-protein nitrogen.

The main cyanogenic product of cassava is linamarin. The removal of cyanide from cassava has always been the main aim in the preparation processes of fermented foods followed by local culture. The normal range of cyanogen content of cassava roots is between 15 and 400 mg HCN kg^{-1} fresh weight (Coursey 1973). The concentration varies greatly between varieties and also depends on the environmental and cultivation conditions.

Globally, and particularly in Africa, the concentration of glycosides may vary widely among cultivars, both for genetic reasons and for environmental reasons (location, soil types, season), and may reach values of up to 2000 mg kg^{-1}. In South America, in general terms, cassava roots have much lower values. In Brazilian varieties, potential cyanide (CN^p) ranges from 20 to 200 mg kg^{-1} and does not generally exceed 100 mg kg^{-1} of CN^p (Brito et al. 2013).

In natural processes the roots are detoxicated by hydrolysis of the cyanogenic glycosides and the subsequent elimination of the liberated HCN. Contact between enzymWes and substrate only occurs when the tissues are mechanically damaged or if there is a loss of physiological integrity, such as during post-harvest deterioration (Conn 1969; Coursey 1973). Most traditional food preparations appear designed to bring about the necessary contact by cell rupture, by grating or pounding, followed by elimination of the HCN by volatilisation or solution in water (Coursey 1973). There is little published information regarding HCN levels in cassava food products and even less that relates those levels to the initial HCN content of the unprocessed roots (Coursey 1973).

Due to increased acidity, fermentation processes are generally not suitable for removing cyanide from products because the low pH prevents the enzymes of the plant itself from acting and hydrolysing the cyanogenic compounds, thereby allowing the HCN to volatilise. This is supported by a study by Bourdoux et al. (1983), who pointed out that the fact that millions of people consume cassava as a staple food without intoxication, suggests that traditional detoxication techniques are generally effective. In order to enable the free cyanide in cassava to be colonised by microorganisms, the most favourable environment is one in which anaerobic or micro-aerobic organisms are present because the CN$^-$ radical inhibits the respiratory chain of most living organisms (Brasil, Cereda and Fioretto, 1982).

The microorganisms that grow in grated cassava mass have little carbohydrate energy for their initial growth and have to hydrolyse the starch that is available. Another important point for the colonisation of the substrate is the acidity of the environment. Cassava has a slightly acidic reaction (pH 5.5 to 6.0), which is typical for healthy plant tissues. However, this reaction to the environment favours moulds, yeasts and some bacteria.

Like all natural starches, cassava starch granules are composed of amylose and amylopectin molecules (Fig. 10.1) in a semi-crystalline state, with different degrees of accessibility for microorganisms.

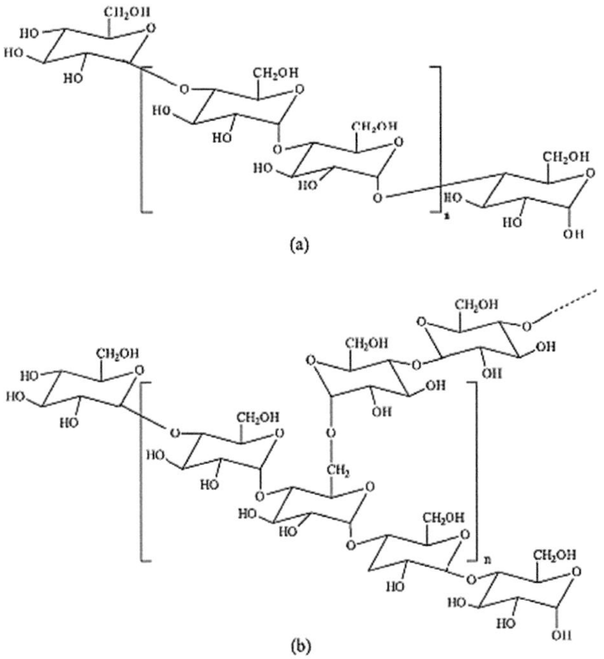

FIGURE 10.1 Components of starch. a) amylose structure (α–1→4 D- glucopyranose). b) amylopectin structure (α–1→4 and α–1→6 D- glucopyranose). Source: Corradini et al. (2005).

Cassava starch contains approximately 18% amylose and 82% amylopectin, expressed as total starch. In the specific case of cassava starch which is fermented and dried in the sun (in case of preparation of *polvilho azedo*), the proportion of amylose is slightly increased due to the partial hydrolysis of the amylopectin during processing, which changes the 20:80 ratio, and which also alters the rheological properties. Once the crystalline areas are in contact with cold water the structure of the granules is maintained, which allows water ingress, and hence swelling. This results in an increase of 10 to 20% by weight of the starch, but this process is reversible by drying (Corradini et al. 2005).

Processing of Fermented Beverages and Foods

To better understand the processing of cassava it is necessary to know a little about cassava roots. As Food and Agriculture Organization (FAO 1998) has pointed out, transversely, cassava roots consist of the following three main areas.

The periderm: comprises the outermost layer of the root. It is composed mostly of dead cork cells, which seal the surface of the root. The periderm is only a few layers of cells thick and as the root continues to increase in diameter, the outermost portions of it are sloughed off and replaced by new cork formations from the inside layers of the periderm. **The cortex:** a layer 1 to 2 mm thick located immediately beneath the periderm. **The starchy flesh:** the central portion of the root, consisting mainly of parenchyma cells packed with starch grains.

Lancaster et al. (1982) classified the types of processes that use fermentation into two categories: those that use spontaneous fermentation; and those that use inoculum from earlier fermentations. These techniques also render the root palatable and, in many cases, storable. These various processing techniques have sequences which are initially similar, but which then diverge, resulting in very different end products. Conversely, very different processes can lead to similar results. Some processes appear to have been developed independently in different countries, often based on methods used to prepare indigenous staple foods (Jones 1959).

In Brazil, some products produced from fermented cassava have evolved from rudimentary technology for commercial products, which have more established technology. When the fermented product is cassava root, as in the preparation of *puba* and *carimã* flour, the root is simply washed, or not at all. The roots are subsequently immersed in water. The water makes the environment either slightly or completely anaerobic and promotes the leaching of sugars and other nutrients. The liquid becomes the medium in which fermentation takes place. The literature overview shows that *Bacillus* spp. are detected during fermentation, which produces amylases, pectinases and cellulases. They also soften the starchy pulp of the roots, which facilitates the removal of the cortex.

In the simplest technology, cassava roots are grated, pressed and then left to dry on a surface over a wood fire to produce what is known in Brazil as *farinha*. The product known as *gari* in Africa is similar to *farinha,* however the latter is fermented. The preparation of *gari* is clearly based on the methods used for preparing *farinha* in Brazil, but usually with a longer fermentation time (Lancaster et al. 1982). For the other fermented products that are obtained from cassava flour it is necessary to wash the roots and often to peel them by removing the periderm and the cortex. The peeling can be done by hand or by using wet-peelers. For fermentation processes the removal of the cortex is not recommended because it contains starch that can be converted into sugars and alcohol.

A widely used technology is the direct fermentation of roots in water (soaked roots) before processing. Microorganisms are involved in the softening of cassava roots through the action of several amylolytic, pectinolytic and cellulolytic enzymes. The roots, either peeled or unpeeled, are soaked for 3–8 days, but sometimes even longer, during which time some fermentation processes occurs. When taken out of the water, the peel is easily removed and the softened roots are either crushed by hand or grated. In addition this process, which is done in unsophisticated conditions, makes it easier. The final product is dried in an oven and has a characteristic and pleasant flavour.

The flour produced in this way is known as *farinha d'agua* or *farinha puba*. Some people prefer the flavour of flour made from slightly fermented yellow cassava roots. The soaking process also softens the roots and makes them easier to grate. For special occasions an extra fine yellow flour known as *carimã* is prepared by crushing dried fermented roots and it is used generally for preparation of cakes (Lancaster et al. 1982).

Many of these products were introduced to Africa and were adapted to the needs and tastes of African countries. In the case of the fermentation of whole cassava roots, it is believed that even in small quantities there are sufficient nutrients for the fermentation process to develop.

In the case of unsoaked roots, they are grated in the form of a moist mass. For traditional products, such as *tiquira*, the grated mass is pressed to remove half of the water contained in the roots, which corresponds to about 30% of the total weight of the cassava root. This pressed liquid, which is decanted off and the starch residue rinsed, is known in Brazil as *manipueira*. In some Brazilians regions, this liquid is not a residue and it can be used to make *tucupi*. *Tucupi* is an example of how popular products can remain within the market. It is a savoury sauce that is fermented by lactic acid bacteria (LAB). It is also found in other regions of Latin America, particularly in Guyana and Surinam. According to Lancaster et al. (1982), *tucupi* is heavily seasoned with peppers, pimento, garlic and local herbs and then boiled down to a thick syrupy consistency. Cassava juice is also known as *yari* in Brazil, *cassareep* in the West Indies, and *kasiripo* in Surinam.

Chisté et al. (2007) reported that the annual per capita consumption of *tucupi* in Belém (the capital of a Para state, in the north of Brazil) was 0.35 kg. As for the free cyanide content of *tucupi*, this was found to range from 9.47 to 46.86 mg HCN.kg^{-1}, while the total cyanide showed wide variations between the samples, ranging from 55.58 to 157.17 mg HCN.kg^{-1}. The concentration of cyanide may be greater than in the cassava root because it is soluble and may become concentrated in the pressed liquid.

Tucupi is consumed all year round and it features in the most famous dishes of the local cuisine. Many local dishes use *tucupi*, including fish and duck, and also a dish known as *tacacá*. The peak of consumption takes place during a religious festival (Círio de Nazaré) when tonnes of this sauce are consumed. In this Brazilian region (Pará state) cassava cultivars are selected for higher humidity. When they are processed to make the sauce, the starch and flour that are obtained are considered as by-products.

In the case of the grated mass of cassava roots, which have lower water content because of pressing, it is easy to form them into balls and flatten them to make *beijus*, which, according to Lancaster et al. (1982) is also called *cassava bread* and are used as the starting points in the preparation of other fermented foods. *Beijus* can be dried or lightly toasted in the ovens used to make cassava flour. Pressing can remove part of the linamarin, but also part of 2% of the soluble sugars. For making fermented products there is no need to peel the roots or to press the grated mass to avoid losing these sugars. In the case of products made by fermentation of cassava, the grating of the roots should be only performed after the removal of the periderm. The grating of the starchy flesh and the cortex destroys the tissues, exposing the starch which facilitates cooking and attack by enzymes.

Fermented Products made from Cassava Roots

Various foods and beverages are prepared by the fermentation of starchy raw materials. In the specific case of cassava, its derivatives form part of Brazilian culture and they are considered as ethnic foods, such as *tiquira* spirit and the products known as *puba, carimã, farinha d'água* and *polvilho azedo*.

Cassava-based beverages

Amerindian people have traditionally prepared a variety of beverages from different raw materials, including fruit and starchy products such as corn and cassava. Most academic studies of beverages based on cassava (either fermented or not) are centred in the Amazon region, e.g. *tiquira*, which is prepared by the indigenous people of Pará and Amazonas states (Brazil).

Gonçalves de Lima (1974) and Lancaster et al. (1982) reported on the origin of several beverages prepared by indigenous communities in the Americas, sometimes making connections with similar beverages produced in different countries. Regarding the beverages which were derived from starchy raw materials and which were in common use by Amerindian communities in Brazil, these authors divided them into two groups: non alcoholic (nutritional and also thirst-quenching) and alcoholic. They further stated that these indigenous people produced various types of wines and spirits; alcoholic drinks which were consumed in abundance at religious ceremonies, parties and banquets.

The custom of preferring the intake of liquids to solid foods is explained by the absence of the habit of drinking water among these people. As they are nomadic, the transport of water is difficult and also the water might be brackish. These beverages are used to simultaneously quench thirst and also as food. They include *chibé, caxirí, chicha, tiquira* and *paiauaru* and they vary from types of gaseous beer to types of non-gaseous wine (Gonçalves de Lima 1974; Lancaster et al. 1982).

Saccharification of Starch with Moulds

Amerindian people had no direct knowledge of enzymes and did not know how to inoculate cassava mass with microorganisms. Consequently, they sought to promote mould growth on the surface in order to take advantage of the enzymes that were produced in order to saccharify the starch into sugars. This technique was reported in historical documents of the time for Amerindian people in Brazil and also Amerindian people living in neighbouring countries and other parts of the world.

The existence of fermented beverages was common in the Americas from Pre-Colombian period; however, only one distilled alcoholic beverage was recorded which remains until the present day - *tiquira*, also known as cassava spirit. Various alcoholic beverages derived from cassava roots are based on *beiju* or *cassava bread* (Gonçalves de Lima 1974; Lancaster et al. 1982).

There is an intriguing fact concerning the origin of the distillation process used in the traditional manufacture of *tiquira*. At the time of Pre-Colombian period, the Brazilian Amerindians were living in Stone Age conditions and did not know how to make metal in order to construct stills. However, several references in the literature cast doubt on this claim. The most intriguing is that by Le Cointe (1922), cited by Gonçalves de Lima (1974), which describes the preparation of *beiju* spirit or *tiquira* by indigenous people in the Amazon region. In addition to the general description of the process, as discussed above, the author points out that the fermented and brewed liquid was distilled in ceramic stills.

Microorganisms Associated with the Fermentation Process

Granular starch is rarely metabolised by microorganisms because the enzymes cannot access the interior of the granules. However, the various types of fermented products that are available in many countries show that some microorganisms can metabolise starch granules in more or less complex sugars due to the availability of different amylolytic enzymes. Moulds and bacteria are known sources of enzymes, and they are the first to colonise raw cassava, metabolising the pre-existing sugar. The most common microorganisms are LAB, as occurs in the fermentation of granular starch in *polvilho azedo* and in constituent water in cassava roots (*manipueira*) in the preparation of *tucupi*.

The traditional methods of preparing food and beverages made from cassava use enzymes produced by moulds and naturally present bacteria, which hydrolyse the starch, producing fermentable sugars, as well as the yeasts that ferment these sugars to ethanol. Among the fermented food and beveragess obtained from cassava in Brazil two particularly stand out: *tiquira*, the distilled alcoholic beverage, and *polvilho azedo*, both for their originality.

Tiquira – cassava spirit

Tiquira is a traditional product from the Brazilian state of Maranhão and among the fermented cassava products it is the least well-known outside Brazil. Cereda and Costa (2008) described the preparation of *tiquira* using the traditional process in small industries and the improvements possible to guarantee a standard quality and yield. After being peeled and grated, the cassava is pressed and the liquid is discarded. According to Chuzel and Cereda (1995), the grated and pressed cassava dough has around 50% humidity. The *beijus* made with this dough are 30 cm in diameter and are 3 to 4 cm thick and weigh about one kilogram.

The resulting dough is used to make large *beijus* (Fig. 10.2), which are toasted on a very hot oven plate in the same oven that is used to make cassava flour.

FIGURE 10.2   Manufacture of *beijus* at traditional small industrial facilities in the Brazilian state of Maranhão. Source: the authors

Fig. 10.2 shows the *beijus* made from grated cassava being cooked on a heated plate. Chuzel and Cereda (1995) observed that *beijus* are occasionally turned in order to cook and bake them on both sides. Once cooled, the *beijus* have moisture content between 30 and 35% with a soluble solids content of 14 °Brix. In order to maintain adequate moisture for the enzymes to have effect, the *beijus* are periodically sprinkled with a little water on both sides, stretched over a bed of banana or palm leaves, and also coated with such leaves. Chuzel and Cereda (1995) state that for the *beijus* to have the best conditions for the growth of microorganisms the storage sites should have high temperature and relative air humidity because such conditions promote the growth of moulds.

After cooking, the *beijus* are placed on a wooden surface and covered by a 2–3 cm thick layer of banana or cassava leaves. This layer receives another layer of leaves of the same thickness as the first. The pile increases and has to be supported so that it does not collapse. After 3–4 days, the layers of leaves are removed and the *beijus*, which are already covered in fungi, are placed in large closed containers (Fig. 10.3). After two days, they are uncovered and the *beijus* are moist, with a clear, yellowish liquid that runs out of them. The colour of the *beijus* varies from "egg yolk" yellow to brown, with a creamy consistency. At this point, the *beijus* are ready for use, but they can be dried and stored for long periods.

FIGURE 10.3   Storage of mouldy *beijus* produced in a traditional company in the Brazilian state of Maranhão. Source: the authors

The hydrolysis phase of the starch is time-consuming and uncertain because it depends on colonisation by the correct moulds, which is well documented. For Gonçaves de Lima (1974), the black colour of the fruiting fungi in *beijus* is charac- teristic of the *Aspergillus* genus, which is capable of hydrolysing starch to glucose.

The moulds that develop on *beijus* have been the subjects of several studies. Park et al. (1982) researched the natural flora of *beijus* that were collected in three *tiquira* manufacturing units in Maranhão state (Brazil). These authors isolated and identi- fied moulds and also measured the α-amylase and amyloglucosidase activity. The results showed that the samples of *Paecilomyces* sp. showed the highest activity for both enzymes, while the lowest values were for *Penicillium* sp. and *Neurospora* sp. Although they presented low values for α-amylase activity, *Aspergillus niger* and *Rhizopus* sp. showed higher values for amyloglucosidase, which is an enzyme that produces glucose. The populations were equivalent in number.

These results show that there may be variation between the microorganisms that develop on cassava mass, which is to be expected in relation to spontaneous pro- cesses that do not use inoculants. In the opinion of Gonçalves de Lima (1943), the agent responsible for starch saccharification was *Neurospora crassa* mould, which was isolated from *beijus* and characterised by the author. In southeastern Brazil, Chuzel and Cereda (1995) isolated strains from *beijus* wrapped in cassava leaves. In addition to *Aspergillus niger* and *Penicillium* spp., both of which have been cited in the literature, these authors identified pink-colored *Manila sitophila*, which showed strong amylolytic activity. The authors also mentioned the pleasant aroma of ripe fruit that was produced.

Regarding the formation of aromas, moulds of the genus *Neurospora*, which are among the many that were isolated from naturally fermented cassava by Park et al. (1982), have a pleasant fruity aroma that is attributed to ethyl hexanoate. Although they are important, the aforementioned results should be viewed with caution because there was no evaluation of bacteria with hydrolytic activity in relation to starch, and the presence of bacteria in the conditions in which the beijus were incubated cannot be ruled out.

When they are stored, *beijus* lose part of their moisture, and mycelia also colonise the interior, as well as the surface. After about 10–12 days, the starch reaction with iodine was negative, indicating that there was no longer any starch present in the product. This does not mean that all the starch had been hydrolysed to fermentable sugars because this type of negative reaction can occur in the presence of dextrins, which are poorly fermented by yeasts (Chuzel and Cereda 1995).

The *beijus* with fungi are crumbled in water. According to Gonçalves de Lima (1943), if the filtrate is not used quickly after 2–3 days it ferments and gives rise to an alcoholic beverage, which when distilled results in the spirit-like beverage called *tiquira*. Chuzel and Cereda (1995) have observed that disintegrated *beijus* look like thick syrup, with °Brix of 14–15. The following day all the sieved suspension is allowed to ferment naturally by native yeasts for at least eight days. The *Saccharomyces* genus was identified among these wild yeasts.

In Brazil, of all the fermented beverages derived from cassava only *tiquira* has specific legislation. The minimum alcohol content (Brasil, 2009) is 38° INPI, which corresponds to the percentage of alcohol by mass expressed as absolute alcohol. Distillation of *tiquira* may maintain the sensory characteristics similar to the natural volatile components contained in the fermented must, or formed during distillation. The level of potentially toxic compounds such as methanol, carbamates, some metal ions and cyanide must be within established limits.

Although, it is generally described as an alcoholic beverage made by Amerindians, *tiquira* has become well-known in the Brazilian state of Maranhão, where many small industries produce it. However, due to its artisan method of production, *tiquira* cannot compete with the low prices of sugarcane spirit produced in the southeast of Brazil.

Innovations proposed in the manufacture of *tiquira*

The flowchart in Fig. 10.4 shows the stages for the preparation of *tiquira* by the traditional process (solid lines), which is part of the processing of cassava flour and which is the basis of the processing of *tiquira*. The proposed innovations (dashed lines) are adaptations to produce *tiquira* in less time without losing the characteristics of the traditional product.

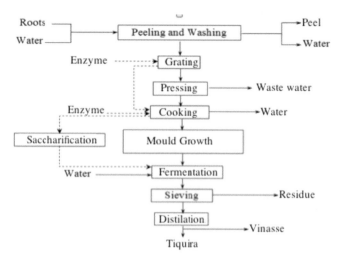

FIGURE 10.4   Flowchart showing the production of *tiquira* by the traditional process (solid lines) and innovations (dashed lines). Source: Adapted from Venturini Filho and Mendes (2003).

The traditional production of *tiquira* is time-consuming and provides low income. To change this pattern improvements have been proposed, with technology adapted for rural communities but without changing the features of this beverage.

The first innovation is to eliminate the peeling of the roots, which are only washed and grated. The second innovation eliminated the need for pressing. These modifications of the traditional process allow the use of about 2% of sugars (sucrose, glucose and fructose) of the constituent water in the cassava roots and with this, avoid a liquid waste. The consistency of the grated mass is favourably changed. To adjust the initial content of sugars to the needs of yeasts, it is necessary to add more water than that afforded by the natural moisture of the cassava roots (average 60%) and thus the consistency of the mass changes from friable to pasty.

To produce ethanol from starch, three steps are necessary: the gelatinisation of the starch, with subsequent saccharification to form sugars and alcoholic fermentation, and distillation. The third innovation is to eliminate the *beijus* and use of cassava flour oven. The gelatinisation of the starch starts with cooking on a direct fire, with temperatures above 60 °C. With gelatinisation, the granules lose their crystallinity, their structure opens; the spiral chains stretch and viscosity increases, making the action of enzymes easier. With elimination of *beijus* it is possible to obtain a viscous liquid that allow the gelatinisation to be performed in closed containers.

The greatest challenge is to replace the tradition of hydrolysis carried out by native moulds. The level of expertise of the producers is not compatible with biotechnological techniques. Additionally, these small businesses are often far from urban centres. Consequently, the use of enzymes was assessed in the laboratory. This showed that the traditional saccharification stage could be favourably replaced by using commercial enzymes. The selection of the enzymes was made bearing in mind the results of research by Surmely et al. (2003) regarding commercial enzymes available in Brazil for the saccharification of cassava starch.

Cereda and Costa (2008) adapted the laboratory procedures for field using droppers to measure the amounts of enzymes, but emphasized the need to use simple equipment to properly measure the alcohol content.

The fourth innovation is the replacement of fungi by commercial enzymes. This ensures the action of two classes of amylases, so that the hydrolysis of the starch can be complete, thereby increasing the yield of fermentable sugars. A standard *beiju* weight is about one kilogram, which contains 200–300 g of starch. To reduce the viscosity of the starch, Thermamyl 120L (3 mL of enzyme per kg starch) was selected. This enzyme is temperature-resistant and therefore hidrolize occurs with gelatinization. This stage does not require an adjustment of the natural pH of cassava or the addition of calcium. The temperature is maintained at 95 °C for one hour and then the suspension is allowed to cool naturally. The pH is adjusted to between 4.00 and 4.50 for the addition of the second enzyme.

Saccharification, which is the second phase of starch hydrolysis, was performed with AMG 300L (2 ml of enzyme per kg of starch) to convert the starch and dextrins to glucose. The temperature was maintained at 60 °C for 2 h. After the action of the enzymes was completed, the grated cassava mass looked like a thin syrup. According to Surmely et al. (2003), this syrup contains oligosaccharides, with 40% dextrins, which are fermentable to lesser degree by the yeasts, followed by glucose and maltose.

The fifth innovation is the replacement of native yeasts by commercial yeast, either in pressed or dried form. Once the mashing of the suspension is completed, the suspension of sugars should be diluted with potable water to 12–14 °Brix, which are values suitable for the fermentation process to occur. Then the must should be boiled to deactivate the enzymes, which greatly reduces contamination and ensures a vigorous, quick fermentation that provides a high yield. As there is no commercial yeast available for the production of alcohol from cassava it is recommended to use the dried yeasts that are used to produce sugar cane spirit at a dose of 1% of must.

The other operations follow the traditional procedures for the production of *tiquira*, where distillation is the most delicate point in the process because of the volatile compounds that are produced during the fermentation process and turn into alcohol. The natural aroma of cassava contains traces of cyanide, which has the smell of bitter almonds.

Copper, or copper with stainless steel equipment was substituted by ceramic stills. The filtering of the must before distillation prevents the deposit burning the bottom of the still, unless the still has a double wall which allows steam to circulate in between. In this case the temperature can be increased and it is possible to reduce the distillation time. Filtering before distillation preserves the aseptic conditions guaranteed by the final boiling.

The distillation of *tiquira* is performed in batches using a simple still. Regardless of the type of heating used to heat the still, distillation must be conducted in a measured manner, without haste. It is necessary to identify and separate the head, heart and tail fractions, which must be added to the must at the next distillation. The best quality part of the distillation is the heart, which is low in impurities, and this is the fraction known as *tiquira* (Venturini Filho and Mendes 2003).

When the process is conducted well, from 100 L of fermented must it is possible to distill 15–20 L of *tiquira* with a minimal alcohol content of 38 INPI and a maximum of 56 INPI, as cited by Venturini Filho and Mendes (2003). The distillate has a pleasant odour and a neutral taste.

There are stills of different capacities, but they must use copper, at least in the ascending part of the still, in order to prevent a cabbage-like smell, which is a characteristic of dimethylsulphide when in higher levels than 8.10^{-7} mol L^{-1} (Polastro et al. 2001). The still should be washed after each distillation to remove the waste from the must that is distilled. The final residue from the distillation is vinasse and the sludge containing the peel, fibres and sediments from the yeasts, and it can be used in animal feed.

Using these innovations means that the entire production process can be carried out within 24 h, not including the aging time, which affects the final value of the spirit and makes it more competitive in the market. These innovations were researched, tested and passed on to small producers. In order to ensure the sustainability of *tiquira* production, production costs must be reduced and the quality needs to be monitored (Cereda and Carneiro 2008).

Furtado et al. (2007) discuss the need for the quality control of *tiquira* and similar beverages in relation to copper and cyanide residues. Brito and Cereda (2009) evaluated the total cyanide content during the processing stages of *tiquira*, after included the above-mentioned innovations. The cyanide content of the cassava roots used in the preparation of *tiquira* was 55.6 mg L^{-1}. After using enzyme Thermamyl this content was reduced to 8.82 mg CN L^{-1} and was reduced further to dropped to 5.41 mg kg^{-1} when the saccharification enzyme AMG was used. After 24 h of fermentation, when the must was distilled in a copper still, the *tiquira* had only 1.98 mg L^{-1} cyanide content. After the bi-distillation stage, which proceeds the aging period, the levels were 0.4 mg L^{-1}. With this reduction, it can be considered that there is no risk of intoxication by the remaining cyanide ions.

Although there is no direct risk of intoxication from the cyanide radical, there are reports that this radical may still be involved in other chemical reactions, particularly in the formation of carbamate. According to Polastro et al. (2001), *tiquira* has average concentrations of amino acids greater than sugar cane spirit, which can lead to the formation of carbamates. According to Cardoso et al. (2003), these carbamates are potentially carcinogenic, which is why it is very important to establish the level of their occurrence in distillates. In addition to issues related to public health, the presence of carbamates in concentrations greater than 0.150 mg L^{-1} also represents a barrier to exports to Europe and North America. The *tiquira* samples analysed by the authors presented an average level of ethyl carbamate of 2.4 mg L^{-1}. This level is above the legally permitted limit, which is worrying and requires investigation.

Polvilho azedo — Fermented Cassava Starch

The main fermented product obtained from cassava starch is *polvilho azedo*. The main attributes of *polvilho azedo* are its characteristic flavour and its ability to expand in the oven without the use of biological or chemical leavening agents. This property occurs after natural fermentation and exposure to the sun.

Starch as a substrate

Brazil is a traditional producer of cassava starch, with 516,000 tonnes per year produced by about 20,000 small and medium-sized industries, as well as about 69 large-scale industries. *Polvilho azedo* is well known in Brazil and throughout Latin

America. In Brazil, it is used to make a very light salty cracker and *pão de queijo* (cheese bread), a round shape baked product with spongy texture, as in bread made from wheat. In Colombia, *polvilho azedo* is known as *almidón agrio*. Cardenas and De Buckle (1980) described the production of *almidón agrio* using natural fermentation. *Almidón agrio* is also used in baked goods, where it is irreplaceable, such as *pan de bono, buñuelo, pan de queso* and *besitos*. In Paraguay *almidón agrio* is used to make *chipas*, similar to *pão de queijo*. In the Brazilian state of Mato Grosso do Sul, which adjoins Paraguay, both *pão de queijo* and *chipas* are commonly consumed. The production of fermented starch by companies in Bolivia is similar to that in Paraguay and Brazil and it is used to prepare *cuñape*, which has a similar formulation to the aforementioned products in Brazil, Paraguay and Colombia (Chuzel and Cereda 1995). The origin of this technology is unknown but Guarani people culture has spread on all these countries.

Polvilho azedo was traditionally extracted from cassava roots by an artisanal process and then fermented. It was used domestically or in small bakeries, but *pão de queijo,* its main by-product, became so popular that it became a type of fast food in Brazil. Cheese bread is a salty product; it is baked in the oven and can be consumed with or without various fillings, salty or sweet either savoury.. Due to its popularity its consumption expanded throughout Brazil. This expansion required increased production, which was transferred to the same industries that produce cassava starch, where it is prepared by hand in a parallel and less technical process (Cereda et al. 2001).

The ability of *polvilho azedo* to expand in the oven without the use of leavening agents of microbial origin (yeasts) or chemical origin (baking soda) is unique. This characteristic of expansion has attracted international researchers, not only because the preparation of snacks with *polvilho azedo* does not require extrusion, which is expensive, but also because it is possible to produce baked goods without gluten. Its expansion corresponds to a volume of 10 $cm^3.g^{-1}$ (Brito and Cereda 2015).

Traditional production of *polvilho azedo*

The production of *polvilho azedo* uses freshly-extracted cassava starch. The starch can be extracted using sophisticated equipment in factories with a minimum daily capacity of 400 tonnes of roots, or in smaller units. The production of *polvilho azedo* follows the traditional method with fermentation and drying in the sun. The starch suspension in water is discharged directly into the fermentation tanks. The water must have good microbial and chemical quality, because if it is rich in iron it can react with cyanide and give a dark blue colour to the *polvilho azedo*. The production of *polvilho azedo* from starch does not generate waste, only a solid by-product, which is generally marketed as adhesive or glue.

Fermentation

The microbial action that produces *polvilho azedo* occurs in granular starch, which is not directly fermentable by microorganisms. The purified starch suspension in water (30%) is transferred to fermentation tanks where it is decanted and forms a layer under the surface of the water. The tanks are either made of wood or masonry, and in the latter case it is common to coat them with black plastic to avoid the acidity releasing sand into *polvilho azedo*.

The starch should remain in the fermentation tanks under a layer of water of 20 cm deep at the outset and dries as time passes. The time needed for complete fermentation varies from 3 to 20 days (Cereda et al. 2001). However, in the traditional producing regions of Minas Gerais state (Brazil), fermentation takes 30–40 days, and up to 60 days at the beginning of the harvest (Cereda and Giaj-Levra 1987).

Most *polvilho azedo* producers did not use inoculum to ensure or accelerate the fermentation process. Fermentation always produces visible evidence within a few days, with the formation of bubbles and froth on the surface, as reported by Leme-Júnior (1967). It is more difficult to establish when fermentation ends. Gas bubbles also appear in the mass of deposited starch; it can be observed when the process occurs in a glass container. The fermentation is characterised by a lowering in pH and a concomitant production of short chain organic acids (Cereda et al. 2001; Chuzel and Cereda 1995).

According to Chuzel and Cereda (1995) the pH in the fermenting starch mass falls from 5.0–6.0 to values of 3.0–3.5 and even 2.5, and this acidity may probably inhibit the natural fermentation process. The fermented starch has different attributes compared with unfermented starch. From being insipid, it begins to show its characteristic flavour. In producing regions it is possible to identify the odour of butyric acid (in warm regions) or lactic acid (in cold regions). The aroma of mature pineapple characterises the *Geotrichum* sp., a yeast-like fungus that grows in the supernatant liquid in the tanks. Fermentation changes the consistency of the starch mass, making it soft and crumbly, like a cheese. Cereda et al. (2001) reported on the fermentation of native cassava starch in the laboratory and established variations in pH, titratable acidity, sugars and organic acids, and related these values to the isolation, enumeration and identification of the microflora that were present.

It is difficult to explain how such a vigorous fermentation can result from a culture medium that is so poor, because the nutrients of the roots are only few. The nitrogenous compounds and vitamins are lost to the water in the starch purification process. The substrate is then restricted to a granular starch suspension in water. However, Chuzel and Cereda (1995) reported an abundant microflora in this liquid from the second day of fermentation. Commercial native starch contains an average of 23×10^3 aerobic and 7.5×10^5 mesophilic aerobic spores per 100 g dry mass, which are sufficient microorganisms to be used as inoculums, as demonstrated in fermentation under aseptic conditions.

Cereda and Lima (1985) found that even in a fermentation performed in an open environment, a decrease occurred in the tension of the oxygen in the supernatant water of the fermentation, which provided microaerobic conditions even in the first days. When fermentations occurred in a closed environment in the laboratory, it was possible to collect the gases that were formed, which were later identified and measured. The initial concentration of O_2 was equivalent to the atmosphere, but as the fermentation progressed there was a decrease of oxygen between the first and third days, rising after the third day to keep in equilibrium with the oxygen content of the air for the remaining period of fermentation.

The final phase of the greatest reduction in the oxygen content coincides with the beginning of the most turbulent phase of the fermentation of the starch, when bubbles are formed inside the deposited starch mass and migrate to the surface, forming a thin layer.

Cereda et al. (2001) reported that fermentation can be divided into well distinguished phases. In the first phase, an undemanding microflora develops, consisting mostly of coliforms and other aerobic mesophilic microorganisms. In the second phase, more demanding microorganisms develop, which are identified as producers of organic acids, many of which are microaerophilic or anaerobic. The third phase is characterised by the presence of yeasts and saprophytic microorganisms. The first phase coincides with a sharp drop in the pH of the supernatant liquid, which stabilises after the second or third day at around pH 3.0. Cardenas and De Buckle (1980) obtained a similar profile in Colombian conditions, where the pH dropped from 6.5 to 3.5 and remained stable until the end. Cereda et al. (2001) observed that titratable acidity showed oscillations until the end of the process, while the pH remained steady, and concluded that pH 3.0 is probably the limit for this type of fermentation.

Carvalho (1994) counted and identified microbial flora in industrial conditions. The results confirmed that the fermentation had already started in the decanting stage in the tank, so the native starch used in small-scale industries can be considered to be semi-fermented. There was no change in the number of microorganisms, but there was a change in the groups that were evaluated. Even when the water that was used was of poor quality, with the presence of total and faecal coliforms, the production process was efficient enough to eliminate them. Carvalho's study also confirmed the presence of microorganisms with the ability to hydrolyse starch, including the yeast, lactic bacteria and *Bacillus* sp. It was not possible to prove the existence of viable lactic acid in *polvilho azedo* that had been dried in the sun, but there was survival only in the genus *Bacillus*.

Differences were observed in the content and composition of acids measured in samples of commercial *polvilho azedo* originating in the producing regions of Minas Gerais, Paraná and São Paulo states (Brazil). These differences could be explained by variations in the microorganisms of the second phase of fermentation due to environmental conditions, especially temperature. The results seem to indicate that *Clostridium butyricum*, which was frequently found, might be one of the key agents in the second phase of fermentation carried out at 30 °C. At temperatures lower than this LAB flora predominates, as noted by Cardenas and De Buckle (1980) in their study, which isolated and identified microorganisms that occur in natural fermentations of Colombian *polvilho azedo*. The latter authors reported that at a temperature of 15 to 25 °C there was a predominance of microaerophilic microorganisms from the lactic group, of which *Lactobacillus plantarum* was the most commonly found bacteria. It was also found that *Lactobacillus casei* and the yeasts *Saccharomyces* spp. and *Geotrichum* spp., as well as Gram-positive sporulating bacteria were present. The fermentation of Colombian cassava starch also confirmed that the microbial flora was composed, in part, by *Lactobacillus* spp., which produces acids that attack starch granules.

Brabet et al. (1994) also found a predominance of microflora producing lactic acid in fermented cassava starch in Colombia; they also found lactic acid to be the main product, as it contributed to the organoleptic characteristics of *almidon agrio*.

The presence of yeasts was also reported by Cereda et al. (2001) in the third stage of fermentation, in which saprophytic microorganisms and contaminants predominated. In addition to consuming the organic acids on the surface of the tanks, these microorganisms may be responsible for the formation of aromatic compounds, which, together with other organic compounds, are responsible for the characteristic flavour of commercial *polvilho azedo*.

According to Cereda et al. (2001), the presence of microorganisms in the first phase of fermentation might be associated with a rapid fall in the concentration of dissolved O_2. The *Escherichia, Alcaligenes, Micrococcus* and *Pseudomonas* genera are capable of consuming oxygen, producing gases (CO_2 and H_2) and organic acids. The predominance of *Bacillus subtilis* was also detected in the first phase. The production of amylolytic enzymes by *Bacillus subtilis* has been highlighted in the literature. It is probable that in this stage the enzymes begin to attack the starch granules, which provides a carbon source for the metabolism of the agents of fermentation.

This hypothesis has been proved through the chromatographic identification of the sugars present in the supernatant liquid throughout fermentation (Cereda et al. 1982). In the aforementioned study, glucose (Gl) was only detected in the first days of fermentation and maltotetroses (G3) was detected on the other days, until the 30th day. This indicated that the sugars, which were produced, were rapidly being consumed and metabolised, mainly in the formation of organic acids, predominantly acetic, butyric and lactic. In a study in Colombia, Cardenas and De Buckle (1980) found a predominance of lactic acid (66% to 82% of the total), followed by mixtures of acetic and butyric acid.

Organic acids, and modifications in the starch that are produced during the fermentation process, impart characteristics of flavour, texture and volume to baked products (Cardenas and De Buckle 1980, Brito and Cereda 2015).

The hypothesis regarding the possible action of amylase on starch granules to release sugars was reinforced by the results of a study by Brabet et al. (1994), in which native starch was compared with *almidón agrio* as well as starch that had only been acidified (20 days at 37 °C) with the organic acids most commonly found in the analysed samples. Under polarised light, the granules of fermented starch showed a partial loss of birefringence and a marked tendency to form aggregates; this characteristic was also reported by Cereda et al. (2001) in samples of commercial *polvilho azedo* in Brazil. Visible evidence of this was found in changes in the surface of the granules, which were smooth and homogeneous in the native starch granules, while coarse and with some slight evidence of perforations in the granules after fermentation (Brito and Cereda 2015). The starch granules that had been treated with acids had a similar smooth surface to native starch. In the opinion of Cardenas and De Buckle (1980), this fact proves that as well as the corrosive action of acids, there is evidence of an enzymatic attack.

That there is a carbon source available, as detected by Cereda et al. (2001) and Cardenas and De Buckle (1980), but it still remains to clarify the source of nitrogen. Cereda et al. (1985) determined the composition of the gases that were given off by revealing the presence of hydrogen, nitrogen, oxygen, argon and carbon dioxide. At first the percentage composition of the gases proved to be very close to the composition of air, and hydrogen was not detected. From the beginning of the visible stage of fermentation a gradual increase in hydrogen and carbon dioxide content occurred. Between the second and fourth days, and corresponding to an intense production of gases, there was an increase in CO_2 as well as nitrogen and oxygen consumption. It appears that the N_2 in the atmosphere within the enclosed system in fermentation was consumed in certain phases, at the expense of the total composition. Thus, the nitrogen required for biomass formation in the early stages of fermentation would have originated from the atmosphere since, according to Chuzel and Cereda (1995) protein content available in native starch is quite low, about 0.15 g per 100 g.

In order to explain this fact, fermentation experiments were performed in a closed system, with counts of microorganisms that were able to grow in a medium lacking a source of nitrogen. The existence of non-symbiotic nitrogen-fixing microorganisms in the starch fermentation was then demonstrated. They were detected from the first day, reaching maximum values after three and four days of fermentation and decreasing thereafter when conditions become adverse. As a result, there was a rising curve of total nitrogen in the fermentation, which reached a maximum on the seventh day and became asymptotic from this point onwards. Although they were not identified, there is a possibility that bacteria of the genus *Bacillus*, such as *Bacillus polymixa, Bacillus macerans, Bacillus circulans, Bacillus cereus* and *Bacillus licheniformis* were present, in addition to the genus *Clostridium*. Bachaman and Gibbons (1975) cite *Clostridium butyricum* as a fixer of atmospheric N_2, and according to Cereda et al. (2001), this occurred in all the laboratory fermentations, as well as 37% of the industrial fermentations that were analysed.

The consumption of oxygen provides conditions for the development of the microorganisms of the second phase, which are microaerophilic, facultative anaerobic, or strictly anaerobes. In this phase, the more demanding microorganisms predominate, which are the producers of organic acids and gas. Groups have been identified which are responsible for butyric, lactic, acetic and propionic fermentations, either separately or concurrently. The predominance of a certain organic acid could therefore be related to the predominance of favourable conditions for certain groups, principally the temperature in the producing region.

Drying or treatment by ultraviolet radiation?

After the fermentation has finished, the surfaces of the tanks are allowed to dry. At this stage the *polvilho azedo* has a humidity of around 30–50%, with the consistency and appearance of cheese. The drying operation starts at dawn, with the spreading of the fermented cassava mass.

Drying is always performed in the sun, in a process that can limit production, but which provides the complementary treatment for ultraviolet action, which confers expansion properties to the *polvilho azedo* (Nunes and Cereda 1994; Brito and Cereda 2015). Artificial drying can only be used as a supplement to drying by sunlight because it does not provide these expansion properties. Drying and UV treatment is usually performed on surfaces such as white cloth or black plastic sheets. According to Plata-Olviedo and Camargo (1998), the maximum effect of the sun was detected after four hours of exposure. Starches, which were previously treated with lactic acid and exposed to the sun or ultraviolet radiation, presented the characteristic of expansion, but with varying degrees of intensity (Nunes and Cereda 1994; Brito and Cereda 2015).

The fermentative process and solar UV radiation shows alteration in the starch granules when compared to untreated granules. The changes that occur during fermentation are such as to change their rheology (Cardenas and De Buckle 1980; Brito and Cereda 2015), so that the rapid visco analyser (RVA) curve begins to show less high peaks that untreated starch shows at the same concentrations.

Polvilho azedo has characteristics that differentiate it from native cassava starch, in terms of its chemical composition, viscographic parameters and also the size of the granules. A decrease in the average diameter of the granules of *polvilho azedo* has been noted, which is reflected in a corresponding reduction in density. The major differences between *polvilho azedo* compared to native cassava starch are lower starch content and higher protein content (crude nitrogen x 6.25). As a result of the fermentative process, acids are formed, which results in higher acidity and lower pH. The profiles of short-chain fatty acids predominate in *polvilho azedo* samples and the literature reports of presence of lactic, acetic and butyric acids. The question is whether it was the effect of these acids or solar radiation that precipitated these changes.

Cardenas and De Buckle (1980) mention that even if there are few differences in appearance, dimensions and gelatinisation temperature, the gelatinisation temperature of *polvilho azedo* is always less than that of unfermented starch. The aforementioned authors also observed an increase in the number of reducing radicals of native starch (1.2) in relation to *polvilho azedo* (8.2), which was probably due to the exposure of the reducing groups to the starch chains by the action of enzymes and acidity. This break in the molecules was confirmed by a reduction in the average molecular weight in the fermented starch samples (30,000) relative to the unfermented samples (215,000); the starch treated with acids showed intermediate values (136,000). The changes caused by the immersion in acids, even after 20 days at 23 °C, were not sufficient to cause an impact similar to that of exposure to the sun.

Complementing this hypothesis, the analysis of more than thirty samples of *polvilho azedo* made it possible to identify changes in the amylose and amylopectin content, which showed higher amylose content than untreated cassava starch. According to the literature, attack by acids and enzymes occurs particularly on the amorphous part of the granule, where the amylopectin is located which is hydrolysed during fermentation (Cereda and Cataneo 1986). Cereda et al. (2001) analysed 25 samples of commercial *polvilho azedo* and concluded that another advantage of the fermentation process of *polvilho azedo* is an increase in protein, which is about 10 times that of non-fermented starch.

Innovations to the production of *polvilho azedo*

The major problem encountered in keeping *polvilho azedo* competitive in the market is the lack of standardisation regarding its expansion properties, which is related to sun exposure. The dependence on the sun requires very large drying areas, where the product is exposed to weather and dirt. In addition, there is a need for periodic turnings, which are performed manually.

The *polvilho azedo* leaves the tank for drying with about 50% humidity. A tank produces about 2 tonnes of wet starch, or a tonne of dry starch. The load used in the drying surface (*jirau*) ranges from 1.0 to 1.5 kg. m². Consequently, in order to dry the equivalent of a tank an area of 1000–1500 square metres is required. In using drying surface, the width of 1.5 metres has been adopted as being appropriate for manual turning, which provides 1.5 m² per meter. The fermented starch mass in one tank would therefore require a drying area of 1000 linear metres. Therefore, in order to increase production it is necessary to increase the drying area, as it depends on the sun. In case of rain it is necessary to collect the entire amount of starch.

Despite the amount of research conducted on the subject of *polvilho azedo*, which has even reached the international level, maintaining a standard of quality is still a challenge. The standards related to food and beverages set out the specifications for starch, which is the raw material for the preparation of *polvilho azedo*, but they do not deal with its characteristic of expansion in the oven without leavening agents (Brito and Cereda 2015). *Polvilho azedo* has a typical granulation and because it is obtained by natural fermentation it has suffered from a lack of uniformity even when it originates from the same source. It is only by improving manufacturing conditions that it will be possible to arrive at a more standardised product.

Bearing in mind the changes caused by fermentation and exposure to the sun, it is noteworthy that lactic acid is always present in commercial samples of *polvilho azedo*. This leads to the hypothesis that it is the reaction of lactic acid and ultraviolet radiation from the sun, which is responsible for the formation of a network that is capable of retaining the gases caused by expansion in the oven (Brito and Cereda 2015), which contradicts reports in the literature that point to the exopolysaccharides (EPS) that are excreted by LAB as the cause (Brabet et al. 1994). This hypothesis was proven (Nunes and Cereda 1994) and it was shown that the process is a natural photochemical reaction, which is induced by ultraviolet radiation on starch that has been pre-treated with lactic acid.

Research conducted by Wang et al. (2012) found that the reaction between starch, lactic acid and gamma radiation arises from the opening of the glucose ring, which forms a vicinal bond by which the outer carbononila is linked to the carbon by a single hydrogen atom. As a consequence of this, more oxygen is added, to form the bond between the amylose or amylopectin chain and the grafted radical. The aforementioned authors also point out that in this reaction, a starch ester is first formed, which later evolves into an ether grouping of starch. This hypothesis has been confirmed for *polvilho azedo* and has led to an application for a patent (Brito and Cereda 2015).

This patent has been developed as an innovation of the process that has been studied, which will allow treatment with ultraviolet radiation in a continuous process. The results that have been obtained show that the reaction with artificial ultraviolet radiation is instantaneous and that its industrial application only depends on the adaptation of the laboratory process to suitable equipment.

Conclusions

The fermented cassava products selected for this study are marketable; they are commercial products and they should continue to be competitive and the subject of research. For this to happen it is necessary not only to research and to clarify by research the basis for their methods of production, but also to develop processes that are compatible with ground reality, and competitive to production companies, and mainly also to develop products compatible with modern life. There are several other similar foods and drinks that will remain unknown unless they receive the exposure that has been granted to *tiquira* and *polvilho azedo*.

References

Bourdoux, P., P. Seghers, M. Mafuta, J. Vanderpas, M. Vanderpas Rivera, F. Delange and M.A. Ermans. (1983). Traditional cassava detoxification process and nutrition education in Zaire. In: F. Delange and R. Aklowalia (eds.), Cassava toxicity and thyroid: research and public health issues. Ottawa, London.

Brabet, C., G. Chuzel, D. Dufour, M. Raimbault and I. Guiraud. (1994). Study of natural fermentation of cassava starch in Colombia. I: characterization of the microflora and the fermentation parameters. In: International meeting on cassava flour and starch. Cirad, Montpellier.

Brasil, O.G., M.P. Cereda and A.M.C. Fioretto. (1982). Indução da respiração resistente ao cianeto em microorganismos pela presença de inibidores de crescimento microbiano. Phiton 42: 49–53.

Brasil. Decreto n. 6.871, de 4 de junho de 2009 que regulamenta a Lei n. 8.918, de 14 de julho de 1994, que dispõe sobre a padronização, a classificação, o registro, a inspeção, a produção e a fiscalização de bebidas. Diário Oficial da União de 20/09/2012. Seção 1. p.4.

Brito, V.H.S. and M.P. Cereda. (2015). Method for the determination of specific volume as a quality standard for fermented cassava starch and substitutes. Brazilian Journal of Food Technology 18: 14–22.

Brito, V.H.S., A.P.M. Rabacow and M.P. Cereda. (2013). Classification of nine month-old cassava cultivars by cyanide levels. Gene Conserve 12: 35–49.

Brito, V.H.S. and M.P. Cereda. (2015), Produção de amido modificado (sucedâneo de polvilho azedo) via tratamento fotoquímico sobre suspensão de amido nativo de qualquer fonte botânica e em qualquer tipo de líquido usando tratamento com radiação ultravioleta artificial (100-400 nm) em reator modular, com adição ou não de reativos ou componentes químicos; B.R. Patent Process # 10 2014 027689 0.

Cereda, M.P., O.F. Vilpoux, O.L.G. Nunes and G. Chuzel. (2001). Polvilho azedo. *In*: Walter Borzani (eds), Biotecnologia Industrial. Blucher, São Paulo, pp.

Cardenas, O.S. and T.S. De Buckle. (1980). Sour cassava starch production: a preliminary study. Journal of Food Science 45: 1509–1528.

Cardoso, D.R., L.G. Andrade-Sobrinho, B.S. Lima-Neto and D.W. Franco. (2003). A rapid and sensitive method for dimethylsulphide analysis in Brazilian sugar cane sugar spirits and other distilled beverages. Journal of the Brazilian Chemical Society 15: 277–281.

Carvalho, E.P. 1994. Determinação da microbiota do polvilho azedo. Ph.D. Thesis. Universidade Estadual de Campinas, Campinas.

Cereda M.P. and M.S.C. Carneiro (2008). Manual de fabricação de Tiquira (Aguardente de Mandioca), por processo tradicional e moderno: Tecnologias e custos de produção. Embrapa Mandioca e Fruticultura Tropical, Cruz das Almas.

Cereda, M.P. (1983). Determinação de viscosidade de fécula fermentada de mandioca (polvilho azedo). Boletim da Sociedade Brasileira de Tecnologia de Alimentos 17: 15-24.

Cereda, M.P. and L.A. Giaj-Levra. (1987). Constatação de bactérias não simbióticas fixadoras de nitrogênio em fermentação natural de fécula de mandioca. Revista Brasileira de Mandioca 6: 29–33.

Cereda, M.P. (1987). Tecnologia e qualidade do polvilho azedo. Informativo Agropecuário 13: 63-68.

Cereda, M.P. and A. Cataneo. (1986). Avaliação de parâmetros de qualidade da fécula fermentada de mandioca. Revista Brasileira de Mandioca 5:55-62.

Cereda, M.P. and U.A. Lima. (1985). Aspectos da fermentação da fécula de mandioca. III-Determinação dos ácidos orgânicos. Turrialba 35: 19–24.

Cereda, M.P., L.A. Bonassi, O.G. Brasil and E. Matsui. (1985). Ensaios de fermentação da fécula de mandioca em diferentes condições de cultivo. Revista Brasileira Mandioca 3:69-81.

Cereda, M.P., O.F. Vilpoux and I.M. Demiate. (2003). Amidos modificados. Tecnologia, uso e potencialidades de tuberosas amiláceas latinoamericanas. Fundação Cargill, São Paulo.

Cereda, M.P., O.F. Vilpoux, O.L.G. Nunes and G. Chuzel. (2001). Polvilho Azedo. In: Walter Borzani (ed.), Biotecnologia Industrial. Blucher, São Paulo.

Chisté, R.C., K.O. Cohen and S.S. Oliveira. (2007). Estudo das propriedades físico-químicas do tucupi. Ciência e Tecnologia de Alimentos 27: 437–440.

Chuzel, G., and M.P. Cereda. (1995). La tiquira: une boisson fermentée à base du manioc. In: E.T. Agbor, A. Brauman, D. Griffon, S. Trèche (eds.), Transformation alimentaire du manioc. Orstom, Paris.

Chuzel, G. and M.P. Cereda and D. Griffon. (1995). L'amidonaigre de manioc. In: E.T. Agbor, A. Brauman, D. Griffon, S. Trèche (eds.),Transformation Alimentaire du Manioc. Orstom, Paris.

Conn, E.E. (1969). Cyanogenic glycosides. Journal Agriculture and Food Chemistry 17:519-526.

Corradini, E., C. Lotti, E.S. Medeiros, A.J.F. Carvalho, A.A.S. Curvelo and L.H.C. Mattoso. (2005). Estudo comparativo de amidos termoplásticos derivados do milho com diferentes teores de amilose. Polímeros: Ciência e Tecnologia 15: 268–273.

Coursey, D.C. (1973). Cassava as food: toxicity and technology. In: B. Nestel and R. MacIntyre (eds.), Chronic cassava toxicity. Proceedings of Interdisciplinary Workshop. Ottawa, London, pp. 2736.

[FAO] Food and Agriculture Organization. (1998). Storage and Processing of Roots and Tubers in the Tropics. D.J.B. Calverley (eds), Rome, pp.1–50.

Furtado, J.L.B., C.W.B. Bezerra, E.P. Marques and A.L.B. Marques. (2007). Cianeto em tiquiras: riscos e metodologia analítica. Ciência e Tecnologia de Alimentos 27: 694–700.

Jones, W.O. (1959). Manioc in Africa. Stanford University Press, Stanford, California.

Lancaster, P.N., J.S. Ingram, H.J. Lin and D.G. Coursey. (1982). Traditional cassava based foods, survey of processing techniques. Economic Botany 36: 12–25.

Leme-Júnior, J. (1967). Amideria e fecularia. *In*: Enciclopédia Delta-Larousse. Delta, São Paulo.

Nunes, O.L.G.S. and M.P. Cereda. (1994). Effect of drying processes on the developement of expansion in cassava starch hydrolyzed by lactic acid. *In*: International Meeting on Cassava Flour and Starch, CIAT, Cali, Colombia.

Park, Y.K., C. Zenin, S. Tueda, C.O. Martins and J.P. Martins-Neto. (1982). Microflora in beiju and their biochemical characteristics. Journal of Fermentation Technology 60: 1–4.

Plata-Oviedo, M. and C. Camargo. (1998). Effect of acid treatments and drying processes on physico-chemical and functional properties of cassava starch. Journal of Science Food Agriculture 77: 103–108.

Polastro, L.R., L.M. Boso, L.G. Andrade-Sobrinho, B.S. Lima-Neto and D.W. Franco. (2001). Compostos nitrogenados em bebidas destiladas: cachaça e tiquira. Ciência e Tecnologia de Alimentos 21: 78–81.

Surmely, R., H. Alvarez, M.P. Cereda and O. Vilpoux. (2003). Hidrólise do amido. *In*: M.P. Cereda and O.F. Vilpoux (eds.), Tecnologia, usos e potencialidades de tuberosas amiláceas sul americanas. Cargill, São Paulo, pp. 377–449.

Venturini-Filho, W.G. and B.P Mendes. (2003). Fermentação alcoólica de raízes tropicais. *In:* M.P. Cereda and O.F. Vilpoux (Eds). Tecnologia, usos e potencialidades de tuberosas amiláceas sul americanas. Cargill, São Paulo, pp. 530–575.

Wang, Q., Y. Hu, J. Hu, Y. Liu, X. Yang and J. Bian. (2012). Convenient synthetic method of starch/lactic acid graft copolymer catalyzed with sodium hydroxide. Bulletin of Materials Science 35: 415–418.

11

Brazilian Indigenous Fermented Food

Rosane Freitas Schwan[a]*, *Cintia Lacerda Ramos*[a],
Euziclei Gonzaga de Almeida[b], *Virgínia Farias Alves*[c]
and Elaine Cristina Pereira De Martinis[d]

Abstract

Not much is known about the fermented food and beverages produced by Brazilian indigenous tribes. This chapter presents a broad review from scientific databases using combined terms such as *fermented food*, *indigenous*, and *Brazil*. The results showed that cassava-based products are the most common fermented foods produced by Brazilian tribes, resulting in food and beverages such as *puba flour, caxiri*, and *cauim*. In addition, maize, nuts, and fruits serve as substrates for the production of fermented food. The microbiota of these products are usually dominated by lactic acid bacteria (LAB), but yeasts can also be considered important for the fermentation process.

Introduction

Fermented foods and beverages, whether based on plant or animal substrates, have been part of the human diet since the dawn of civilization and represent important methods of food preservation. Fermentation processes are also key to obtaining exquisite sensory attributes (flavor and texture), in addition to the enhancement of the nutritional quality of food products due to higher protein solubility and the availability of nutrients (Anukan and Reid 2009; Das and Deka 2012; Nehal 2013). Fermented food products are defined as those that have been submitted to the action of microorganisms (or their enzymes) in order to cause desirable biochemical changes (Blandino et al. 2003). During the natural fermentation process, microbial communities (bacteria, yeasts, and molds) undergo ecological succession within the fermentation container, since the community structure changes in response to the environmental conditions created by the previous species (Chavez-López et al. 2014;

---

[a] Universidade Federal de Lavras, Departamento de Biologia, Setor de Microbiologia, Lavras, MG, Brazil

[b] Universidade Federal de Mato Grosso, Instituto de Biociências, Departamento de Botânica e Ecologia. Cuiabá, MT, Brazil

[c] Universidade Federal de Goiás, Faculdade de Farmácia, Goiânia, GO, Brazil

[d] Universidade de São Paulo, Faculdade de Ciências Farmacêuticas de Ribeirão Preto, Av. do Café s/n, Monte Alegre, 14098-558, Ribeirão Preto (SP), Brazil. E-mail: rschwan@dbi.ufla.br

* Corresponding Author: rschwan@dbi.ufla.br

Colehour et al. 2014). Thus, the acidic environment created by bacteria favors yeast growth and metabolism, while the yeasts may provide growth factors, vitamins, and soluble nitrogen compounds that, in turn, stimulate bacterial growth (Nout and Sarkar 1999; Ramos et al. 2011).

Indigenous people around the world have prepared and consumed fermented foods for thousands of years. The nature of the raw materials and fermented products vary from tribe to tribe; they are not only strongly linked to the region of production but are also highly influenced by folk traditions (Sekar and Mariappan 2007; Nehal 2013). Although many studies on indigenous fermented food in different regions of the world (e.g. Africa and Asia) have been published, data on fermented foods produced by native American tribes are scarce (Santos et al. 2012).

Currently, about 12.5% of the Brazilian territory is represented by indigenous lands, home to approximately 510,000 Indians living mainly in the Amazon region (Brasil 2012). Several of these communities use artisanal small-scale fermentation processes to produce food and beverages with a nutritional, medicinal, and even religious value. Nevertheless, most of these products are unknown outside the region where they are produced, which is an obstacle to their study (Schwan et al. 2007; Santos et al. 2012; Puerari et al. 2015). They are prepared according to the local weather conditions and the ethnicities of the region, and there are variations in the available raw materials and in the fermentation time; therefore, the sensory attributes and quality are variable (Chavez-López et al. 2014). The raw materials used by Brazilian Indians as substrates for the production of fermented foods and beverages include cassava, maize, rice, cotton seed, sweet potato, peanuts, and fruits such as pineapple and banana (Lacerda et al. 2005; Almeida et al. 2007; Ramos et al. 2011; Miguel et al. 2014; Puerari et al. 2015). Table 11.1 gives examples of indigenous fermented foods and the substrates used for their preparation.

TABLE 11.1 Examples of indigenous fermented foods and their respective substrates

Fermented food	Substrate	Region	Reference
Puba flour	Cassava	Brazil	Silveira et al. 1999; Lacerda et al. 2005
Caxiri	Cassava, corn	Brazil	Santos et al. 2012; Miguel et al. 2015
Cauim	Cassava, rice, peanuts, cotton seed	Brazil	Almeida et al. 2007; Schwan et al. 2007; Ramos et al. 2010; Ramos et al. 2011
Calugi	Cassava, corn, rice	Brazil	Miguel et al. 2012; Miguel et al. 2015
Chicha	Corn, rice	Brazil, Argentina, Peru	Vallejo et al. 2013; Elizaquível et al. 2015; Puerari et al. 2015
Tarubá	Cassava	Brazil	Ramos et al. 2015
Yakupa	Cassava	Brazil	Freire et al. 2014
Tejuino	Corn	Mexico	

Substrates

There are numerous roots and tubers that have a high content of carbohydrates (starch and other polysaccharides) and can be fermented by microorganisms producing metabolites such as ethanol and organic acids. In tropical regions of Africa and South and Central America, there are several native root species such as *Ipomoea batatas* (sweet potato), *Dioscorea* spp. (yam), *Polymnia sonchifolia* (yacón), and *Manihot esculenta* (cassava) which have a potential for the production of fermented products (Venturini Filho and Mendes 2003).

Tubers and starchy grains are highly caloric; they are considered subsistence foods providing energy to under served populations. Among the cereals, corn is the crop that provides the most energy. Among tubers, potatoes and yams are emphasized due to easy growth. Cereal grains are the most important sources of starch, as their starch content is 40–90% (dry weight). The starch content of legume grains is 30–70%, while that of tubers is 65–85% (dry weight; Guilbot and Mercier 1985). Corn, wheat, rice, potato and cassava are the five major worldwide crops containing starch that are used as commercial sources,

Cassava (*Manihot esculenta*, Crantz), also known as manioc, mandioca, tapioca, or *yuca*, is a nutty flavored starchy tuber native from tropical America; it grows in poor and acidic soil and is quite tolerant of a alack of water (Kimaryo et al. 2000; Almeida et al. 2007). Several varieties of cassava exist, but it is possible to divide them into two groups: sweet and bitter. Both types can be used to prepare a variety of foods. Despite their low protein content, cassava tubers are very rich in carbohydrates and are widely consumed by Amerindians due to their energy value (Boonnop et al. 2009; Montagnac and Tanumihardjo 2009). However, fresh cassava roots are highly perishable, with a storage life of less than 48 h. In addition, cassava may be rich in toxiccyanogen compounds and contains substances that can diminish its nutrient bioavailability, such as phytate and polyphenols (Kimaryo et al. 2000; Oguntoyinbo and Dodd 2009; Montagnac and Tanumihardjo 2009). Therefore, prior to consumption, cassava tubers are processed by several methods (peeling, boiling, steaming, pounding, slicing, grating, roasting, soaking, pressing, and fermentation) to avoid their postharvest deterioration, reduce their toxicity, and improve the sensorial qualities of the derived products (Padonou et al. 2010).

Fermentation is an important step employed by the Indians for cassava processing in order to produce fermented foods. Lactic acid bacteria (LAB) have been described as predominant microorganisms among indigenous fermented food. Species belonging to several genera (*Lactobacillus*, *Pediococcus*, *Leuconostoc*, *Weissella*, *Streptococcus*, and others) have been described in various indigenous fermented products produced in Africa, South America, and Asia (Blandino et al. 2003; Almeida et al. 2007; Schwan et al. 2007; Ramos et al. 2010; 2011, Miguel et al. 2012; Owusu-Kwarteng et al. 2012; Santos et al. 2012, Adimpong et al. 2012, 2013; Akabanda et al. 2013; Freire et al. 2013). LAB initiate rapid and adequate acidification in the raw materials through the production of various organic acids from carbohydrates, of which lactic acid is the most abundant, followed by acetic and butyric acids. In addition, LAB can produce ethanol, bacteriocins, aroma compounds, exopolysaccharides, and certain enzymes. The common occurrence of LAB in food products, coupled with their long historical use, contributes to their identification as generally recognized as safe (GRAS) for human consumption (Silva et al. 2002). LAB contributes to food preservation and flavor promotion; furthermore, they might increase functional properties. In carbohydrate-rich environments, LAB frequently grows in association with yeasts, which are often present in lower populations. The fermenting yeasts may proliferate and spontaneously promote alcoholic fermentation.

Types of Indigenous Fermented Foods

Puba Flour

Retting (*pubagem*) is the natural submerged fermentation of fresh cassava roots to produce pub a flour or *carimã*, a food of indigenous origin, that is made in a manner similar to some fermented foods from West Africa, such as *lafun* and *foo-foo* (Padonou et al. 2010; Adebayou-Oyetoro et al. 2013; Fayemy and Ojokoh 2014). The manufacturing of puba is based on ancient empirical knowledge of indigenous Brazilian tribes, whose traditional methods of preparation have hardly changed over time (Crispin et al. 2013). Currently, puba flour also serves as a staple food for much of the Brazilian population, particularly for inhabitants of the north and northeast regions.

To prepare puba, unpeeled manioc roots are directly steeped in stagnant water or are placed in a bag kept in running water at room temperature for a period of 3–7 days. This is done to soften the roots and begin the skin relieving (Di Tanno 2001; Brauman et al. 2003; Crispin et al. 2013). After this period, the softened roots are removed from the water and strained to remove fibrous materials. They are then washed, sieved, and allowed to stand in order to break apart in the water. The mass is placed in a cloth bag and the excess water is removed by squeezing, thereby producing the wet pub a (50% moisture). Otherwise, the roots may be left in the sun or in ovens to produce dry puba flour, which has a moisture content of approximately 13% (Di Tanno 2001; Crisp in et al. 2013). Either the wet or dry puba may be used to prepare couscous, cassava paste porridge, and a variety of puba cakes.

During the *pubagem* (retting) the spontaneous fermentation process starts, acidification and softening of the roots take place, and degradation of cyanogenic compounds, synthesis of aromatic compounds, and phytate removal occur (Kimaryo et al. 2000; Boonnop et al. 2009; Montagnac and Tanumihardjo 2009; Anyogu et al. 2014; Colehour et al. 2014). The spontaneous fermentation process is the result of the competitive activity of different microorganisms, and each stage of the process is dominated by the best-adapted strains so that, during this process, considerable changes occur in the microbial population (Holzapfel 1997). Although there is little available data on the microbiota involved in the fermentation of puba in Brazil, it has been well documented that microorganisms associated with submerged fermentation of cassava include LAB (particularly *Lactobacillus* and *Leuconostoc* sp.), Enterobacteriaceae, *Corynebacterium* sp., *Bacillus* sp., *Clostridium* sp., yeasts, and molds (Padonou et al. 2010; Ahaoto et al. 2013; Anyogu et al. 2014; Fayemy and Ojokoh 2014). During all phases of cassava fermentation, the LAB predominate and their metabolic activity contributes to the production of organic acids, the development of desirable sensory characteristics (flavor, visual appearance, texture), and the extension of the shelf life and safety of the final product by delaying microbial spoilage or inhibiting microbial pathogens (Lacerda et al. 2005; Anyogu et al. 2014; Fayemy and Ojokoh 2014).

Studies on African fermented cassava products using molecular methods showed that, during *lafun* production, *Lactobacillus fermentum* was the ruling LAB, followed by *Lactobacillus plantarum* and *Weissella confusa*, whereas during the *foo-foo* fermentation process, *Lactobacillus delbrueckii*, *Lactobacillus fermentum*, *Lactococcus lactis*, and *Leuconostoc mesenteroides* were recovered, but *Lactobacillus plantarum* was the dominant species at the end of the process (Brauman et al. 2003; Padonou et al. 2010). In Brazil, using amplified ribosomal DNA restriction analysis (ARDRA), Crispin et al. (2013) verified that the LAB most frequently isolated from commercial puba samples were *Lactobacillus fermentum*, followed by *Lactobacillus delbrueckii* and other LAB from *Lactobacillus plantarum* groups. Also in Brazil, Lacerda et al. (2005) identified that during the production of sour cassava starch, *Lactobacillus plantarum* and *Lactobacillus fermentum* were the dominant LAB, whereas *Lactobacillus perolans* and *Lactobacillus brevis* were identified at lower levels. These studies indicate that LAB are usually found in two to four species associations, which suggests a role of synergistic phenomena.

Caxiri

Cassava constitutes the main ingredient in caxiri, an ancient alcoholic fermented beverage consumed by many indigenous tribes from the Amazonian region. Its alcohol content is about 10/11°GL, similar to the alcoholic content of some fermented fruit products (Pinelli et al. 2009). In addition to cassava, sweet potato, yam, and/or sugar cane juice can be added to prepare the beverage. Caxiri is prepared exclusively by women and its preparation is both a culinary and a religious task, because the drink plays an important role in tribal religious ceremonies, agricultural festivities, and other social events (Assis 2001; Lima 2005).

In Brazil, to prepare the caxiri, the Juruna tribe starts by immersing cassava roots for 2 days in running water to produce a softened and fermented product, the puba (Santos et al. 2012). After the retting of cassava, the tubers are peeled, cut into small pieces, pressed to remove excess water, grated, sieved, and roasted to produce a flour. The roasting process is important because it contributes significantly to the detoxification of cassava roots (Montagnac and Tanumihardjo 2009). The flour is then diluted with water, mixed with grated sweet potatoes and placed into barrels to initiate the fermentation process. The mixture is kept in the barrels for about 24–48 h at approximately 30 °C. The beverage is consumed within 48 h after preparation (Santos et al. 2012). Santos et al. (2012) determined that the fermentation of caxiri results from the action of a complex microbial ecosystem. In their study, yeasts represented the predominant microbial group and *Saccharomyces cerevisiae* was the predominant species. *Rhodotorula mucilaginosa*, *Pichia membranifaciens*, *Pichia guilliermondii*, and *Cryptococcus luteolus* were also found. Regarding bacteria, caxiri demonstrated the major involvement of spore-forming bacteria belonging to the genus *Bacillus*. The *Bacillus cereus* group, *Bacillus subtilis*, and *Bacillus pumilus* were the predominant species, while *Bacillus megaterium*, *Bacillus amyloliquefaciens*, and *Bacillus simplex* were found at low levels (Santos et al. 2012). Caxiri produced from cassava and corn showed a predominance of *Saccharomyces cerevisae*, *Rhodotorula mucilaginosa*, *Lactobacillus fermentum*, *Bacillus subtilis*, and *Lactobacillus helveticus* (Miguel et al. 2015).

Cauim

Cauim is a nonalcoholic fermented beverage produced by some Amerindian tribes from a variety of substrates, including cassava, rice, corn, peanuts, cotton seed, sweet potatoes, and fermentable fruits. This beverage constitutes an important staple food for Indigenous people, from young children to the elderly.

Brazilian Indians make nonalcoholic beverages that are fermented using different techniques and processes, resulting in a beverage called "cauim" (*kawí*; Lima 2005). The beverage is produced via a very rudimentary and traditional process and may be "cooked cauim" and/or "fermented cauim." For preparation of cooked cauim, the substrates are boiled for approximately 2 h and then cooled to room temperature. For fermented cauim, when the porridge is cold, an inoculum is added to initiate the fermentation process, which usually takes 24–48 h. The inoculum is obtained from the fluid generated by female members of the tribe chewing sweet potato (Schwan et al. 2007; Ramos et al. 2010). The nonalcoholic beverage is a staple food for both adults and children. It is produced during the *Cauinagem* festival, which is held to celebrate the arrival of the time of harvesting due to rains and subsequent land productivity (Wagley 1988).

The inoculation starts the fermentation process because saliva contains the amylase enzyme, which acts on starch, the main carbohydrate present in the substrates. Amylase metabolizes this complex carbohydrate, releasing simple sugars that are used by microorganisms. Studies have shown that there is a predominance of LAB (*Lactobacillus* genus), which are responsible for conferring a slightly acidic flavor due to the presence of lactic acid (Almeida et al. 2007; Ramos et al. 2010, 2011). Yeast species belonging to the *Candida*, *Pichia*, *Clavispora*, *Kluyveromyces*, and *Saccharomyces* genera have been described in the fermentation of cauim from different substrates (Schwan et al. 2007; Ramos et al. 2010, 2011).

The importance and role of microorganisms during cauim fermentation are still unclear. It is known that LAB are the microorganisms most commonly found in indigenous fermented foods. Their crucial importance is associated mainly with their physiological features, such as substrate utilization, metabolic capabilities, and functional properties. A recent work studying the probiotic characteristics of microorganisms isolated from different fermented foods provided a preliminary selection of some potential probiotic strains isolated from cauim, such as *Lactobacillus brevis* UFLA FFC199. This isolate showed a tolerance to low pH and bile salts, these are conditions imposed by the GIT (Gastrointestinal Tract) environment. In addition, the isolate was able to adhere to a human intestinal epithelial cell line (Caco-2) and then increased the TEER (Trans-Epithelial Electrical Resistance) of the Caco-2 cell line (Ramos et al. 2013).

Calugi

To produce calugi, a nonalcoholic food, the Brazilian Indians of the Javaé tribe use corn, cassava, and rice as the raw materials (Miguel et al. 2012; 2014). Calugi is a porridge consumed by adults and children. For its preparation, dried corn and rice are left to soak in water for approximately 30 min and then grinded with a pestle. The obtained flour is diluted with water and sieved for husk removal. Water is added to the mass and cooked for about 2 h. The mixture is then allowed to cool to room temperature. An inoculum from the fluid generated by chewing sweet potato as described for cauim is added to the mass (Miguel et al. 2012). The microbial communities were studied by Miguel et al. (2012) using the PCR-DGGE independent-culture method, and the researchers described the presence of *Lactobacillus plantarum*, *Streptococcus salivarius*, *Streptococcus parasanguis*, *Weissella confusa,* *Enterobacter cloacae*, *Bacillus cereus*, and other *Bacillus* sp., as well as the yeasts

Saccharomyces cerevisiae, *Pichia fermentans*, and *Candida* sp. A pH decrease of 6.0 to 3.5 after 48 h of fermentation was also observed (Miguel et al. 2012).

Chicha

Chicha is a generic Spanish term for a traditional fermented or non-fermented beverage that is widely consumed by indigenous groups prevalent from the Amazon to the Andes (Blandino et al. 2003, Chavez-López et al. 2014). It plays an important role in the Amerindians' life, not only as a nutritional source, but also in rituals and religious practices. Chicha's alcoholic content is generally low, but this can vary between 1 and 12% (v/v) depending on the production method (Jennings 2005; Chavez-López et al. 2014). There is a wide range of recipes for chicha's production, but the beverage is made primarily from maize (therefore, it is also called maize beer). Other starchy plant crops, such as cassava and millet, can be used as raw materials for chicha's preparation (Jennings 2005; Colehour et al. 2014). In Brazil, the Umutina tribe has replaced the original corn chicha with a rice-based one (Puerari et al. 2015).

During the production of traditional chicha, saliva is used as a source of amylase for the conversion of starch into fermentable sugars (Blandino et al. 2003). After chewing and sun drying, corn is diluted with water for the beginning of the fermentation process. Microbiologically, the fermentation is a spontaneous process carried out by the indigenous microbiota at room temperature. The fermentative microorganisms involved in chicha's fermentation can vary according to the environment and fermentation vessels, but they generally include cooperation among yeasts such as *Saccharomyces cerevisiae*, LAB (*Lactobacillus* sp. and *Leuconostoc* sp., among others), and molds (*Aspergillus* sp. *Penicillium* sp. Blandino et al. 2003; Chavez-López et al. 2014).

For the production of the rice-based chicha in the Brazilian Umutina tribe, an artisanal process is used. The non-fermented beverage is called "Katazeru", while the fermented one is called "Ulumuty." According to Puerari et al. (2015), for the preparation of the fermented beverage, the rice (*Oryza sativa*) undergoes spontaneous and uncontrolled fermentation initiated by its endemic microbiota, as well as those microorganisms associated with the utensils used during beverage preparation, individuals' hands, and the local environment. First, some fine rice flour is prepared. Then, the flour is stirred into boiling water to prepare the beverage. Separately, a little rice flour is roasted, chewed, and used as the inoculum during cooking. Sugar and additional water are added to the mixture and the drink undergoes fermentation for about 36 h at room temperature. Puerari et al. (2015) found that a combination of LAB (*Enterococcus* sp., *Leuconostoc* sp., *Lactobacillus* sp.) and *Bacillus* spp. (*Bacillus subtilis* and *Bacillus cereus*) dominated throughout the rice-based chicha fermentation process, and unlike the results of other studies on fermented beverages (Blandino et al. 2003; Chavez-López et al. 2014), yeasts were not detected.

Tarubá

Tarubá is a beverage produced from cassava by Indians living in the Brazilian state of Amazonas. Although it is made using cassava, the tarubá preparation uses a solid substrate fermentation, unlike the submerged fermentation employed for other

Brazilian indigenous cassava products described in this chapter (Almeida et al. 2007; Schwan et al. 2007; Miguel et al. 2012; Miguel et al. 2014; Santos et al. 2012; Freire et al. 2014). After solid fermentation, the substrate is diluted in water for consumption. To prepare the beverage, cassava roots (*Manihot utilissima*) are peeled and washed in running water. The cleaned cassava is crushed, pressed, and sieved to remove excess water using a handmade straw press called a "*tipiti*." Then, the cassava mass is toasted and covered with *Trema micrantha* (L.) *Blume* and *Musa* spp. (banana) leaves. This mass is left to ferment for 12 days. After fermentation, the cassava mass is diluted in water and filtered in order to remove the solids. The liquid obtained is the beverage, and ready to be consumed (Ramos et al. 2015). Microbial studies on this beverage have also shown an association of bacteria and yeast. According to Ramos et al. (2015), in the tarubá prepared by the Saterê-Mawé tribe, LAB (mainly *Lactobacillus plantarum, Lactobacillus brevis, Leuconostoc mesenteroides,* and *Leuconostoc lactis*) and *Bacillus* species were predominant during fermentation. The yeast *Torulaspora delbrueckii* was the dominant species found during fermentation and in the final tarubá beverage. A combination of culture-dependent and independent methods (PCR-DGGE) also resulted in the detection of *Candida tropicalis, Candida rugosa, Candida ethanolica, Hanseniaspora uvarum, Pichia exigua, Pichia kudriavzevii, Pichia manshurica, Rhodotorula mucilaginosa,* and *Wickerhamomyces anomalus* (Ramos et al. 2015).

Although studies regarding the microorganisms involved in indigenous fermented foods have been performed, knowledge about microbial metabolism and the role of microbes in these fermentations is still limited. Some isolates of *Torulospora delbrueckii, Pichia exigua*, and *Bacillus* species identified by Ramos et al. (2015) were able to grow at 45 °C and showed amylolytic and proteolytic activities. They probably make an important contribution to the tarubá beverage and may be potential starter cultures for a better control of the fermentation process.

Yakupa

Yakupa is a spontaneously fermented non-alcoholic beverage consumed daily by children and adults and is classified as a refreshing cauim (Lima 2005). Indigenous Juruna people, who inhabit the Indigenous Xingu Park (Mato Grosso, Brazil), prepare yakupa using cassava roots, which are first left to soak in water for about 2–3 days. This step allows for the pre-fermentation of the substrate. Next, the pre-fermented roots, called puba, are sundried and mashed, water is added, the mixture is sieved, and a white liquid is obtained. This is then cooked and mixed with grated sweet potatoes (Freire et al. 2014). The beverage is consumed as soon as it is prepared and after fermentation for 24–48 h. The final product presents a liquid consistency and yellowish color; it is slightly acidic with a sour taste.

Studying the microbiological and physicochemical parameters of yakupa, Freire et al. (2014) attributed the beverage's flavor to the presence of microbial metabolism, particularly in the production of volatile compounds (mainly esters) in the middle to final periods of the fermentation process. Esters are described as important flavor compounds found in different fermented foods, and they are associated with fruity flavor notes. Yeast may produce a wide variety of volatile compounds, including esters, which contribute to the flavor of fermented foods. According to Freire et al. (2014), the culture-dependent and independent methods used served to identify the yeasts *Saccharomyces cerevisiae* and *Pichia kudriavzevii*. The bacteria *Lactobacillus fermentum*, *Lactobacillus plantarum*, *Weissella cibaria*, and *Weissella confusa* were also found. The dynamic of yakupa's spontaneous fermentation demonstrated the dominance of heterolactic LAB species, mainly *Lactobacillus fermentum*, which may be responsible for the lactic acid production, and thus, the beverage acidification. The beverage has a pH of around 3.0 and lactic acid (around 8 g/L) is the main organic acid.

A recent study was performed using a cassava-based substrate for fermentation by LAB and yeast isolated from yakupa and other fermented foods (Freire et al. 2015). The authors demonstrated that these co-cultures have the ability to improve the safety of the product by lowering pH and producing organic acids (e.g., lactic and tartaric acids). In addition to this, certain volatile compounds related to pleasant odors (e.g., 2-phenylethyl alcohol [odor of roses], 1-butanol, 3-methyl, or isoamyl alcohol [banana and pear-like aroma], and acetoin [buttery odor]) were produced by these microorganisms.

Tejuino

Tejuino is a traditional Mexican beverage prepared by fermentation of nixtamalized corn mass. The nixtamalization process is performed by cooking corn with calcium oxide (CaO) and this process can be applied in combination with other processes to improve the processing and use of corn. Sefa-Dedeh et al. (2004) reported that nixtamalized corn subjected to solid-state fermentation showed a reduction in pH and viscosity. The corn nixtamalization process may improve several nutritional parameters, enhancing free calcium levels, free nicotinic acid, and the availability of niacin and amino acids, while reducing tannin levels and the bioavailability of iron and other minerals (Bharati and Vaidehi 1989; Sefa-Dedeh et al. 2004).

Tejuino is prepared using the same mass of processed corn used to produce tortillas and tamales, typical dishes in Mexican cuisine. The term *tejuino* comes from different states in Mexico, for example, Jalisco, Oaxaca, Nayarit, and Chihuahua (Wacher-Rodarte, 1995). The preparation may vary from region to region and even from one producer to another, however mixing the mass of nixtamalized corn with water and sugar usually produces the beverage. The mass is then boiled and the liquid is allowed to ferment at room temperature for about 24 h. The beverage is consumed cold with lemon juice, a pinch of salt, and a shaved ice in shell or lemon sorbet. It is usually sold on the street in small plastic cups. It has a bittersweet taste and low alcohol content. Although some corn foods have been described, there is no data and literature about the microorganisms involved in tejuino fermentations.

Conclusions

The data and literature on beverages and food technologies used by indigenous people are of great importance for preserving their cultural identity and to facilitate a better dissemination of their culture in the world. Although the number of publications on indigenous fermented foods has increased, there are very few studies related to nutritional and microbial composition of these foods. Researchers are challenged to uncover more information and increase the understanding of these rudimentary fermentation processes, which will represent an important empirical basis for the knowledge about the current technologies used.

References

Adebayou-Oyetoro, A.O., O.B. Oyewole, A.O. Obadina and M.A. Omemu. (2013). Microbiological safety assessment of fermented cassava flour "Lafun" available in Ogunand Oyo states of Nigeria. International Journal of Food Science 2013: 1–5. Article ID 845324, doi: 10.1155/2013/845324

Adimpong, D.B., D.S. Nielsen, K.I. Sørensen, P.M.F. Derkx and L. Jespersen. (2012). Genotypic characterization and safety assessment of lactic acid bacteria from indigenous African fermented food products. BMC Microbiology 12: 1–12.

Adimpong, D.B., D.S. Nielsen, K.I. Sørensen, F.K. Vogensen, H. Sawadogo-Lingani, P.M.F. Derkx and L. Jespersen. (2013). *Lactobacillus delbrueckii* subsp. *jakobsenii* subsp. nov., isolated from dolo wort, an alcoholic fermented beverage in Burkina Faso. International Journal of Systematic and Evolutionary Microbiology 63: 3720–3726.

Ahaoto, I., C.C. Ogueke, C.I. Owuamanam, N.N. Ahaotu and J.N. Nwosu. (2013). Fermentation of undewatered cassava pulp by linamarase producing microorganisms: effect on nutritional composition and residual cyanide. American Journal of Food and Nutrition 3: 1–8.

Akabanda, F., J. Owusu-Kwarteng, K. Tano-Debrah, R.L.K. Glover, D.S. Nielsen and L. Jespersen. (2013). Taxonomic and molecular characterization of lactic acid bacteria and yeasts in nunu, a Ghanaian fermented milk product. Food Microbiology 34: 277–283.

Almeida, E.G., C.C.T.C. Rachid and R.F. Schwan. (2007). Microbial population present in fermented beverage "cauim" produced by Brazilian Amerindians. International Journal of Food Microbiology 120: 146–151.

Anukan, K.C. and G. Reid, African traditional fermented foods and probiotics. (2009). Journal of Medicinal Food 12: 1177–1184.

Anyogu, A., B. Awamaria, J.P. Sutherland and L.I.I. Ouoba. (2014). Molecular characterization and antimicrobial activity of bacteria associated with submerged lactic acid cassava fermentation. Food Control 39: 119–127.

Assis, L.P.S. (2007). Da cachaça a libertação: mudanças nos hábitos de beber do povo Dâw no alto do Rio Negro. Revista Antropos 1: 1–73.

Bharati, P. and M.P. Vaidehi. (1989). Calcium enrichment of sorghum in grains. Food and Nutrition Bulletin 11: 53–55.

Blandino, A., M.E. Al-Aseeri, S.S. Pandiella, D. Cantero and C. Webb. (2003). Cereal-based fermented foods and beverages. Food Research International 36: 527–543.

Boonnop, K., M. Wanapat, N. Ngarmnit and S. Wanapat. (2009). Enriching nutritive value of cassava root by yeast fermentation. Scientia Agricola 66: 629–633.

Brasil, 2012. Instituto Brasileiro de Geografia e Estatística (IBGE). (2012). Os indígenas no Censo Demográfico 2010: primeiras considerações com base no quesito cor ou raça. Acessed in 15 March 2015. Available at: http://www.ibge.gov.br/indigenas/indigena_censo2010.pdf.

Brauman, A., S. Keleke, M. Malonga, E. Miambi and F. Ampe. (2003). Microbiological and biochemical characterization of cassava retting, a traditional lactic acid fermentation for Foo-Foo (Cassava Flour) production. Applied and Environmental Microbiology 62: 2854–2858.

Chavez-López, C., A. Serio, C.D. Grande-Tovar, R. Cuervo-Mulet, J. Delgado-Ospina and A. Paparella. (2014). Traditional fermented foods and beverages from a microbiological and nutritional perspective: The Colombian Heritage. Comprehensive Reviews in Food Science and Food Safety 13: 1031–1048.

Colehour, A.M., J.F. Meadow, M.A. Liebert, T.J. Cepon-Robins, T.E. Gildner, S.S. Urlacher, B.J.M. Bohannan, J.J. Snodgrass and L.S. Sugiyama. (2014). Local domestication of lactic acid bacteria via cassava beer fermentation. Peer Journal 2: e479. DOI 10.7717/peerj. 479.

Crispin, S.M., A.M.A. Nascimento, P.S. Costa, J.L.S. Moreira, A.C. Nunes, J.R. Nicoli, F.L. Lima, V.T. Mota and R.M.D. Nardi. (2013). Molecular identification of *Lactobacillus* spp. associated with puba, a Brazilian fermented cassava food. Brazilian Journal of Microbiology 44: 15–21.

Das, A.J. and S.C. Deka. (2012). Fermented foods and beverages of the North-East India. International Food Research Journal 19: 377–392.

Di Tanno, M.F.P. (2001). Influência da temperatura, tempo e concentração de pectinase na textura, rendimento e características físico-químicas da mandioca (*Manihot esculenta* C.) durante fermentação. M.S. Thesis, University of São Paulo, Piracicaba, São Paulo.

Elizaquível, P., A. Pérez-Cataluña, A. Yépez, Aristimuño, C. E. Jiménez, P.S. Cocconcelli, G. Vignolo, R. Aznar. (2015). Pyrosequencing vs. culture-dependent approaches to analyze lactic acid bacteria associated to chicha, a traditional maize-based fermented beverage from Northwestern Argentina. International Journal of Food Microbiology 198: 9–18.

Fayemy, O.E. and A.O. Ojokoh. (2014). The effect of different fermentation techniques on the nutritional quality on the nutritional quality of the cassava product (fufu). Journal of Food Processing and Preservation 38: 183–192.

Freire, A.L., C.L. Ramos and R.F. Schwan. (2015). Microbiological and chemical parameters during cassava based-substrate fermentation using potential starter cultures of lactic acid bacteria and yeast. Food Research International 76(3): 787-795.

Freire, A.L., C.L. Ramos, E.G. Almeida, W.F. Duarte and R.F. Schwan. (2014). Study of the physicochemical parameters and spontaneous fermentation during the traditional production of yakupa, an indigenous beverage produced by Brazilian Amerindians. World Journal of Microbiology and Biotechnology 10: 567–577.

Guilbot, A. and C. Mercier. (1985). Starch. *In*: G.O. Aspinall (eds.), The polysaccharides. Academic Press Inc., Orlando, pp. 209–282.

Holzapfel, W. (1997). Use of starter cultures in fermentation on a household scale. Food Control 8(5-6): 241-258.

Jennings, J. (2005). La chichera y El Patrón: chicha and the energetics of feasting in the prehistoric Andes. Archaeological Papers of the American Anthropological Association 14: 241–259.

Jennings, J., K.L. Antrobus, S.J. Atencio, E. Glavich, R. Johnson, G. Loffler, and C. Luu. (2005) "Drinkin beer in a blissful mood": Alcohol production, operational chains, and feasting in the Ancient World. Current Anthropology 46(2):275–294.

Kimaryo, V.M., G.A. Massawe, N.A. Olasupo and W.H. Holzapfel. (2000). The use of a starter culture in the fermentation of cassava for the production of "kivunde", a traditional Tanzanian food product. International Journal of Food Microbiology 56: 179– 190.

Lacerda, I.C.A., R.L. Miranda, B.M. Borelli, A.C. Nunes, R.M.D. Nardi, M. Lachance and C.A. Rosa. (2005). Lactic acid bacteria and yeasts associated with spontaneous fermentations during the production of sour cassava starch in Brazil. International Journal of Food Microbiology 105: 213–219.

Lima, T.S. (2005). Um peixe olhou para mim: o povo Yudjá e a perspectiva. Editora UNESP, São Paulo.

Miguel, M.G.C.P., C.F. Collela, E.G. Almeida, D.R. Dias and R.F. Schwan. (2015). Physicochemical and microbiological description of caxiri a cassava and corn alcoholic beverage. International Journal of Food Science & Technology 50(12): 2537-2544.

Miguel, M.G.C.P., C.C.A.A. Santos, M.R.R.M. Santos, W.F. Duarte and R.F. Schwan. (2014). Bacterial dynamics and chemical changes during the spontaneous production of the fermented porridge (Calugi) from cassava and corn. African Journal of Microbiology 8: 839–849.

Miguel, M.G.C.P., M.R.R.M. Santos, W.F. Duarte, E.G. Almeida and R.F. Schwan. (2012). Physico-chemical and microbiological characterization of corn and rice calugi produced by Brazilian Amerindian people. Food Research International 49: 524–532.

Montagnac, J.A. and S.A. Tanumihardjo. (2009). Processing techniques to reduce toxicity and antinutrients of cassava for use as a staple food. Comprehensive Reviews in Food Science and Food Safety 8: 17–27.

Nehal, N. (2013). Knowledge of traditional fermented food products harbored by the tribal folks of the Indian Himalayan Belt. International Journal of Agriculture and Food Science Technology 4: 401–414.

Nout, M.J.R. and P.K. Sarkar. (1999). Lactic acid fermentation in tropical climates. Antonie van Leeuwenhoek 76: 395–401.

Oguntoyinbo, F.A and C.E.R. Dodd. (2009). Bacterial dynamics during the spontaneous fermentation of cassava dough in gari production. Food Control 21: 306–312.

Owusu-Kwarteng, J., F. Akabanda, D.S. Nielsen, K. Tano-Debrah, R.L.K. Glover and L. Jespersen. (2012). Identification of lactic acid bacteria isolated during traditional fura processing in Ghana. Food Microbiology 32: 72–78.

Padonou, W., D.S. Nielsen, N.H. Akissoe, J.S. Hounhouigan, M.C. Nago and M. Jakobsen. (2010). Development of starter culture for improved processing of Lafun, an African fermented cassava food product. Journal of Applied Microbiology 109: 1402–1410.

Pinelli, L.L.O, V.C. Ginani and R.N. Xavier. (2009). Caxiri. *In*: W.G. Venturini Filho (ed.), Bebidas alcoólicas: Ciência e Tecnologia. Editora Blucher, São Paulo, pp.1–13

Puerari, C., K.T. Magalhães-Guedes and R.F. Schwan (2015). Physicochemical and microbiological characterization of chicha, a rice-based fermented beverage produced by Umutina Brazilian Amerindians. Food Microbiology 46: 210–217.

Ramos, C.L., E.G. Almeida, A.L. Freire and R.F. Schwan, (2011). Diversity of bacteria and yeast in the naturally fermented cotton seed and rice beverage produced by Brazilian Amerindians. Food Microbiology 28: 1380–1386.

Ramos, C.L., E.G. Almeida, G.V.M. Pereira, P.G. Cardoso, E.S. Dias and R.F. Schwan. (2010). Determination of dynamic characteristics of microbiota in a fermented beverage produced by Brazilian Amerindians using culture-dependent and culture-independent methods. International Journal of Food Microbiology 140: 225–231.

Ramos, C.L., E.S.O. Sousa, J. Ribeiro, T.M.M. Almeida, C.C.A. Santos, M.A. Abegg and R.F. Schwan. (2015). Microbiological and chemical characteristics of tarubá, an indigenous beverage produced from solid cassava fermentation. Food Microbiology 49: 182–188.

Ramos, C.L., L. Thorsen, R.F. Schwan and L. Jespersen. (2013). Strain-specific probiotics properties of *Lactobacillus fermentum, L. plantarum* and *L. brevis* isolates from Brazilian food products. Food Microbiology 36: 22–29.

Santos, C.C.A.A., E.G. Almeida, G.V.P. Melo and R.F. Schwan. (2012). Microbiological and physicochemical characterisation of caxiri, an alcoholic beverage produced by the indigenous Juruna people of Brazil. International Journal of Food Microbiology 156: 112–121.

Schwan, R.F., E.G. Almeida and L. Jerpersen. (2007). Yeasts diversity in rice-cassava fermentations produced by indigenous Tapirapé people of Brazil. FEMS Yeast Research 7: 966–972.

Sefa-Dedeh, S., B. Cornelius, W. Amoa-Awua, E. Sakyi-Dawson and E.O. Afoakwa. (2004). The microflora of fermented nixtamalized corn. International Journal of Food Microbiology 96: 97–102.

Sekar, S. and S. Mariappan. (2007). Usage of traditional fermented products by Indian rural folks and IPR. Indian Journal of Traditional Knowledge 6: 111–120.

Silva, J., A.S. Carvalho, P. Teixeira and P.A. Gibbs. (2002). Bacteriocin production by spray-dried lactic acid bacteria. Letters in Applied Microbiology 34: 77–81.

Silveira, I. A. ; E. P. Carvalho; L. Pilon. (1999). Alimentos e bebidas obtidos pela fermentação natural da fécula de mandioca. Revista Higiene alimentar 13: (66/67): 44-47.

Vallejo, J., P. Miranda, J. Flores, F. Sánchez, J. Ageitos, J. González, E. Velásquez, T. Villa. (2013). Atypical yeasts identified as Saccharomyces cerevisiae by MALDI - TOF MS and gene sequencing are the main responsible of fermentation of chicha, a traditional beverage from Peru. Systematic and Applied Microbiology 36: 560 - 564.

Venturini Filho, W.G. and B.P. Mendes. (2003). Fermentação alcoólica de raízes tropicais. *In*: M.P. Cereda et al. (eds.), Tecnologias, usos e potencialidades de tuberosas latino americanas. Série culturas de tuberosas amiláceas Latino Americanas. Fundação Cargill, São Paulo, pp. 530–575.

Wacher-Rodarte, C. (1995). Alimentos y bebidas fermentados tradicionales. *In*: M. García-Garibay, R. Quintero and A. López-Munguía (Eds.), Biotecnología Alimentaria. Limusa, México, pp. 313–349.

Wagley, C. (1988). Lágrimas de boas vindas: Os índios Tapirapé do Brasil Central. Editora da Universidade de São Paulo, São Paulo.

12

Andean Fermented Beverages

Julio Rueda, Eugenia Jimenez, Manuel Lobo and Norma Sammán*

Abstract

Latin American indigenous tribes used the fermentation processes as means of food processing and preservation. In the Andean regions of Latin America where the Inca empire had its civilization, the production of beverages was among the major practices that involved a natural fermentation processes. For the Inca's population, the production of alcoholic beverages was important because they played an essential role in their spiritual beliefs., These drinks were considered to have a high symbolism because of the Incan cosmovision. The present chapter describes the traditional fermented beverages of pre-Columbian cultures of the Andean region of South America. Spontaneous fermentation processes were used as a way of alcohol production by using native (in pre-Colombian time) and Old World crops (in post-Colombian time). By allowing an aqueous mixture of fruits or vegetables to ferment, a great variety of alcoholic beverages with a low to a moderate alcoholic content (2–12%) were obtained. Such substrates are rich in fermentable sugars and/ or polysaccharides. In some cases preliminary steps such as chewing, germination or roasting were carried out to allow the process to easily occur and also to obtain assorted drinks. This chapter elaborates on the raw materials used, including: Manioc or Çassava, Banana, Sugar Cane, Sweet Potato, Rice, Cacao, Carob Beans, Maize and others. Maize production was of great importance, and for this reason, food crop also became an ingredient beverage manufacturing and this elaboration method was spread along the entire Andean region. Thus the most important and well known drink in this region was a Maize beer or maize-based beverage, called *chicha*. In Andean Latin America all chicha-like beverages with similar production methodology were called *chicha* as well. Nowadays many of these beverages are still being artisanally produced and consumed although they may have undergone changes in their composition and chemical characteristics due to the incorporation of new raw materials and elaboration methodologies. The main objective of this chapter is to introduce an improved knowledge of these traditional Andean fermented beverages, highlighting those coming from the Northwest region of Argentina. A revision of the processes involved in their manufacture will be useful in understanding

Facultad de Ingeniería, Universidad Nacional de Jujuy, Avenida Italia esq. Martiarena, 4600 Jujuy, Argentina. E-mail: julioruedafca@gmail.com; eugeniajimenez81@hotmail.com; molobo@ arnet.com.ar; nsamman@arnet.com.ar

* Corresponding Author: nsamman@arnet.com.ar, Phone: +54 3884 493160; fax: +54 3884 493141

biochemical changes associated to the action of native microorganisms on raw materials. Traditional and new technological procedures involved in their manufacture are discussed. By increasing and updating the available knowledge from a technical and scientific point of view it is expected to make a helpful contribution for those who are interested in traditional fermented foods.

Introduction

Food fermentation techniques are one of the oldest methods discovered by humanity for the purpose of food modification and preservation. These techniques probably emerged as a spontaneous phenomenon and were subsequently improved and developed as a complex practice applied by native people to many of the raw materials for food production and preservation. Cheese, bread, wine and beer are some of the oldest fermented food products produced by earlier civilizations and these foods are clear examples of the high grade of development that these techniques reached in the past. These products have also been strongly linked to culture, tradition and to the evolvement of the human race (Battcock and Azam-Ali 1998).

Fermented beverages like beer and wine are evidence of this association as they were very frequently related to power and religious acts. Empires from the cradle of civilizations all over the world like Sumerians, Babylonians and Assyrians in Asia, the Chinese empire in Asia and the Inca and pre Inca civilizations in America have a developed knowledge of brewery techniques. Mead from honey and beer from barley (*Hordeum vulgare* L.) were the most common fermented beverages in the Old World and were the most common fermented beverages and they had their counterparts in the highlands of the Andean mountain range. A considerable variety of drinks were produced from fruits, tubers, seeds, grains and cereals in the fertile valleys of Andean region. Most of these were alcoholic beverages and they played a key role in the organization and expansion of the Inca Empire.

Nowadays, most South American traditional fermented beverages are still artisanally produced. The production process has remained unchanged, but with notable variations in some regions. These modifications are related to the purpose of the drink, the available raw materials, depending on the geographical location, climate conditions and the skills of the people involved in the production of these traditional beverages. Many of these practices have fallen by the wayside and many others have persisted but have changed due to the introduction of food crops from the old world along with modifications introduced by Spaniards. This chapter introduces improved and systematic knowledge of alcoholic fermented beverages from the Andean region which have been in existence since pre Inca and pre Colombian times, with special mention of some from the Northwest region of Argentina. A revision of processes involved in their manufacture will be useful to understand changes associated with the action of native microorganisms on raw materials. Traditional and new technological procedures involved in their manufacture are commented upon. By increasing and updating the available knowledge it is hoped that this will be a helpful contribution for those who are interested in traditional fermented foods from South America.

Developmnet of Fermented Beverages Production: An Ovwerview

Fruits and vegetables are thought to be the first foods consumed as fermented products. Early hunter-gatherers could have collected and eaten fresh fruits during times of abundance but they could also have consumed rotten and fermented fruits at times of scarcity. This habit led to a developed taste for certain fermented fruits and also for the alcohol produced by spontaneous fermentation (Battcock and Azam-Ali 1998). China is believed to be the birthplace of fermented vegetables. These food products were prepared by the action of molds such as *Aspergillus* sp. and *Rhizopus* sp. to make fermented foods (Battcock and Azam-Ali 1998). This included the development of an amylolysis fermentation method, in which fungi broke down the polysaccharides in rice (*Oryza sativa* L.) and millet (McGovern et al. 2004). Archaeochemical, archaeobotanical, and archaeological evidence support the fact that the preparation of fermented beverages dates back to nearly 9,000 years ago in China (McGovern et al. 2004). There is also reliable information supporting the theory that fermented drinks were being produced over 7,000 years ago in Babylonia (now Republic of Iraq), 5,000 years ago in Egypt, 4,000 years ago in Mexico and 3,500 years ago in Sudan (Dirar 1993; Battcock and Azam-Ali 1998).

Mead, wine and beer were the first alcoholic fermented beverages produced by indigenous people in human history. Production of wine and mead was simple in comparison to beer making. Certainly, mead was the first alcoholic fermented drink because honey (the most antique fermentative substrate) and water were readily available. The process was easily feasible by just allowing a mixture of these ingredients to stand and ferment (Haehn 1956).

Inhabitants of *Mesopotamia* already knew the manufacturing process of beer according to cuneiform findings (2,700 years ago) which clearly describe the process. In the 5th century A.D. (1,500 years ago) there existed in Babylonia skilled brewers who prepared malted grains from barley. This process was described on cuneiform inscriptions as follows: barley is soaked, allowed to germinate, dried and ground to obtain flour and make bread which is baked in a clay oven. Separately, a portion of malted grains is dried under the sunlight and milled. Then, the bread together with the malt and water are blended, and allowed to macerate for several days. Finally, a clear liquid is filtered and transferred to pots for the final fermentation. This process was found to be very similar to the elaboration of fermented beverages in Latin American Andean countries.

The literature describes the existence of at least twenty different types of Babylonian beers and an export trade of beer to Egypt. Egyptians implemented the browning techniques for producing a stout or dark beer and they flavored such beers it with fruits. In the 7th Century A.D. the Arabs introduced beer to Spain and prepared with malt bread along with the addition of cinnamon (*Cinnamomum verum* J. Presl), rue (*Ruta graveolens* L.) and marjoram (*Origanum majorana* L.). Greeks and Romans had wine as their preferred beverage but during the Germanic expansion, beer became an important beverage in Middle and Western Europe after the Germanic conquest of Greece and Rome 1,600 years ago (Haehn 1956).

The early inhabitants of Latin America had their own brewing techniques. Findings of microscopic archeological remains of starch and phytoliths from maize (*Zea mays* L.) in ancient pottery in Northwest Argentina, reinforces the evidence that brewery practices could have been carried out by South American natives 4,000 years ago

(Lantos et al. 2015). Brewing tools and vessels were found on the north coast of Peru where pre Inca civilizations like Moche (1,900 years ago) and Chimu (1,000 years ago) had their empires (Hayashida 2008). There is no accurate data to exactly when fermented drinks began to be produced in Andean Latin American lands but archaeological studies support the theory that even prior to the Inca expansion and establishment in South America, aboriginal peoples from Northwest Argentina already had knowledge of complex processes for the production of drinks in large amounts (Leibowicz, 2013). Identification has been based on the presence of brewing tools and vessels (for milling, cooking, fermentation), by products (e.g., strained mash), and features (e.g., hearths, pits for germinating grain) (Hayashida 2008).

Fermented beverages constituted a fundamental part of indigenous meals, which were produced by native American peoples. With the Incan civilization, technologies such as drying and fermentation achieved a exceptance for food conservation and modification. It is remarkable that these techniques were especially developed in the South American highlands along the Andean mountain range. The Indigenous peoples of South America were skilled farmers and owners of an advanced knowledge in practices for soil cultivation. The aboriginal agricultural systems were noteworthy and allowed them to produce a large quantity of diverse fruits and vegetables which were the main constituents of the Inca diet (Browne 1935).

The production of different meals and dried and fermented products was possible due to the fertile and cultivable soils of the lowlands and highlands of South America, which resulted in the production of an astonishing variety of vegetables. Different varieties of maize and potatoes (*Solanum tuberosum* L.) were domesticated in this region and they became an important staple food along with tubers and other crops such as the now acclaimed quinoa (*Chenopodium quinoa* Willd.) and amaranth (*Amaranthus* spp. L.). Such richness in the food crops available resulted in the development of indigenous techniques for food conservation such as the freezing and fermentation.

Fermentation became a complex process that was applied to many vegetables in the Andean lands which in turn led to the manufacture of different fermented beverages in Latin America that received the common or general name of *chicha*. This term in discussed later in this chapter. Andean Aboriginals fermented most of what they grew in the Andes. In the Andean regions of Latin America that were ruled by the Inca Empire, the manufacturing of beverages was among the main practices involving spontaneous or wild fermentation processes. The production of alcoholic drinks played an important role in their spiritual beliefs and they were considered to have a level of high symbolism because of the Incan beliefs. The spontaneous fermentation was used as a mean of alcohol production on the basis of native (in pre Colombian times) and Old World crops (during post Colombian times). In certain cases this process was refined and became a complex technique. These processes led to the obtainment of alcoholic beverages with a low to moderate alcoholic content (2–12%). Substrates rich in sugar and/or starch such as manioc or cassava (*Manihot esculenta* Crantz), banana (*Musa paradisiaca* L.), sugar cane (*Saccharum officinarum* L.), sweet potato (*Ipomoea batatas* (L.) Lam.), rice, cacao carob beans (*Theobroma cacao* L.) and maize were and are still being used for the elaboration of traditional fermented beverages. In some cases, preliminary steps such as chewing, germination or roasting can be carried out to allow the processes to occur easily and to create assorted drinks. Considering that maize production was of great importance, this food crop also gained importance due to the fact it was the main ingredient in beverage manufacturing. Along the Andean Region the most important and well known drink was produced: *maize chicha*. There existed other chicha-like beverages in Andean Latin America whose production processes followed a similar methodology and they also received the same name.

Indigenous fermented beverages played key roles in human culture towards the development of fermentation technologies. These products were always considered to be healthy because of their nutritional values and sensorial attributes. They helped in the establishment of agriculture and food processing techniques (McGovern et al. 2004). These indigenous technologies have passed from parents to progeny for thousands of years and they are still in use. Certainly, some fermented products and practices not only survived the passage of time but were also tested for different conditions and changes in the region they belong to (Battcock and Azam-Ali 1998).

Vegetable Material Employed For Fermented Beverages Production in Andean Latin America

Beverages from Andean Latin America were prepared from a considerable variety of edible fruits and vegetables. Alcoholic drinks could be obtained from roots, tubers, seeds, fruits, leaves and even animal body parts, in a few cases (León and Hare 2008). Watery vegetables and fruits served as main substrates in the lowlands of Andean countries due to their proximity with the Amazonian jungle. Cereals, Andean crops, grains and seeds were used in the highlands. Spanish colonizers introduced new food crops to Latin America which were also included in the production of fermented beverages. The diversity of drinks found in Latin America is also attributed to the different ways of preparation and the intended purposes for which it is to be used, since meals were considered foods to feed the body and soul.

Aboriginal peoples from Andean Latin America take advantage of the broad biodiversity of edible food plants available in Andean lands for the production of alcoholic beverages. In the highlands, grains and tubers were used while in the tropical Andean regions sugary fruits and starchy tubers were added to common preparations.

The variety of beverages was immense, though the most popular was an alcoholic drink made on the basis of *Jora*, germinated or sprouted maize. There is another important drink called *masato,* prepared from vegetables, such as the tuber cassava. This beverage is well known in tropical areas and in low-lying regions of Andean countries where the growing of this tuber is feasible. This was the first drink that the Spaniards learnt about from the Andean tribes after arriving in Latin America. Fermented beverages in Andean Latin America were commonly prepared from cereal-like grains such as quinoa and kañiwa (*Chenopodium pallidicaule* Aellen.). Fruits from *molle* (*Schinus molle* L.) and *algarrobo* (*Prosopis* sp. L.) trees were employed for the production of fermented and non-fermented drinks, especially in the southern countries of the Andean region. The tuber oca (*Oxalis tuberose* Molina) was also an ingredient in the manufacturing of chicha. Another well-known beverage was one made with peanuts (*Arachis hypogaea* L.) and it was probably the second most important beverage after maize beer. In the tropical Andes fermented beverages included mainly apples (*Malus domestica* Borkh.), loquat (*Eriobotrya japonica* (Thunb.) Lindl.), pineapple (*Ananas comosus* (L.) Merr.) and cactus pear (*Opuntia* sp. Mill.,) among others (de Leon Pinelo 1636).

The Spanish arrival in America brought several plant materials that were incorporated later in the traditional production methodologies along with the native Andean crops. Most of the new ingredients for drinks were almost all spices such as cloves (*Syzygium aromaticum* (L.) Merr & Perry), cinnamon, citrus (*Citrus* L.), anise and sugar cane, among others. Additionally, cereals like wheat (*Triticum* L.), rice and barley were incorporated passively, but Andean crops already had a strong tradition of use. Sugar cane became an important ingredient in the production of several fermented beverages since it allowed to produce sweet drinks and permitted the production of highly alcoholic drinks and to accelerate the process of fermentation. *Panela* is a solid bar of unrefined sugar cane and can be found in the literature of fermented beverages production with Spanish names like *rapadura*, *chancaca*, *papelon* or *piloncillo*. Nowadays, the production of Andean drinks includes traditional ingredients from Andean America and those that were introduced from the Old World.

Diversity of Beverages in Andean Lands

As previously mentioned, the Spanish term *chicha* is a general name given to traditional or aboriginal Andean Latin American beverages. These may be alcoholic or non-alcoholic drinks. In most cases this term is associated with the popular maize beer which was found to be the most important alcoholic beverage of the region, since pre Inca times and even after the Spanish conquest. The production process of South America's beer is reviewed and discussed in the coming sections of this chapter. *Aloja* was another term for beverages used in Latin America and they were normally prepared from fruits. *Aloja* was the common name of certain beverages ranging from non-alcoholic to mildly alcoholic. In northern South America and Andean deserts of Peru this term especially alludes to the beverage made from *algarroba* or *molle* fruits. Comparatively, *aloja* does not have a complex production process and it requires few ingredients and less time in processing. It is a weak drink; chicha, on the other hand, needs more preparation time, more ingredients and entails a complex process for its manufacture. Chicha can be thick in consistency, but still aqueous. This generalization needs to be taken as a general concept since in Latin American fermented beverages were and are still being produced in similar ways.

Native people in Andean countries prepared a beverage based on the bright red or pink fruits from a Peruvian native tree called molle. This tree is an evergreen, and tolerant of the dryness which is endemic in the arid zones from the northern South America to the central Argentinian and Chilean deserts. The beverage prepared with molle fruits was simple and only fruits and water were necessary. It was described as "sweet, agreeable and excellent in the treatment of oedema". For the production of this drink around 250 g of fresh seeds yielded 20 L of *chicha de molle* (Goldstein and Coleman 2004). Technically, fruits of *Schinus molle* are drupes and are ready for chicha production when the resin pockets are fleshy and sweet due to the high level of concentrated sugars. For preparing the beverage, the fruits are boiled with water in a ceramic pot for about half an hour. Cinnamon sticks and cloves are added to flavoring the water. Then the clay pot containing the water is removed from the heat. This decoction will have turned sweet but once cooled, some tablespoons of sugarcane can be added. The overnight mixture is strained through a sieve into a ceramic pot. The pot is covered with a wet cloth and the liquid The pot is covered with a wet cloth and the liquid is allowed to ferment for 10 days in a cool and shaded room (Goldstein and Coleman 2004; de Leon Pinelo 1693).

In Peru another chicha is prepared with assorted fruits. This drink includeds apples, loquat, pineapple, cactus pear and others. It was only considered a refreshing drink if it was not fermented. The preparation of this drink is made from fruits and also included rice, grated potatoes and carrots (*Daucus carota* L.), black grapes (*Vitis* L.), sugar cane, loquat and pineapple husks. All these ingredients were grated and well mixed. The mixture was then allowed to macerate for about 20 days and an acidic, slightly sweet liquid was obtained through straining the mixture through a sieve. This sweet liquid could be consumed immediately or an alcoholic version could be obtained through fermenting the drink for an additional number of days. At certain intervals, unrefined sugar cane was added until the drink reached the desirable alcoholic content (León and Hare 2008)

Chicha de manguey or chicha de cabuya a beverage made from *manguey* (*Agave americana* L.) and produced from the sap of manguey. To produce this drink a stalk of this plant is perforated and the draining liquid was collected and allowed to ferment spontaneously, it was normally consumed without the fermentation step; in certain cases the sap could be boiled prior to drinking. This beverage was regarded as having medical attributes. In Mexico this beverage was called *Pulque* (de Leon Pinelo 1636).

Peanuts were also employed by the native Andean peoples to prepare a beverage called *chicha de mani* which was quite popular. In its original version, this beverage was prepared only with peanuts, but after the Spanish conquest more ingredients such as rice, sesame seeds (*Sesamum indicum* L.), cinnamon, anise, cloves and digestive or aromatic herbs were added to the traditional recipe. To prepare this drink, peanuts were roasted and ground. Then, the spices were boiled in water and the peanut flour was added. This mixture was simmered and the decoction obtained was cooled. The broth obtained was allowed to sediment, the fermentation was spontaneous and it could take up to three or four days. It was rich in fat from the peanuts (de Leon Pinelo 1636).

In the Andean highlands, ground barley was also employed to obtain a fermented beverage. To prepare this alcoholic drink barley flour was browned and then boiled in water, was cooled and allowed to ferment. Sediment from other *chichas* was used as a starter culture to provide yeasts to the new drink and thus, accelerate the process.

Cassava was also used for the elaboration of a non-alcoholic beverage, a popular drink in the Amazonia region called *massato*, which was different from the alcoholic version. Cassava roots were peeled, washed and cut. It was parboiled in water and more water and honey were added. It was consumed as a refreshing beverage (de Leon Pinelo 1636).

Indigenous peoples in Latin America had no written language before the Spaniards arrival to South America. *Quechua* and *Aimara* were the most important and widespread spoken dialects and fermented beverages were named with different terms aside from *chicha*. In manuscripts of Spanish explorers the voices *aqa*, *acca* or *akka* referred to the fermented beverages of indigenous peoples and their pronunciation belongs to the *Quechua* language. In a similar way the voice *kufa* was attributed to the *Aimara* language (Nicholson 1960; Gastineau et al. 1979).

After the Spaniards arrived and colonized the region from Central to South America fermented beverages were collectively named as *chicha*. Some theories claim the explorers named the beverage *chicha* as an abbreviation of the expression *chicha co-pa*, in which the part *chichah* means maize, and *co-pah* means drink (Pardo and Pizarro 2005). Another hypothesis about the origin and meaning of the word *chicha* refers to the preparation of this beverage. Thus, the name *chichia-tl* could mean "to ferment" and "water" and another interpretation says that *chichial* means "with saliva" (Nicholson 1960). The Spanish advance towards South America spread the name *chicha* which replaced local denominations and even the aboriginal terms of *mudai* and *palcu* for Chilean fermented beverages which were falling into disuse. Thus, the term *chicha* was assigned to every fermented beverage prepared in a similar way, but became popular and assigned to the *chicha* made from maize and cassava which were heavy and popular drinks. The Spaniard word *aloja* was attributed also to fermented and non-fermented beverages prepared mostly from fruits, (*algarrobo* mainly) and with a weaker consistency (Pardo and Pizarro 2005; León and Hare 2008).

The South American Beer: The Manufacturing of *Chicha De Maiz*

The manufacturing of maize beer is a long-lasting and complex process that requires several steps and days to obtain the alcoholic beverage. It is not a single-step spontaneous fermentation but involves the degradation and consecutive extraction of starch and sugars, which are then worked on separately and reunited again in the last step. The production of chicha also results in the acquisition of several edible by-products. Some of these are very useful and appreciated because of their nutritive value. While some products are employed for animal feeding, others can still be used to obtain other beverages which may be alcoholic or non-alcoholic (Cutler and Cardenas 1947). The Fig. 12.1 shows the production procedure for *chicha de maiz* in the Northwest Argentina.

The first stage for this procedure is the selection of the maize kernels. The quality and variety of maize beers depends on several factors among which the maize cultivar employed in production is one of the most important. Traditionally, this step was carried out by women skilled in the production of chicha which are known as *chicheras*. Several versions of *chicha* can be obtained according to the variety of maize utilized in the production of this drink. Since in the Andean region many cultivars of maize can be grown, a considerable diversity of maize beers can be produced.

Rigid kernels from a local variety of maize called *quellu sara* were used to produce an alcoholic drink in Peru. The cultivar *kulli* which had a deep purple colour was employed to elaborate a dark chicha. Another chicha was prepared with leaden-coloured grains from the variety *oqe* (León and Hare 2008). Bolivian *chicheras* preferred kernels from the *chuspillo* cultivar to prepare their alcoholic drink. This variety is cultivated with the only purpose of beer production because it is very sweet and with a many-rowed cob. The *kulli* cultivar is also used in Bolivia for the production of a burgundy coloured chicha. The kernels of this maize are appreciated for their colour, which can vary from reddish to a dark purple and they have a high content of anthocyanins (Cutler and Cardenas 1947). In the north of Argentina two local varieties are utilized in the production of chicha: *maíz criollo* and *maíz abajeño* (Otero and Cremonte 2009).

Once the maize kernels were selected they can be treated in different ways (grinding, browning, sprouting, etc.) and thus, even more varieties of maize chicha can be produced.

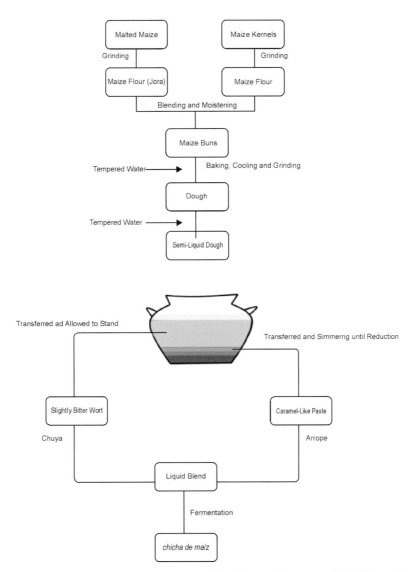

FIGURE 12.1 Flow Chart of *chicha de maiz* making in the northwest Argentina. This procedure is the confluent of both Spaniards and indigenous cultures.

At the beginning of the process maize kernels can be malted, browned and even chewed (with a previous slow baking). Some of these steps are preliminary to obtaining a flour (León and Hare 2008).

Malted cereals have been traditionally used in beer manufacturing because they have a high content of enzymes for starch degradation and hydrolysis into fermentable sugars. Malt does have a good enzymatic activity and can still be useful if even more starchy cereals are added (Ward 1989). In the Andean highlands of Peru malted kernels of maize were called *Jora* or *sora* and the beverage produced with the malted grains of maize was called *chicha de Jora* or *chicha de sora*. The germination is interrupted when the radicle reaches approximately 2.5 cm in length (León and Hare 2008; Cutler and Cardenas 1947). The flour obtained after grinding this maize is called *pachuc-cho* and is sold in some markets. It can be stored for six months (Nicholson 1960). In Bolivia, both the germinated maize kernels and the flour obtained are called *hui-ñapu* or *wiñapu*. The enzyme α-amylase is synthesized during germination or malt-ing which produces the hydrolysis of starch by breaking the α-1,4 glycosidic bonds. Another enzyme, the β-amylase, is already preformed in the kernel before the sprout-ing and contributes to the degradation of maltose into single units of glucose. The resultant hydrolysis products of starch are sugars like maltose and glucose which are available for subsequent fermentation (Bewlyseed 2001; Biazus et al 2009).

The browning technique in maize kernels produces a partial release of single sug-ars from starch (Tonroy and Perry 1975). Additionally, this reduces the level of toxic and anti-nutritional compounds such as the hydroxycinnamic acid and tannins pres-ent in maize (Ayatse et al. 1983).

The practice of chewing/salivating and spitting out food as a preliminary step in the production of beer was very common among Latin American brewers. In the Andean highlands this custom was called *mukeado* and the alcoholic drink obtained through this tradition received the name *chicha mukeada*. The purpose of chewing and/or salivating maize kernels or maize flour was to produce a partial hydrolysis of carbohydrates. This action was attributed to the amylase enzyme found in saliva. Currently, the chewing practice has fallen into disuse but it is still implemented by minor Andean tribes. In the following sections, viewpoints on this ancestral practice are given. Nowadays, *chicha* tend to be prepared from a dough from maize flour instead of *mukeado* or the chewing of kernels. The dough mentioned is called *moy-apo* and the chicha produced by using this substrate is known as *chicha de masa* (Otero and Cremonte 2009).

The next step after the selection and treatment of maize kernels is the grinding for the production of maize flour. In ancient times a half-moon shaped stone over a flat stone were used as a mortar and was employed for the grinding and the production of maize flour (Cutler and Cardenas 1947). Nowadays, this method may be still in use but mechanical or electrical mills are the most widespread devices used to grind the grain (Otero and Cremonte 2009). Some *chicheras* do not ground the grains until they are reduced to a fine powder or flour, instead they grind the kernels into small particles of maize. They argue that *chicha* will be clear and more palatable with this degree of grinding (de Mayolo 1981).

In order to elaborate *chicha de masa,* a firm dough is prepared with at least two varieties of flour and the addition of previously boiled water. Then, the dough is baked in a clay oven at a high temperature (Otero and Cremonte 2009), during this process starch is partially hydrolyzed. The reducing sugars produced take part in the Maillard and browning reactions. The browning phenomenon is important because it induces changes in colour, flavour and in the nutritive value of baked food products (Purlis 2010). Maize flour has been characterized as having a high reactivity toward the Maillard reaction and to the formation of favourable compounds (Rufian-Henares et al. 2009). This step is important in the manufacture of *chicha de masa* because it initiates the desirable compounds in flavour and colour. The Maillard reaction is responsible for the typical brown colour of chicha (Elizaquivel et al. 2015).

After the obtention of a baked bun, the next step is the extraction of fermentable substrates. The extraction of carbohydrates is carried out in special ceramic potteries

(Fig. 12.2). Carbohydrates and others nutrients become soluble by macerating the cooked maize bread in water in a clay pot called *virque* (Fig 12.2A). This stage can be performed by different methods including the soaking of the bread, by decoction or by combining both methods (Ward 1989). Preliminary, the baked dough can be dis aggregated by hands or ground (Otero and Cremonte 2009).

The soaking step needs warm or tempered water (70 ºC) that has been previously boiled. Boiling water is not used since it produces an undesirable viscous paste (Cutler and Cardenas 1947), additionally, water at the boiling temperature inativates the amylolytic enzymes. Temperatures between 63–65 ºC are favourable for the acti-vation of enzymes found in cereals and for starch degradation (Ward 1989). The substrate in maceration is continuously stirred with a large wood spoon (Fig. 12.2B) until a homogeneous slurry is obtained. Then the semi-liquid preparation is allowed to decant for few hours after which a multilayer wort can be noted. The disposition of these layers inside the *virque* is due to differences in density, solubility and grind-ing level of the initial substrate compounds (Fig. 12.1). Each layer has utility and none of these are discarded. Two layers are the most important for the production of chicha. One is the maize wort, locally known as *chuya* and the other is a jelly-like layer called *arrope* which is composed mainly of starch and a minor portion of single sugars. Each of these two layers is transferred into separate clay pots and given a different treatment after which they will be blended again to form the *chicha*. First, the upper layer *chuya* is transferred to a narrow-mouth ceramic vessel or *cántaro chichero* (Fig. 12.2C) and is allowed to stand for the duration of three or four days until it becomes acid (Otero and Cremonte 2009). In some cases this maize wort can be boiled for about three hours and then is cooled (Cutler and Cardenas 1947). In this vessel the lactic acid fermentation is carried out and the pH value drops up to 3.7; this step, the wort acidification, is optional and some brewers may not allow the wort to become acidic (Cutler and Cardenas 1947; Elazequivel et al. 2015). On the other hand, the jelly-like layer called *arrope* is placed in a wide-mouthed pot or *olla arropera* (Fig. 12.2D) and boiled down until it becomes thick and with a caramel-like or brown colour. At this point a two-fold increase in starch, maltose, sucrose and glu-cose occurs (Elazequivel et al 2015). Finally, the two most important components of *chicha de masa* are blended. The naturally acidified wort and the caramel-like paste of starch are mixed at different proportions to turn from a weak to highly alcoholic *chicha*. Immediately after the two layers are blended, *chicha* starts to bubble and when this phenomenon ceases the production gas declines and the process is consid-ered to be complete (Cutler and Cardenas 1947). As the environmental temperature may vary according to the geographical region and season, the time of fermentation may also vary. In some circumstances, the fermentation is completed in just one or two days (Cutler and Cardenas 1947) or it can last up to one month or more (León and Hare 2008). This last stage of the process is called *the maturing of chicha*. If it is necessary to accelerate the process, sugar cane bars (locally know as *chancaca* or *panela*) can be added to the vessel pot or hot coals can be placed under the ceramic vessel to warm up the fermentation. Thus, sucrose and warmth will help the yeasts to start the fermentation and the production of alcohol.

The final pH of this fermented beverage has been recorded as 3.5 (Elazequivel et al. 2015) and the alcoholic content of the *chicha de masa* can reach values of 4–6% (Aguado et al. 2006). This amount varies amongst the diversity of *chichas* found in the Andean lands from 2% in a weak chicha up to 12%. The average percentage of alcohol for *chichas* prepared in Cochabamba in Bolivia has been recorded as 5% or less. A well made *chicha* is an attractive drink, clear in color, effervescent and similar in taste to cider. According to Antunez de Mayolo (1981) a half-cup of a heady *chicha* is enough to produce exhilarating effects.

FIGURE 12.2   Chicha pottery employed in Northwestern Argentina. A: *virque*; B: *large wood spoon*; C: *cántaro chichero*; D: *olla arropera*.

In certain cases, superficial foam taken from a mature *chicha* or the sediment inside the vessel is used to accelerate the process of fermentation in a new *chicha*. Both, the superficial foam and the remnant sediment, are rich in yeasts and can be used in the produccion of maize bread. The yeasts with a high capacity of ethanol production (15–22 °GL) ascend to the top layer in the pot and can be separated with the foam. In turn, yeasts with a moderate capacity of alcohol production (8–15 °GL) descend to the bottom layer of the vessel (Ward 1989). It is well known that even the yeasts found in the pores of ceramic vessels were enough to initiate the process of fermentation with no addition of foam or sediment (Cutler and Cardenas 1947).

Microbial fermentation in Indigenous Beverages

Indigenous fermented food products are the result of fermentative processes that might be carried out by the microorganisms already present in the raw material that allows the fermentation to begin spontaneously, or it can be induced by the inoculation of starter cultures taken from previous traditional fermented foods or drinks. Normally, these fermented products are produced by more than one type of autochthonous microorganisms, that work together or in a sequenced mode leading to what is known as the microbial succession.

The first species of microorganisms that colonize the substrate grow until they are self-inhibited by their own by-products. At this point, other organisms present begin to get active until they cannot tolerate the new environmental conditions created by their own metabolism. Bacteria will initiate the work and yeasts will continue until molds take over the substrate, but normally molds do not play a significant role in the desirable fermentation of fruit and vegetable products. The genera *Leuconostoc* and *Streptococcus* grow first and ferment the substrate at an increased rate to other close related species. Often, these genera are the first group of bacteria to colonize a substrate of fermented fruits and vegetables (Battcock and Azam-Ali 1998).

Many Latin American fermented beverages have been characterized as having a sour, spicy and sweet taste with a low to moderate alcoholic content and this is the result of two main types of fermentation: the one that produces the acidic taste because of the action of lactic acid bacteria (LAB) and the other that produces alcohol. Ethyl alcohol is produced mostly by yeasts of the genus *Saccharomyces*, particularly *Saccharomyces cerevisiae* and it also can be produced in small amounts by some LAB. If present, some *Acetobacter* species can synthesize acetic acid from ethanol oxidation. These main products of the microbial metabolism are produced on the basis of available simple sugars (Battcock and Azam-Ali 1998).

A common factor in the production of most fermented beverages is the use of yeasts to convert sugars into ethanol. Approximately, 96% of the fermentation is carried out by strains of *Saccharomyces cerevisiae* or related species. Ethyl alcohol is synthesized through the Embden-Meyerhof-Parnas pathway (EMP) by which the piruvate produced through glycosylation is metabolized into acetaldehyde and ethanol. The global reaction is:

$$\text{Glucose} + 2 \text{ ADP} \rightarrow 2 \text{ Ethanol} + 2 \text{ CO}_2 + 2 \text{ ATP}$$

During the process of beer manufacturing yeasts metabolize sucrose, fructose, maltose and maltotriose, in this order. Sucrose is hydrolyzed by a microbial extracellular invertase. Maltose and maltotriose are hydrolyzed inside the yeast by the enzyme α-glucosidase. Most of *Saccharomyces* strains are unable to hydrolyze starch and dextrin. For this reason, the use of starchy raw materials requires the action of enzymes like α and β–amylases found in malt or produced during malting. Many strains of *Saccharomyces cerevisiae* can produce ethanol up to 12–14%. At these high values of alcoholic content the rate of growth is strongly affected and the process of ethanol production is inhibited by itself. Often, values of pH between 3.0 and 6.0 are favourable for the growing of yeasts and for the fermentative process. At pH values lower than 3.0 the growing rate decreases remarkably. Temperatures between 15 °C and 35 °C increase the rate of fermentation and the level of compounds like glycerol, acetone, acetaldehyde, pyruvate and ketoglutarate in the wort (Ward 1989).

The Role of Saliva in the Production of Andean Drinks

The chewing practice was a common tradition of Inca and pre-Inca civilizations for the production of foods. It was performed for the production of natural sweeteners from maize stalks. Maize stalks were chewed for a time and the resultant mixture of sap with saliva inside the mouth was spat out to produce a sweetener (Gade 1975; Browne 1935). Peruvian brewers chewed cassava as a preliminary step in the production of the beverage *masato* (de Leon Pinelo 1636). The chewing of maize was a common practice of Andean and Amazonian tribes of South America. Indeed, it was a paid activity when large amounts of *chicha* were necessary to be produced for feasts and religious acts (León and Hare 2008).

The custom of chewing roots, fruits and grains for the production of fermented beverages was a widespread technique among the indigenous peoples from the Andes. Explorers found this method being practiced in both small tribes and in more structured communities. Several edible vegetables were chewed for the production of fermented beverages in South America. Maize was the most important food crop chewed in the highlands of Peru, cassava in the lowlands of Ecuador, cassava and maize in central Brazil, maize and sweet potato in the Brazilian coast, carobs from *algarroba* and *tusca* (*Acacia aroma* Hook. & Arn.) and fruits from *chañar* (*Geoffroea decorticans* (Gillies ex Hook. & Arn.) Burkart), in the semi-arid lowlands of Bolivia, Paraguay and Argentina. This practice is well described by Culter and Cardenas 1947 and, in the authors opinion the teeth played a minor role in this practice. The researchers, in their study of chicha-making in Bolivia, describe the process as follows: Once the maize flour is obtained from ground kernels a small portion is moistened with water and rolled into a medium–sized ball which is placed into the mouth. The flour ball is kept inside the mouth and is thoroughly worked with saliva until a uniform paste is formed. After this, the salivated paste in moved forward with the tongue and taken out of the mouth with the fingers. The salivated flour ball is called *muko* and can be dried under the sunlight and stored. The production of *muko* and *chicha mukeada* was abolished by Spaniards because they considered this practice to be unacceptable (León and Hare 2008).

Culter and Cardenas (1947) preferred to call this technique salivation instead of mastication or chewing because they considered these last terms should be used only when fibrous vegetables such as algarroba carobs, cassava or sweet potatoes were employed in the production of fermented beverages and the fibrous pulp needs to be disaggregated to produce the drink.

This quirky methodology, which was a distinctive feature of Andean tribes, was supposed to help the fermentation process. It was believed that the action of saliva is to produce a partial hydrolysis of carbohydrates found in starchy vegetables through the salivary amylase enzyme, but according to Pedersen et al (2002), this enzyme has an important but minor role in the gastrointestinal digestion because of its low concentration in saliva.

The Microbial Succession in Chicha

Indigenous beverages are a varied group of traditional fermented foods that still need a deep and comprehensive study of their microbial ecosystems. There are few studies accurately describing the microbiomes present in these indigenous foods. The microbial composition is the result of different factors such as the autochthonous microorganisms present on raw materials, the environmental microorganisms, human mediated factors and the ancestral practices which have led to the local domestication of certain microorganisms.

While most studies on the microbial composition of traditional drinks relied on the traditional culture method of the final products (Cox et al. 1987), recent studies have adopted a methodology based on the study of each step in the production of fermented drinks. The study of microbial changes with a stage by stage methodology has resulted in the knowledge of common patterns of fermentation along the procedure to produce indigenous drinks.

Maize chicha was studied by Eliazequível et al. (2015) at different stages of production for the evolution of LAB. At the beginning, plate counts for lactic acid bacteria (LAB) were recorded to be of 10^4–10^6 colony forming units (CFU) g^{-1} in flour and 10^8 during the buns making. Since maize chicha was produced by procedures that included cooking or baking, most of the initial microbial community was lost at this stage. Hence, the remaining microorganisms in the substrate, the environmental microorganisms and human factors played an important role in the reconstitution of the microbial ecosystem. After this reconstitution, the acid fermentation began and counts for LAB rose up to values of 10^9 CFU g^{-1}. Some LAB produced small amounts of ethanol, but the main production was carried out by the yeasts that initiated the second phase in this fermentation. Thus, the level of LAB decreases again (10^3–10^5 CFU per mL) due to the inhibition caused mainly by ethanol, the production of lactic acid (2.0–4.5 g%) and the reduction in available single sugars.

The fermentative phase that produces alcohol involves the metabolism of yeasts. Lopez-Arboleda et al. (2010), by using PCR-restriction fragment length polymorphism (RFLP) it was found that the most abundant species in maize chicha were *Sacharomyces cerevisiae* (15%), *Pichia guilliermondii* (18%) and *Candida tropicalis* (18%). Before the fermentative phase, isolates belonged to the species *Candida tropicalis, Pichia guilliermondii* and *Pichia kluyveri.* When the production of alcohol and carbon dioxide took place, these species along with *Pichia fermentans, Sacharomyces cerevisiae* and *Hanseniaspora gulliermondi* metabolized sugars to produce more alcohol. After the bubbling phase, yeasts found were *Candida tropicalis, Sacharomyces cerevisiae, Pichia fermentans* and *Hanseniaspora gulliermondi*.

The bacterial diversity associated with acid fermentation has been investigated by Elizaquível et al. (2015) in maize *chicha* in Northwest Argentina. Pyro-sequencing methodology was able to detect a higher number of species than those found by traditional culture methods. Species from the genera *Enterococcus, Lactococcus, Streptococcus, Weissella, Leuconostoc* and *Lactobacillus* were the most predominant in maize *chichas*. The genus *Lactobacillus* was found as the group with high diversity in species, while the genera *Enterococcus* and *Leuconostoc* were the second group in importance.

In samples of chicha from one village the diversity of species included mainly *Lactobacillus plantarum, Lactobacillus rossiae, Leuconostoc lactis* and *Weissella viridescens*. On the other hand, samples from the village at a distance of five kilometers away from the first evidenced the presence of species like *Enterococcus hirae, Enterococcus faecium, Leuconostoc mesenteroides* and *Weissella confusa* as being predominant. The microbial ecosystem of chicha from both villages at a very close distance was different. Similar conclusion was reached by Colehour et al. (2014), who found that there existed differences in the microbiome associated to cassava chicha from different villages, even in those that were separated by short geographical distances. Both authors attributed this difference in genera of LAB to the elaboration procedure and human mediated factors, since the same beverages were prepared in a similar way but with slight variations.

Benefitial Aspects Related to Andean Fermented Beverages

Many indigenous fermented beverages are still poorly studied and they are thought to own beneficial properties, which have not been deeply studied from a scientific point of view (Marsh et al. 2014). This is the case of many fermented drinks from Latin America which lack of factual information about the beneficial properties attributed to its consumption. Some of these beneficial traits are known because they were orally transmitted from parents to progeny and they were always perceived as good for health since pre Incan times.

From the nutritional point of view it is well known that beneficial effects attributed to the microbial activity on the nutritional value of fermented foods are related to the conversion of single sugars into acids, alcoholic compounds, the reduction of anti-nutritional factors and the presence of potential probiotic bacteria (van Hylckama Vlieg et al. 2011). Some maize varieties may have different content of anti-nutritional factors such as phytates, condensed tannins, α-amylase inhibitors and lectins (Ejigui et al. 2005). Additionally, some metabolites that inhibit or limit the growing of pathogenic bacteria in food products are produced during the LAB fermentation. During the fermentative process the gross composition of fermented food products does not change significantly. However, the water-soluble fraction of nutritional compounds improves remarkably. During the fermentation single proteins, polypeptides and aminoacids are released from conjugated proteins. Bacterial activity increases notably the level of free aminoacids and water soluble carbohydrates during fermentation (Gaden 1992).

Fermentation improves the nutritional value of products produced from maize. There is an increase in the content of Vitamin A, Vitamin B_{12} and amino acids like arginine and methionine (Steinkraus 1997; Chelule et al. 2010). Steinkraus (1997) reported that maize *chicha* is rich in vitamins from the B complex with an increase in the levels of riboflavin.

The consumption of *chicha* was always considered as safe and was a common habit in Andean lands because water sources employed for consumption were always associated with illnesses and to parasitism. It is believed that the Incan never drank from water sources but in the form of *chicha* after but, and the regular intake could have reached up to a litre and half per day. Maize *chicha* was an everyday beverage since most of the Andean meals were produced on the basis of dehydrated vegetables. There existed beverages with different alcoholic content, from the non-alcoholic maize *chicha* to the highly alcoholic drinks prepared for celebration. *Chicha* was not recommended to be consumed by people suffering from skin disorders and was known as being effective against respiratory diseases when it was consumed warm and with the addition of leaves of coca (*Erythroxilum coca* Lam), especially for throat and bronchial affections. Due its acidic trait, *chicha* was employed by some Andean physicians for the treatment of parasitism and tuberculosis (Leon and Hare 2008). Another property assigned to this drink was the diuretic effect, and as a stimulant effect for the supply of breast milk and the anti-inflammatory activity of *chicha* sediment. Numbing properties in rituals were attributed to maize *chicha* if barks of the tree ishpingo [*Ocotea quixos* (Lam.) Kosterm] or sap from succulent plants (Cactaceae) were added. (Cactaceae). Molle *chicha* was considered as a diuretic beverage as long as it was not fermented (de Mayolo 1981).

According to Marsh et al. (2014), on several native communities, the reason for consuming fermented beverages is related to their beneficial effects. It is believed that these effects are due to the synergistic action of several factors such as the raw materials used, the microbial community and the metabolites produced by them. However, further research is necessary to clarify this topic.

Latin America Fermented Beverages Today

The term *chicha* in Latin America has always been associated to a traditional or indigenous beverage which may be an alcoholic or non-alcoholic drink. It can be produced from a single ingredient or by combining several ones. Whether roots, cereals, tubers, nuts or watery fruits were used, beverages are found to be very diverse. Spices can be added to give a special flavor and sugarcane is always employed in the elaboration. Thus bars of unrefined sugar cane became an important ingredient. The proportion and type of raw materials used in the elaboration varies according the geographical region. In Venezuela the expression *chicha* is commonly used for a beverage made of rice and milk that does not involve a fermentative process. It is a sweet and refreshing beverage sold in public markets. In the Andean states Tachira, Merida and Trujillo in Venezuela, *chicha* is made from maize flour, pineapple's juice and skins, which are fermented with the maize wort. Cloves, cinnamon, and guava (*Psidium guajava* L.) are normally added. In Colombia a similar *chicha* is still traditionally made and sold in bars and restaurants. It can be an alcoholic drink or not. The spices employed in Colombian fermented beverages are peppermint leaves, marjoram and orange tree leaves. In the tropical regions of Ecuador *chicha* de yucca or cassava is a popular beverage prepared in rural communities and by aboriginal peoples. Cassava is harvested by hands, washed, peeled and crushed. Then, a sweet carrot's puree (*Arracacia xanthorrhiza Bancr.* 1826) is added with the intention to accelerate the natural fermentation process. This last step can be done at different degrees and thus, a slightly sweet and acidic beverage can be obtained. At the beginning of the process the drink is sweet but acquires a yeasty smell as the days pass by. The beverage has a milky appearance. In the Ecuadorean hills *chicha de jora* is elaborated and sweetened with *panela*. In the Andean region of Ecuador a *chicha* made from quinoa, panela and pineapple juice is still traditionally produced. Cloves and allspice (*Pimienta dioica* (L.) Merr.) are also ingredients for this beverage. Peru has a vast diversity of *chichas.* At present, a non-alcoholic beverage is sold as *chicha morada*, which is produced artisanally and the process of its production has been industrialized for export overseas. This drink is found in markets as a ready to drink beverage or can be prepared from a powdered formula. Alcoholic *chichas* in Peru are prepared mainly with maize but they also can include other fruits and vegetables, especially in rural regions. With the black or purple maize cultivar, called *guiñapo,* a purple and sweet *chicha* is elaborated. The white *chicha* is made with several grains such as maize, cereals and spices. *Chicha* de cacao is elaborated during cacao harvesting season. *Chica de jora* can be the base for the elaboration of other kind of drinks. Smashed strawberries can be added to *Chicha de jora* and the drink *frutillada* is obtained. It can be added also with *chicha de maní* which is rich in oil. Bean flour, browned barley and quinoa can become ingredients in *chicha de jora*. In Peru *jora* is produced, packaged and commercialized overseas. Companies such as Inca's Food and Arezzo produce and export bottled chicha. These products can be found even on the Amazon.com web site. Peru is the main and most important producer of *chicha*. Markets for *chicha* are found in Europe and in the United States in which this preparation is sold as a ready to drink beverage or as a seasoning due to its acidic taste. In Bolivia the most traditional and well known *chicha* is the maize *chicha* made from *jora*. Bolivia has an export trade of bottled *chicha* to the United States, Spain, Italy and Sweden. The production is still artisanal but in few cases the process has been fully industrialized. There are regulations that standardize physical-chemical and microbiological parameters in bottled *chicha*. In Argentina *chicha* is still an artisanal and traditional drink prepared mainly by northwestern communities. The production is carried out for traditional celebrations such as the carnival, the day of Dead and in the veneration of goddess Pachamama. *Chicha mukeada* is not allowed by regulatory law but is allowed to be prepared only with beer yeasts. The *chicha* prepared in northwest of Argentina has modifications introduced by Spaniards in some stages in the process. In this region the most traditional *chicha* is prepared on the basis of maize *criollo* and *jora*. Another well-known traditional fermented beverage found in the Northwest of Argentina is the *aloja*. *Aloja* is prepared from carobs from the tree called *algarrobo*. In central Chile, *chichas* are prepared from fruits, especially grapes and apples.

Acknowledgments

We would like to thank Ana Lúcia Barretto Penna, Luis Augusto Nero and Svetoslav Dimitrov Todorov for the invitation to participate in this book and for promoting the knowledge of traditional foods from Latin America.

References

Aguado L. and M. Cabana. (2006). Obtención y caracterización de una bebida alcohólica destilada a partir de chicha. Thesis, Universidad Nacional de Jujuy, San Salvador de Jujuy.

Ayatse, J.O., O.U. Eka and E.T. Ifon. (1983). Chemical evaluation of the effect of roasting on the nutritive value of maize (*Zea mays*, Linn). Food Chemistry 12: 135–147.

Battcock, M. and S. Azam-Ali. (1998). Fermented fruits and vegetables: A global perspective. Food and Agriculture Organization of the United Nations Rome. FAO Agricultural Services Bulletin, 134.

Bewley, J.D. (2001). Seed germination and reserve mobilization. Encyclopedia of life sciences. Available at http://rubisco.ugr.es/fisiofar/pagwebinmalcb/contenidos/Tema27/seed_germination.pdf

Biazus, J.P.M., R.R.D. Souza, J.E. Márquez, T.T. Franco, J.C.C. Santana and E.B. Tambourgi. (2009). Production and characterization of amylases from Zea May malt. Brazilian Archives of Biology and Technology 52(4): 991–1000.

Browne, C.A. (1935). The chemical industries of the American aborigines. Isis,23(2), 406–424.

Chelule, P.K., H.P. Mbongwa, S. Carries and N. Gqaleni. (2010). Lactic acid fermentation improves the quality of amahewu, a traditional South African maize-based porridge. Food Chemistry 122(3): 656–661.

Colehour, A.M., J.F. Meadow, M.A. Liebert, T.J. Cepon-Robins, T.E. Gildner, S.S. Urlacher and L.S. Sugiyama. (2014). Local domestication of lactic acid bacteria via cassava beer fermentation. Peer J, 2:e479; DOI 10.7717/peerj.479.

Cox, L.J., B. Caicedo, V. Vanos, E. Heck, S. Hofstaetter and J.L. Cordier. (1987). A catalogue of some Ecuadorean fermented beverages, with notes on their microflora. MIRCEN Journal of Applied Microbiology and Biotechnology 3(2): 143–153.

Cutler, H.C. and M. Cardenas. (1947). Chicha, a native South American beer. Botanical Museum Leaflets, 13: 33–60.

de Leon Pinelo, A. (1636). Question moral: si el chocolate quebranta el ayuno eclesiástico: tratase de otras bebidas i confecciones que usan en varias provincias. Por la viuda de Iuan Gonçalez.

de Mayolo, S.E.A. (1981) La nutrición en el antiguo Perú. Sociedad Geografica de Lima, Lima.

Dirar, H.A. (1993). The indigenous fermented foods of the Sudan: a study in African food and nutrition. CAB International, Wallington.

Ejigui, J., L. Savoie, J. Marin, T. Desrosiers. (2005). Beneficial changes and drawbacks of a traditional fermentation process on chemical composition and antinutritional factors of yellow maize (*Zea mays*). Journal of Biological Science 5(5): 590-596.

Elizaquível, P., A. Pérez-Cataluña, A. Yépez, C. Aristimuño, E. Jiménez, P.S. Cocconcelli, and R. Aznar. (2015). Pyrosequencing vs. culture-dependent approaches to analyze lactic acid bacteria associated to chicha, a traditional maize-based fermented beverage from Northwestern Argentina. International Journal of Food Microbiology 198: 9–18.

Gade W.D. (1975). Plants, Man and the Land in the Vilcanota Valley of Peru. Dr. W. Junk B.V., Publishers, The Hague.

Gaden, E.L. (1992). Applications of Biotechnology in Traditional Fermented Food. Panel on the Applications of Biotechnology to Traditional Fermented Foods. National Academic Press, Washington D.C.

Gastineau, G.F., W.F. Darby and T.B. Turner. (1979). Fermented food beverages in Nutrition, Academic Press. Washington D.C.

Goldstein, D.J. and R.C. Coleman. (2004). *Schinus molle L. (Anacardiaceae)* chicha production in the central Andes. Economic Botany 58: 523–529.

Haehn, H. (1956). Bioquímica de las fermentaciones. Aguilar Madrid.

Hayashida, F.M. (2008). Ancient beer and modern brewers: Ethnoarchaeological observations of chicha production in two regions of the North Coast of Peru. Journal of Anthropological Archaeology 27: 161–174.

Lantos, I., J.E. Spangenberg. M.A. Giovannetti. N. Ratto and M.S. Maier. (2015). Maize consumption in pre-Hispanic south-central Andes: chemical and microscopic evidence from organic residues in archaeological pottery from western Tinogasta (Catamarca, Argentina). Journal of Archaeological Science 55: 83–99.

Leibowicz, I. (2013). "Una chichería en la Quebrada de Humahuaca" El caso de Juella, Jujuy, Argentina. Intersecciones en Antropología 14: 409–422.

León, R., and B. Hare. (2008). Chicha Peruana: una bebida, una cultura. Universidad de San Martín de Porres, Fondo Editorial, Lima.

López-Arboleda, W.A. and M. Ramírez-Castrillón. (2010). Diversidad de levaduras asociadas a chichas tradicionales de Colombia. Revista Colombiana de Biotecnología 12(2): 176-186.

Marsh, A.L., C. Hill, R.P. Ross and P.D. Cotter. (2014). Fermented beverages with health promoting potential: Past and Future perspectives. Trends in Food Science and Technology 38: 113–124.

McGovern, P.E., J. Zhang, J. Tang, Z. Zhang, G.R. Hall, R.A. Moreau, A. Nuñez, E.D. Butrym, M.P. Richards, C. Wang, G. Cheng, Z. Zhao and C. Wang. (2004). Fermented beverages of pre-and proto-historic China. Proceedings of the National Academy of Sciences of the United States of America 101: 17593–17598.

Nicholson, G.E. (1960). Chicha maize types and Chicha manufacture in Peru. Economic Botany 14: 290–299.

Otero, C. and M.B. Cremonte. (2009). Desde el maíz a la chicha. Preparación de la bebida ancestral andina en Villa El Perchel, quebrada de Humahuaca, Jujuy. Editorial de la Facultad de Filosofía y Letras Universidad de Buenos Aires, Buenos Aires.

Pardo, O. and J.L. Pizarro. (2005). La chicha en el Chile precolombino. Mare Nostrum, Santiago de Chile.

Pedersen, A.M., A. Bardow, S.B. Jensen and B. Nauntofte. (2002). Saliva and gastrointestinal functions of taste, mastication, swallowing and digestion. Oral Diseases 8(3), 117–129.

Purlis, E. (2010). Browning development in bakery products–A review. Journal of Food Engineering 99: 239–249.

Rufián-Henares, J.A., C. Delgado-Andrade, and F.J. Morales. (2009). Assessing the Maillard reaction development during the toasting process of common flours employed by the cereal products industry. Food Chemistry 114: 93–99.

Steinkraus, K.H. (1997). Bio-enrichment: production of vitamins in fermented foods. In: Wood BJB (eds.). Microbiology of fermented foods. Thomson Science. New York, pp. 603–621.

Tonroy, B.R. and T.W. Perry. (1975). Effect of roasting corn at different temperatures on grain characteristics and in-vitro starch digestibility. Journal of Dairy Science 68: 566

Van Hylckama Vlieg J.E.T., P. Veiga, C. Zhang, M. Derrien, L. Zhao. (2011). Impact of microbial transformation of food on health – from fermented foods to fermentation in the gastro-intestinal tract. Current Oppinion in Biotechnology 22(2): 211-219.

Ward, O.P. (1989). Fermentation biotechnology: principles, processes and products. Open University Press, Milton Keynes.

13

Increasing Folate Content Through the Use of Lactic Acid Bacteria in Novel Fermented Foods

Marcela Albuquerque Cavalcanti de Albuquerque[a]*, Raquel Bedani*[a]*,
Susana Marta Isay Saad*[a] *and Jean Guy Joseph LeBlanc*[b]*

Abstract

Folate is an essential B-group vitamin that plays a key role in numerous metabolic reactions such as energy usage and the biosynthesis of DNA, RNA, and some amino acids. Since humans cannot synthesize folate, an exogenous supply of this vitamin is necessary to prevent nutritional deficiency. For this reason, many countries possess mandatory folic acid enrichment programs in foods of mass consumption; however, there is evidence that high intakes of folic acid, the synthetic form of folate, but not natural folates, can cause adverse effects in some individuals such as the masking of the hematological manifestations of vitamin B_{12} deficiency. Currently, many researcher groups are evaluating novel alternatives to increase concentrations of natural folates in foods. Lactic acid bacteria (LAB), widely used as starter cultures for the fermentation of a large variety of foods, can improve the safety, shelf life, nutritional value, flavor, and overall quality of the fermented products. Although most LAB are auxotrophic for several vitamins, it is now known that certain strains have the capability to synthesize some B-group vitamins. In this Chapter, the use of specific strains of folate producing LAB for the design of novel fermented food products will be discussed as well their use as an important strategy to help in the prevention of folate deficiency and as a safer alternative to mandatory folic acid fortification programs.

Introduction

Folic acid or vitamin B_9, is an essential component of the human diet and is involved in many metabolic pathways (Rossi et al. 2011; LeBlanc et al. 2013). This micronutrient is a water-soluble vitamin and is part of the B-group vitamins. As it is not

[a] Departamento de Tecnologia Bioquímico-Farmacêutica, Faculdade de Ciências Farmacêuticas, Universidade de São Paulo, 05508-000, São Paulo, SP, Brasil.
[b] CERELA-CONICET, C.P. T4000ILC, San Miguel de Tucumán, Argentina.
* Corresponding author: leblanc@cerela.org.ar or leblancjeanguy@gmail.com

synthesized by mammals, this vitamin must be obtained through food ingestion (Lin and Young 2000; Sybesma et al. 2003; Padalinno et al. 2007; Kariluoto et al. 2010).

Folate may be synthesized by various plants and some microorganisms. Dairy and non-dairy products are considered important sources of folate (Sybesma et al. 2003; Padalino et al. 2012; Laiño et al. 2013b). The main forms of folate are tetrahydro-folate (THF), 5-formyltetrahydrofolate (5-FmTHF), and 5-methyltetrahydrofolate (5-MeTHF). According to Lin and Young (2000) and Uehara and Rosa (2010), the latter is the principal form that is transported and stored in the human body and, therefore, folates derived from the diet needs to be converted into smaller residues called monoglutamates in order to be properly absorbed.

Various activities of our body are related to folate, such as: DNA replication, repair and methylation, biosynthesis of nucleic acids and amino acids protect against certain types of cancers and decreased risk of cardiovascular disease (Kariluoto et al. 2010).

Folate deficiency is a public health problem (Dantas et al. 2010) with great impact for pregnant women and consequently on the development of the fetuses (Santos et al. 2007). This deficiency, among other factors, may cause defects in neural tube formation, which is a congenital malformation resulting from the failure of the embryonic neural tube closure. This phenomenon may lead to anencephaly and spina bifida (Laiño et al. 2012; Fujimore et al. 2013). As a result, recommended doses of daily intake were proposed by agencies of various countries in order to reduce the problems caused by folate deficiency in individuals.

To avoid problems caused by folate deficiency, many people consume vitamin supplements. However, these supplements are usually developed from the synthetic folate form (folic acid), which is chemically produced. Besides having a high cost of production, it is known that synthetic folate may cause harm to human health (Wyk et al. 2011; Hugenschimidt et al. 2011). Folic acid may, among other things, mask vitamin B_{12} deficiency. On the other hand, folate in its natural form (as found in foods or through the production of certain microorganisms) does not cause adverse effects in health (Laiño et al. 2013a). Thereby, the inclusion of bioenriched foods in the diet, in which folate is produced by non-synthetic technologies can be an alternative for the lower daily intake of folate (Uehara and Rosa 2010).

Certain strains of LAB possess, among other properties, the ability to produce folate, which is dependent on species, strain, growth time, and growth conditions (Sybesma et al. 2003; Pompei et al. 2007; Kariluoto et al. 2010; D'Aimmo et al. 2012; Laiño et al. 2012; Padalino et al. 2012; Laiño et al. 2013a). The use of vitamin-producing microorganisms in food may represent a more natural supplement alternative with increased acceptability by consumers. The production of foods with higher concentrations of natural vitamins produced by LAB would not cause harm to human health (LeBlanc et al. 2013).

Folate

Chemical structure, bioavailability, and functions

Folate, also known as vitamin B_9, is a critical molecule in cellular metabolism. Folate is the generic term of the naturally occurring folates and folic acid (FA), which is the fully oxidized synthetic form of folate added to foods as fortifier and used in dietary supplements (Laiño et al. 2013a; Fajardo et al. 2015). Folic acid or pteroyl glutamic acid (PGA) is comprised of *p*-aminobenzoic acid flanked by a pteridine ring and L-glutamic acid (Laiño et al. 2013a; Walkey et al. 2015). On the other hand, the naturally occurring folates differ in the extent of the reduction state of the pteroyl group, the nature of the substituents on the pteridine ring and the number of glutamyl residues attached to the pteroyl group (Laiño et al. 2013a). The natural folates include 5-methyltetrahydrofolate (5-MTHF), 5-formyltetrahydrofolate (5-formyl-THF), 10-formyltetrahydrofolate (10-formyl-THF), 5,10-methylenetetrahydrofolate (5,10-methylene-THF), 5,10-methenyltetrahydrofolate (5,10-methenyl-THF), 5-formiminotetrahydrofolate (5-formimino-THF), 5,6,7,8-tetrahydrofolate (THF), and dihydrofolate (DHF). The most naturally occurring folates are pteroylglutamates with two or seven glutamates joined in amide (peptides) linkages to the g-carboxyl of glutamate. The main intracellular folates are peteroylpentaglutamates and the major extracellular folates are pteroylmonoglutamates. pteroylpentaglutamates with up to 11 glutamic acid residues occur naturally (LeBlanc et al. 2007).

In general, bioavailability may be defined as the proportion of a nutrient ingested that becomes available to the organism for metabolic process or storage (LeBlanc et al. 2007). It is important to point out that the dietary folate bioavailability may be impaired by the polyglutamate chain that most natural folate have (McNulty and Pentieva 2004; LeBlanc et al. 2007). Natural food folates or pteroylpolyglutamates are hydrolyzed by the enzymes γ-glutamyl hydrolase or human conjugase to pteroylmonoglutamate forms prior to absorption in small intestine, mainly in duodenum and jejunum. Therefore, the dietary polyglutamates are hydrolyzed and then reduced and methylated in the enterocyte before being absorbed (Silva and Davis 2013). The monoglutamates forms of folate, including FA, are transported across the proximal intestine through a saturable pH-dependent process (Iyer and Tomar 2009). Higher doses of folic acid are absorbed via a non-saturable passive diffusion process (Hendler and Rorvik 2001; Iyer and Tomar 2009). Folate may be absorbed and transported as monoglutamates into the liver portal vein (LeBlanc et al. 2007). Folate enters the portal circulation as methyl-THF, which is the predominant form of the vitamin in the plasma (Silva and Davis 2013). The main transporter of folate in plasma is the reduced folate transporter (RFC), which delivers systemic folate to the tissues. The high affinity folate receptors (FRs) are expressed on various epithelia and are another family of folate transporters (Zhao et al. 2011).

Approximately 0.3 to 0.8% of the body folate pool is excreted daily, in both urine and feces. Renal excretion increases at higher folate intakes (Ohrvik and Witthof 2011; Silva and Davis 2013), which is normal since the excess of all water soluble vitamins (including the B group vitamins) is excreted.

Evidences suggest that the polyglutamate form is 60 to 80% bioavailable compared to the monoglutamate form (Gregory 1995; Melse-Boonstra et al. 2004; LeBlanc et al. 2007). Although there are controversies, the absorption efficiency of natural folates is approximately half of that of synthetic FA and the relative bioavailability of dietary folates is estimated to be only 50% in comparison with synthetic folic acid (Sauberlich et al. 1987; Gregory 1995; Forssen et al. 2000; McNulty and Pentieva 2004; Iyer and Tomar 2009). Additionally, folic acid absorption in an empty stomach is twice as available as food folate, and folic acid taken with food is 1.7 times as available as food folate (Hendler and Rorvik 2001). In this line, according to Shuabi et al. (2009), the assessment of folate nutritional status is incomplete if content values in the food composition database do not account the differences in bioavailability between naturally occurring folate and synthetic FA as a food fortificant, and folate supplement usage.

Folate-binding proteins from milk may increase the efficiency of folate absorption by protecting dietary folates from uptake by intestinal bacteria, thus leading to increased absorption in the small intestine (Eitenmiller and Landen 1999). Other dietary factors that may influence the folate bioavailability include: the effects of foods on intestinal pH with potential modification of conjugase activity, presence of folate antagonists, intestinal changes influenced by dietary factors (for example, alcoholism), chelation, and factors that influence the rate of gastric emptying (LeBlanc et al. 2007).

A study developed by Pounis et al. (2014) assessed the possible differences in folate status in two European Union countries and their possible association with dietary patterns and/or other lifestyles. These researchers reported that both inadequate dietary folate intake and serum levels were observed in Italian participants of their study, whereas in individuals from southwest London, folate status seemed slightly better. According to the authors, differences between countries in food group consumption as good sources of folate might explain this result. Additionally, non-smoking habits and physical activity were the two non-dietary, lifestyle characteristics positively associated with folate serum levels.

Evidences suggest that serum folate concentrations express recent folate intake, while red cell folate has a tendency to provide a better reflection of tissue folate status. Thus, serum folate may be considered a good predictor of recent dietary intake (Truswell and Kounnavong 1997; Shuabi et al. 2009).

Folate is an essential micronutrient that plays an important role in the human metabolism. It acts as a cofactor in several biosynthetic reactions, serving primarily as a one-carbon donor. It is involved in the methylations and formylations that occur as part of nucleotide biosynthesis; therefore, the folate deficiency may cause defect in DNA synthesis in tissue with rapidly replicating cells (Chiang et al. 2014; Walkey et al. 2015). Thereby, the most remarkable consequence of folate deficiency occurs during pregnancy (Walkey et al. 2015). In general, folates are involved in a wide number of key metabolic functions including DNA replication, repair, and methylation and biosynthesis of nucleic acid, some amino acids, pantothenate, and other vitamins (Laiño et al. 2013).

Owing to the role of folate in nucleotide biosynthesis, its privation impairs DNA synthesis in embryonic tissue, resulting in a reduction in the rate of cellular division and congenital malformations of the brain and spinal cord, such as neural tube defects (NTDs). Chronic alcoholism may lead to folate deficiency that may result in megaloblastic anemia. Additionally, the folate deficiency leads to elevated plasma homocysteine levels, a risk factor for cardiovascular diseases (Cravo et al. 2000).

Food rich in folate and folate requirements

As human beings do not synthesize folates, and it is necessary to assimilate this vitamin from exogenous sources. Folate occurs in most foods, with at least 50% being in the polyglutamate form. FA is thermolabile and thus may be destroyed by cooking (Silva and Davis 2013).

In general, folates are present in most foods such as legumes (beans, nuts, peas, and others), leafy greens (spinach, asparagus, and Brussels sprouts), citrus, some fruits, grains, vegetables (broccoli, cauliflower), liver, and dairy products (milk and fermented dairy products) (Eitenmiller and Landen 1999; Walkey et al. 2015). Fermented milk products, for example yogurt, may contain higher concentration of folates (around 100 µg/L) compared to non-fermented milk (around 20 µg/L) (Lin and Young 2000; Iyer and Tomar 2009). Scientific evidences suggest that certain strains of LAB may synthesize natural folates (Lin and Young 2000; Aryana, 2003;, Laiño et al. 2014). In this sense, the use of LAB in fermentative process may represent an interesting biotechnological approach to increase folate levels in milk (Laiño et al. 2013b; Laiño et al 2014). It is noteworthy that the ability of microorganisms to produce folate is considered strain-dependent (D'Aimmo et al. 2012; LeBlanc et al. 2013, 2014). Nevertheless, these folate sources are unstable and large losses occur as a result of heat exposure, typical of many food and cooking procedures (Liu et al. 2012). This phenomenon may hamper to estimate the intake of total dietary folates by consumers (Walkey et al. 2015).

The lack of folates in the diet is one of the most common nutritional deficiencies in the world and has severe consequences on health (Herbison et al. 2012). Traditionally, folate deficiency in humans has been related to macrocytic or megaloblastic anemia. However, this deficiency is also associated with health disorders such as cancer, cardiovascular diseases, and neural tube defects in newborns (Wang et al. 2007; Ohrvik and Witthoft 2011). In this line, to reduce the risk of neural tube defects in newborns, the increased folate consumption for woman in the periconceptional period is crucial to keep an optimal folate status by considering the relationship between folate intake and blood folate concentration (Stamm and Houghton 2013).

Regarding the mean dietary reference intakes, epidemiological evidences indicate that a suboptimal folate intake may be widespread in the population in both developing and developed countries (Hermann and Obeid 2011; Fajardo et al. 2015). The increase in folate intake in the population may be achieved by the consumption of foods naturally rich in folates; use of FA supplements; and consumption of foods fortified with synthetic FA or natural folates (López-Nicolás et al. 2014). It is worth mentioning that the potential adverse effects of synthetic FA, such as masking symptoms of vitamin B_{12} deficiency and promoting certain types of cancer, have inhibited mandatory fortification in some countries (Fajardo and Varela-Moreiras 2012). According to Fajardo et al. (2015), the promotion of folate intake from natural food sources continues to be a health strategy to reach a safe and adequate nutritional status.

Regarding the promoting of certain kind of cancers, a case-control design study with 408 volunteers developed by Chiang et al. (2014) evaluated the association between serum folate and the risk of colorectal cancer (CRC) in subjects with CRC or colorectal adenomatous polyps (AP, a precursor of CRC), and healthy subjects. The authors concluded that higher serum folate concentration ($\geq$13.55 ng/mL) appeared to be associated with increased risk in subjects with AP while serum folate had no effect on CRC risk in healthy controls. Additionally, these researchers speculate that serum folate might play a dual role regarding CRC risk.

The daily recommended intake (DRI) of folate in the European Union (EU) is 200 and 600 µg/day for adults and women in periconceptional period, respectively (IOM, 2006). The recommended dietary allowance (RDA) of folate in adults is also

200–400 mg/day (FAO/WHO 2002) and the body stores around 10–20 mg which is usually sufficient for only about 4 months. Pregnancy significantly increases the folate requirement, particularly throughout periods of rapid fetal growth (for example, in the second and third trimester). In addition, during the lactation, losses of folate in milk also increase the folate requirement of mothers (McPartlin et al. 1993; FAO/WHO 2002). According to the Brazilian legislation, the daily recommended intake of folate is also 0.4 mg/day for adults, 0.6 mg/day for pregnant, and 0.5 mg/day for women who are breastfeeding (ANVISA 2005).

Fortification programs

Several countries have attempted to ensure adequate folate intake and prevent the disorders related to folate deficiency by mandatory FA fortification of cereal products. Procedures for mandatory fortification of wheat flour with FA have been in place in several countries; however, in many cases, these regulations have not been implemented or lack independent controls (Crider et al. 2011). As a result, in 2006, the World Health Organization and the Food and Agricultural Organization of the United Nations issued guidelines to help countries to set the Target Fortification Level, the Minimum Fortification Level, the Maximum Fortification Level and the Legal Minimum Level of folic acid to be used to fortify flour with FA (FAO/WHO 2006). In the United States of America and Canada, the implantation of mandatory fortification of cereal grain products with FA occurred in 1998. The United States of America program adds 140 µg of FA per 100 g of enriched cereal grain product and has been estimated to provide 100–200 µg of folic acid per day to women of childbearing age (Rader et al. 2000; Quinlivan and Gregory 2007; Yang et al. 2007; Crider et al. 2011). In countries such as Argentina and Brazil, the flour fortification with FA became obligatory in 2002 and 2004, respectively. The Brazilian legislation recommends the addition of 150 µg of FA per 100 g of wheat and maize flours (ANVISA 2002). Other countries with mandatory FA fortification programs include Canada (150 µg/100 g), Costa Rica (180 µg/100 g), Chile (220 µg/100 g), and South Africa (150 µg/100 g) (Crider et al. 2011). Evidences suggest that NTDs has declined (19 to 55%) in Canada, South Africa, Costa Rica, Chile, Argentina, and Brazil since the introduction of folic acid fortification practices (Crider et al. 2011).

On the other hand, many countries have not adopted a National Fortification Program with FA due to its potential undesirable adverse effects (Laiño et al. 2013a). In several European countries, the fortification is not obligatory mainly due to a concern that FA fortification may damage individuals with undiagnosed vitamin B_{12} deficiency (Smith et al. 2008). In Italy, for example, FA fortification is not mandatory and the supplementation of women of childbearing age or health promotion strategies aim at increasing intake of dietary sources (Pounis et al. 2014). Similarly, Finland does not allow mandatory fortification of staple foods with FA. In this country, a balanced diet, rich in folate, is recommended for all women planning a pregnancy or in early pregnancy, to obtain at least 400 µg of folate daily. Additionally, a daily supplement of 400 µg of FA is recommended for all women planning a pregnancy or in early pregnancy and voluntary fortification of certain food products is allowed (Samaniego-Vaesken et al. 2012; Kariluoto et al. 2014).

Due to the potential risks of the fortification with FA, there has been growing interest in the fortification of foodstuffs with natural form folates (Scott 1999; LeBlanc et al. 2007; Iyer and Tomar 2009). In such cases, natural folates, such as 5-MTHF, that are usually found in foods and produced by microorganisms do not mask vitamin B_{12} deficiency; therefore they might be considered a promising alternative for fortification with FA (Scott 1999; Kariluoto et al. 2014; Laiño et al. 2014).

Folate Production by Microorganisms

Nowadays, there is an increased demand by consumers to acquire healthy diets through the intake of natural foods without or at least with lower amounts of chemical preservatives. Folate deficiency occurs in several countries and the main reasons are the insufficient intake of food, restricted diets, and low purchasing power to obtain fortified foods. In order to prevent this deficiency problem, the governments of some countries adopted mandatory FA fortification programs as discussed previously. However, the supplementation of foods with synthetic FA is considered by some researchers as potentially dangerous for human health because people who keep normal or elevated folate ingestion from normal diets would be exposed to a higher FA intake which could mask hematological manifestations of B_{12} vitamin, as also mentioned previously (LeBlanc et al. 2007). For these reasons, it is necessary to develop alternatives to the use of synthetic FA. In some counties, mandatory folate fortification is not allowed and, in these cases, natural folate enhancement of foods appears as a promising alternative.

LAB are a very important group of microorganisms for the food industry since they are used to ferment a large variety of foods, such as dairy, vegetables, and various types of bread (Capozzi et al. 2012). This bacterial group can improve the safety, shelf-life, nutritional value, flavor, and overall quality of fermented products through the production of many beneficial compounds in foods. Among these compounds, some strains have the ability of producing, releasing and/or increasing B-group vitamins. It is known that some strains used as starter cultures in the fermentation process are able to synthesize folate as reported by many studies (Lin and Young 2000; Crittenden et al. 2003; Iyer et al. 2010; Laiño et al. 2012). Folates produced by microorganisms are natural (especially 5-MTHF are produced) and do not cause adverse effects to the human body. Therefore, studying and selecting folate-producing strains and using them to develop folate bio-enriched food would be beneficial to the food industry since it is a very cheap process that provides a high value-add to their products and to consumers who demand more natural foods.

Some studies have focused on the screening of several strains of LAB for their ability to produce folate, which can be produced intracellularly and/or extracellularly by selected microorganisms (Table 13.1). Sybesma et al. (2003) evaluated the effects of cultivation conditions on folate production by LAB strains and they observed that the intracellular and extracellular folate produced by *Streptococcus thermophillus* was influenced by the medium pH values. At lower pH, this microorganism produced more extracellular folate than those that grew in higher pH values. A possible explanation is that at low intracellular pH, folate is protonated and becomes electrically neutral. Thus, folate would enhance transport across the membrane, increasing the amount of this vitamin in the extracellular medium. However, for *Lactococcus lactis*, these authors identified no difference in the intra- and extracellular folate distribution when the microorganism grew in low or high pH. Kariluoto et al. (2010) identified some folate producer microorganisms which presented higher intracellular folate content when they grew in high pH values.

TABLE 13.1 Reports on folate production by microorganisms in folate-free medium

Microbial species	Extracellular content	Intracellular content	Total content	Reference
Lactococcus species				
Lactococcus lactis subsp. *cremoris* MG1363	46 µg/L	69 µg/L	115 µg/L	Sybesma et al. (2003)
Lactococcus lactis subsp. *lactis* NZ9000	11 µg/L	245 µg/L	256 µg/L	Sybesma et al. (2003)
Lactobacillus species				
Lactococcus amylovoros CRL887	68.3 ± 3.4 µg/L	12.9 ± 1.3 µg/L	81.2 ± 5.4 µg/L	Laiño et al. (2014)
Lactobacillus plantarum CRL103	16.7 ± 3.4 µg/L	40.5 ± 4.2 µg/L	57.2 ± 5.2 µg/L	Laiño et al. (2014)
Lactobacillus acidophilus CRL1064	21.9 ± 2.3 µg/L	15.3 ± 1.4 µg/L	37.2 ± 3.1 µg/L	Laiño et al. (2014)
Lactobacillus delbrueckii subsp. *bulgaricus* CRL8711	86.2 ± 0.3 µg/L	8.6 ± 0.1 µg/L		Laiño et al. (2012)
Lactobacillus helveticus	1 ng/mL	90 ng/mL	89 µg/L	Sybesma et al. (2003)
Streptococcus species				
Streptococcus thermophilus CRL803	76.6 ± 7.0 µg/L	15.9 ± 0.2 µg/L		Laiño et al. (2012)
Bifidobacterium species				
Bifidobacterium adolescentes ATCC 15703			8865 ± 355 µg/100g DM[1]	D'Aimmo et al. (2012)
Bifidobacterium catenulatum ATCC 27539			9295 ± 750 µg/100g DM[1]	D'Aimmo et al. (2012)
Propionibacterium species				
Propionibacterium freudenreichii	25 ± 3 ng/mL			Hugenschimidt et al (2011)

[1]Dry matter

Lactococcus lactis and *Streptococcus thermophillus* are two industrially important microorganisms that have the ability to produce folate, but other LAB such *Lactobacillus delbrueckii* subsp. *bulgaricus*, *Lactobacillus acidophilus*, *Lactobacillus plantarum*, *Lactobacillus reuteri*, *Leuconostoc lactis*, *Propionibacterium* spp., and *Bifidobacterium* spp. also have this ability (Lin and Young 2000; Crittenden et al. 2003; Pompei et al. 2007; Santos et al. 2008; Hugenschmidt at al. 2011; D'Aimmo et al. 2012; Laiño et al. 2014). These species are normally involved in the fermentation of dairy products. However, natural microbiota from raw materials used by the food industry may also affect the folate level of some cereal products (Kariluoto et al. 2010).

LAB starter cultures isolated from artisanal Argentinean yogurts were tested by Laiño et al. (2012). In this study certain strains of *Streptococcus thermophillus* and *Lactobacillus delbrueckii* subsp. *bulgaricus* were reported to be able to increase folate in folate-free culture medium. It is important to mention that not all lactobacilli strains are able to produce folates and it has historically been believed that, in yogurt production, *S. thermophilus* produces folates whereas *Lactobacillus bulgaricus* consumes this vitamin in a symbiotic relationship. It is now known that some strains of *Lactobacillus bulgaricus* can in fact produce folates (Laiño et al. 2012). Therefore, the screening of different lactobacilli strains within different species should be conducted since the production of folate by microorganism is strain-dependent (Table 13.1). Sybesma et al. (2003) identified a *Lactobacillus plantarum* strain that was able to produce folate. In this context, Nor et al. (2010) verified that the use of *Lactobacillus plantarum* I-UL4 led to an increase in folate content from 36.36 to 60.39 µg/L using an optimized medium formulation compared to Man Rogosa Sharp (MRS) broth.

Masuda et al (2012) isolated 180 LAB strains from a Japanese food named *nukazuke,* a traditional Japanese pickle made of salt and vegetables in a fermented rice bran bed. From these 180 isolated strains, only 96 grew in a free-folate medium. Since 58.4% of the strains belonged to the *Lactobacillus* genus, a significant number of strains did not grow in the folate-free medium clearly demonstrating that not all lactobacilli strains produce this important vitamin. However, three lactobacilli strains (*Lactobacillus sakei* CN-3, *Lactobacillus sakei* CN-28 and *Lactobacillus plantarum* CN-49) were shown to produce extracellulary high levels of folate 101 ± 10 µg/L, 106 ± 6 µg/L, and 108 ± 9 µg/L, respectively.

Different studies have evaluated the effect of *p*ABA (*para*-aminobenzoic acid), a precursor of folate, on folate production by LAB. Not all strains possess the genes necessary for *p*ABA biosynthesis and this could be a limiting factor for folate production in some microorganisms (Rossi et al. 2011). In a review about production of folate by probiotic bacteria, Rossi et al. (2011) showed the presence or absence of genes and enzymes necessary for the biosynthesis of DHPPP (6-hydroxymethyl-7,8-dihydropterin pyrophosphate), tetrahydrofolate-polyglutamate, chorismate, and *p*ABA from the sequenced genomes of some *Lactobacillus* spp., *Bifidobacterium* spp., and other LAB. According to these authors, *Lactobacillus plantarum* is able to produce folate only in the presence of *p*ABA, since the ability to synthesize *p*ABA *de novo* does not appear in several members of the genus *Lactobacillus*. This could explain why many lactobacilli do not produce this vitamin.

Other strains of lactobacilli, such as *Lactobacillus amylovorus, Lactobacillus acidophilus, Lactobacillus casei, Lactobacillus fermentum, Lactobacillus paracasei,* and *Lactobacillus plantarum*, isolated from a wide range of artisanal Argentinean dairy products, were tested for the ability to produce folate in a folate-free synthetic medium. Folate amounts were found in the supernatant of some strains belonging to each of these bacterial species (Laiño et al. 2014).

In addition to *Lactobacillus*, another important probiotic genus is *Bifidobacterium*. This group has a relevant impact on human health, due to its association to beneficial effects by the gut microbiota. D'Aimmo et al. (2012) investigated a total of 19 strains of *Bifidobacterium* for their capacity to produce folate in free-folate medium. The results showed that the highest value of folate was found for *Bifidobacterium catenulatum* ATCC 27539 (9,295 µg per 100 g of dry matter). On the other hand, the lowest value was found for *Bifidobacterium animalis* subsp. *animalis* ATCC 25527 (220 µg per 100 g of dry matter).

Pompei et al. (2007) administered three bifidobacteria strains (*Bifidobacterium adolescentis* MB 227, *Bifidobacterium adolescentis* MB 239, and *Bifidobacterium pseudocatenulatum* MB 116) to folate-depleted Wistar rats. These deficient rats were positively affected by the administration of bifidobacteria strains, since these bacteria produced folate *in vivo* and could thus be consider probiotic microorganisms.

Folate-producing probiotic strain could be used to develop new functional foods without the need of recurring to fermentation, since these microorganisms could produce vitamins directly in the GIT. Probiotics are "live microorganisms which when administered in adequate amounts confer a health benefits on the host" (FAO/WHO 2002). The most commonly studied probiotics have been associated with strains from *Lactobacillus* and *Bifidobacterium*.

Folates are very sensitive to heat treatments and the amount of this vitamin in vegetables, for example, could decrease after the cooking process (Delchier et al. 2014). Pasteurization and ultra high temperature (UHT) processes of raw milk also reduce folate levels (Lin and Young 2000). Since lower pH levels have been shown to protect folates from heat-destruction, fermentation can be used to produce folate bio-enriched foods by lowering pH values and preventing vitamin losses and, in this way, avoiding the need to supplement/fortify foods with synthetic folic acid.

Besides LAB, other microorganisms are also able to produce folate. Kariluoto et al. (2010) isolated 20 strains of bacteria from three commercial oat bran products and tested them for their ability to produce folate. *Bacillus subtilis* ON5, *Chryseobacterium* sp. NR7, *Curtobacterium* sp. ON7, *Enterococcus durans* ON9, *Janthinobacterium* sp. RB4, *Paenibacillus* sp. ON11, *Propionibacterium* sp. RB9 and *Staphylococcus kloosii* RB7 were the best folate producers in culture medium. Kariluoto et al. (2006) also evaluated the potential of three sourdough yeasts, *Candida milleri* CBS 8195, *Saccharomyces cerevisiae* TS 146, and *Torulaspora delbrueckii* TS 207 to produce folate. A baker's yeast *Saccharomyces cerevisiae* ALKO 743 and four *Lactobacillus* strains from rye sourdough were also examined. The strains did not produce significant amounts of extracellular folates in yeast extract-peptone-D glucose medium but others should be tested in order to identify new folate producing strains that could be useful in the production of bakery products.

Bio-enriched Foods with Folate Produced by Microorganisms

As previously mentioned, an alternative to FA supplementation is the development of new food products bio-enriched with natural folates produced by microorganisms, using fermentative process. This strategy might be an innovative and cheap way to increase this vitamin in different products. However, it is very important to identify more microorganisms as folate producers, especially LAB, since these bacteria are widely used in fermentative process of dairy products. As discussed above, this group of bacteria is extensively used by the food industry, mainly in dairy products like fermented milk or yogurts. It is known that the folate content in milk is not high, especially after the application of pasteurization or UHT processes (Lin and Young

2000), and thus the fermentation of this product by folate-producing microorganisms could increase the levels of this vitamin. Natural folates produced by microorganisms, such as 5-methyltetrahydrofolate, are usually found in foods (Laiño et al. 2014).

Several studies have evaluated the production of folate by different LAB strains in milk environments. In this line, Laiño et al (2013a) tested starter cultures, including *Streptococcus thermophilus* and *Lactobacillus delbrueckii* subsp. *bulgaricus*, for folate production in milk. In the study, the authors observed that the strains used in co-culture increased folate levels significantly (180 ± 10 µg/L of folate) compared to unfermented milk (250% increase) and to commercial yogurts (125% increase). Also, the folate amount showed no significant changes during the product shelf-life (28 days of storage at 4 °C) making this product interesting from a technological point of view.

Gangadharan and Nampoothiri (2011) evaluated a fermented skimmed milk using a strain of *Lactococcus lactis* subsp. *cremoris*. The authors obtained 187 ng/mL of folate using a 5 L bioreactor. The effect of sorbitol and mannitol on folate content was evaluated. Mannitol promoted an increased folate production compared to sorbitol. This strain also enriched cucumber (10 ± 0.2 to 60 ± 1.9 ng/mL) and melon juice (18 ± 0.9 to 26 ± 1.6 ng/mL) in folates. Folate binding proteins from milk scavenger the vitamin from blood plasma, protecting it, thus preventing folate losses as well as improving its bioavailability and stability when consuming dairy products (Nygren-Babol and Jägerstad 2012).

Holasová et al. (2005) investigated folate increase in fermented milk by the fermentation process and through the addition of fruit components, such as pineapple, sour cherry, kiwi, apricot, peach, apple, strawberry, blueberry, and raspberry. After 12 hrs at 37 °C, the researchers observed that milk sample inoculated with butter starter cultures of *Streptococcus thermophilus* and *Bifidobacterium longum* achieved 3.39 µg/100 g while the milk sample inoculated with butter starter cultures of *Streptococcus thermophilus* and *Propionibacterium* spp. achieved 4.23 µg/100 g of 5-methyltetrahydrofolate. Moreover, the incorporation of strawberry led to the highest amount of folate. The authors concluded that the addition of this fruit component to the fermented milks may increase the product's natural folate content.

Kefir grains are a kind of natural immobilized culture and the beverage fermented by these grains is recognized as a probiotic dairy product. In this way, the milk fermentation process may increase vitamin content using kefir and a *Propioni bacterium* culture. Wyk et al. (2011) included *Propionibacterium freudenreichii* strains into kefir grains and observed that the best treatment delivered 19% Recommended Dietary Allowance of folate per 200 mL of product.

The applicability of *Lactobacillus amylovorus* strain in co-culture with yogurt starter cultures (*Lactobacillus bulgaricus* CRL871, *Streptococcus thermophilus* CRL803 and CRL415) to produce folate bio-enriched fermented milk was evaluated by Laiño et al. (2014). In the study, a yogurt containing high folate content (263.1 ± 2.4 µg/L) was obtained. Divya and Nampoothirii (2015) identified two strains of *Lactococcus lactis* (CM22 and CM28) isolated from cow milk and checked if these strains, when encapsulated, might be used to fortify milk and ice cream with natural folate through the product fermentation. The resulting fermented products showed an enhancement of the folate content; however, the vitamin production by *Lactococcus lactis* CM22 was higher than by *Lactococcus lactis* CM28 demonstrating once again that folate production is a strain-dependent trait.

An interesting study conducted by Iyer and Tomar (2011) assessed the effect of folate-rich fermented milk produced by two strains of *Streptococcus thermophilus* (RD 102 and RD 104) on hemoglobin level using mice model. Four groups of eight mice 30 ± 10 days old each were fed with four formulations (group 1, a basal diet with a synthetic anemic diet; group 2, a basal diet with skim milk; group 3, a basal diet with fermented skim milk produced by RD 102 and group 4, a basal diet with fermented skim milk produced by RD 104). The groups of animals that received the milks fermented with the folate producing strains (group 3 and 4) showed significant increases in mice hemoglobin level compared with the control groups.

Several studies have shown that some LAB strains may exert, beyond their sensorial attributes to the food, beneficial properties to the host. Certain strains of LAB can produce significant amounts of folate and are able to survive to the gastrointestinal tract (GIT) passage. Therefore, the identification and selection of possible probiotic folate producers would be very important to the development of probiotic foods with increased nutritional value (Laiño et al. 2013b). In this context, Crittenden et al. (2003) investigated the potential of probiotic cultures regarding the synthesis and utilization of folate in milk. The authors concluded that the combination of strains of *Streptococcus thermophilus* and *Bifidobacterium animalis* increased more than six fold (72 ng/g) the folate content of milk. The researchers also showed that *Streptococcus thermophilus* and all probiotic bifidobacteria strains were the best folate producers in the study, whereas *Lactobacillus* strains depleted folate in the skimmed milk. In addition, milk fermentation by *Enterococcus faecium* also increased folate content.

In another study, the production of natural folates was evaluated using *Lactococcus lactis* subsp. *cremoris* to fortify skimed milk and fruit juices (Gangdharan and Nampoothiri 2014). The results showed that this microorganism was able to produce folate and enhance the vitamin content in skimmed milk and in cucumber and water melon juice. To test different food matrices, not only milk, in order to enrich food with natural folate, it is important to develop new bio-enriched products since, due to lactose intolerance or milk proteins allergy, not everybody can consume dairy products. Fermented folate enriched fruit juices could be an alternative for this kind of consumers, as well as for vegetarians.

Pompei et al. (2007) evaluated the production of folate by some *Bifidobacterium* strains with potentially probiotic properties. According to the results obtained, the best folate producers were *Bifidobacterium adolescentis* MB 115 (65 ng/mL) and *Bifidobacterium pseudocatenulatum* MB 116 (82 ng/mL). Even though the strains were cultivated in folate-free synthetic medium, their use as folate producers ought to be evaluated since probiotic bifidobacteria strains are commonly used in different food matrices and these might affect folate production. All this information is very significant since it has been suggested that the microbiota in the small and large intestine, which contains LAB and bifidobacteria, is able to produce folate that can be assimilated by the host (Camilo et al. 1996; D'Aimmo et al. 2014). Prebiotics are defined as "selectively fermentable ingredients that allow specific changes in the composition and/or activity of gastrointestinal microbiota that allow benefits to the host" (Gibson et al. 2004, 2010). Padalino et al. (2012) studied the effect of galactooligosaccharides (GOS) and fructooligosaccharides (FOS) on folate production by some bifidobacteria, lactobacilli, and streptococci strains in milk. The authors observed that the milk containing fructooligosaccharides (FOS) and fermented by *Bifidobacterium catenulatum* (23.5 µg/100 mL) and *Lactobacillus plantarum* (11.21 µg/100 mL), after 10 hrs of fermentation, showed the highest folate levels. On the other hand, when milk was supplemented with galactooligosaccharides (GOS), *Streptococcus thermophilus*, *Bifidobacterium adolescentis*, and *Lactobacillus delbrueckii* were able to produce higher concentrations of folate than in milk supplemented with FOS. According to these results, the production of folate depends on the strain and the growth media, as mentioned previously. These results suggest that the consumption of prebiotics could selectively be used to increase folate production in the GIT.

Besides LAB, other microorganisms may produce folate and increase this vitamin content in foods. The fermentative process of rye dough may promote an increase in folate content. *Candida milleri* CBS8195, *Saccharomuces cerevisiae* TS 146, and *Torulaspora delbrueckii* TS 207 are sourdough yeasts evaluated by Kariluoto et al. (2006) for their abilities to produce or consume folate. The researchers verified that folate content was increased by yeasts after sterilized rye flour fermentation. Since most studies using food-grade microorganisms involve milk fermentation, the development of new products using different food matrices fortified with natural folate is an additional challenge in this area. In this sense, the applicability of yeast strain for bio-fortification of folates in white wheat bread was investigated by Hjortmo et al. (2008). White wheat bread is usually produced with commercial baker's yeast that is able to produce natural folate (27-43 µg/100g). However, when *Sacchacaromyces cerevisiae* CBS7764 was used by the authors, folate levels in white wheat bread were 3 to 5-fold higher. In this way, according to Kariluoto et al. (2006) and Hjortmo et al. (2008) it is possible increase folate content in yeast fermented foods using specific yeast strains. However, it is also important to determine an efficient cultivation procedure to allow the maximum development and activity of the selected strain.

Folate Analysis in Food

Microbiological assay and tri-enzyme treatment

Folate quantification in food may be conducted using several methods. Nevertheless, the microbiological assay seems to be the only official method according to American Association of Analytical Chemists (AOAC). In this method (AOAC 2006), the strain *Lactobacillus rhamnosus* ATCC 7469 is used as the indicator strain to estimate total folate in food. For this purpose, bacterial growth in a 96-well microtiter plate is compared through turbidity given by optical density values of different samples after incubation. This technique is able to detect most of the folate natural forms. However, the response decreases when the number of glutamyl residues linked to the pteroyl group increases and the measurement of folates is complicated since there are many different forms of the vitamin. In order to measure all the polyglutamated forms it is important to enzymatically deconjugate them prior to analysis.

In this sense, the use of tri-enzyme treatment before folates measurement is essential for obtaining the maximum values of food folate since this vitamin, in food, is possibly trapped by carbohydrate and protein matrices (Aiso and Tamura 1998; Iyer et al. 2009; Tomar et al. 2009; Chew et al. 2012). The treatment includes the use of α-amylase and protease, besides the traditional treatment that uses pteroylpoly-γ-glutamyl hydrolase (AACC 2000; Chew et al. 2012).

After the use of tri-enzyme treatment, the amount of folate usually increases when compared to the traditional microbiological assay. Iyer et al. (2009) evaluated the use of tri-enzyme method followed by the microbiological assay to determine the folate content of different Indian milk species. Buffalo milk showed the highest amount of folate (60 μg/L) when compared to goat, cow, and sheep milk (10, 44, and 56 μg/L, respectively). According to Tamura et al. (1997), the use of tri-enzyme treatment seems to show an essential rule to determine food folate content and the food folate tables should be updated after using this tri-enzyme methodology to accurately establish the dietary folate requirements in human. The instability of several folate forms (for example, tetrahydrofolate) promotes underestimated folate values in data-banks and antioxidants like ascorbic acid are important as folate protectors during analysis (Strandler et al. 2015). Composition of foods and analytical procedures are difficulties faced by researchers to perform international folate content comparisons and to estimate the real intake of this vitamin (Fajardo et al. 2012).

The use of commercial enzymes still shows an important barrier: their very high cost. Alternatives have been developed to make these assays cheaper. In this way, an in-house folate conjugase from chicken pancreas was prepared and tested to quantify the folate content present in several foods (Soongsongkiat et al. 2010). The authors observed that single-enzyme treatment, using folate conjugase from chicken pancreas, may be used to deconjugate folate in some food matrices (i.e. soybean and asparagus); however, the tri-enzyme treatment was necessary to quantify total folate content in egg and whole milk powder. Total folate may be 20–30% higher after tri-enzyme extraction than after treatment with conjugase alone (Rader et al. 1998). The researchers also quantified the folate values after cooking and observed that cooked (boiled) soybean and asparagus retained about 75% and 82% of total folate. In this line, Maharaj et al. (2015) investigated the effect of boiling and frying on the retention of folate in some Fijian vegetables using microbiological assay and tri-enzyme treatment. The authors concluded that the boiling process promoted higher folate loss (10–64%) and that this fact might have been favored by water solubility of this vitamin.

As folate values may be underestimated due to the methods employed, Yon and Hyun (2003) measured the folate content in several foods consumed by Koreans by microbiological assay, comparing the two extraction methods (single and tri-enzymatic). The values obtained by the authors are presented in Table 13.2. Folate contents obtained after tri-enzymatic treatment are apparently higher than those which did not receive this treatment. This observation shows the importance of using amylase and protease plus conjugase to recover higher values of the vitamin.

Divya and Nampoothiri (2014) used *Lactococcus lactis* CM28 as probiotic strain to ferment and fortify skimmed milk with natural folate. The addition of folate precursors, prebiotics, and reducing agents was performed to optimize the medium and, thus, the extracellular folate was increased four folds. After deconjugation, the total folate value achieved 129.53 ± 1.2 μL. After 15 days of cold storage of fermented milk, about 90% of the folate produced was retained in the active form.

The folate content of six common food samples of Bangladesh (lentil, Bengal gram, spinach, basil, milk, and topa boro rice) were measured by microbiological assay using tri-enzyme extraction method (protease, α-amylase, and chicken pancreases as deconjugase). The highest folate contents were recorded for spinach (195 μg/100 g) and the lowest for milk (10 μg/100 g) (Rahman et al. 2015). Iwatani et al. (2003) also evaluated the folate content of vegetables commonly consumed in Australia. However, tri-enzyme treatment was not as efficient as single-enzyme extraction in the study since the vegetables samples investigated contain low amounts of starch and protein and the highest reported folate level was 425 μg/100 g.

TABLE 13.2 Measurement of folate content in food after conjugase (CT) and tri-enzyme treatment (TT)

Food	Folate (μg/100g)		%increase[1]
	CT[2]	TT[2]	
Corn	100 ± 10	129 ± 5	29
Rice	5 ± 1	18 ± 3	260
Rice (cooked)	3 ± 1	8 ± 0	167
Wheat flour	6 ± 1	16 ± 2	167
Soybean	176 ± 31	318 ± 62	81
Soybean milk	16 ± 7	34 ±7	113
Potatoes	14 ± 2	27 ± 3	93
Cabbage	71 ± 20	135 ± 68	90
Carrot	29 ± 19	31 ± 19	7
Lettuce	46 ± 21	57 ± 19	24
Tomato	34 ± 4	52 ± 8	53
Apple (red)	5 ± 3	7 ± 6	40
Banana	16 ± 14	16 ± 7	0
Orange	47 ± 9	51 ± 4	9
Orange juice	31 ± 3	58 ± 6	87
Chicken's egg	36 ± 9	115 ± 18	219
Milk	6 ± 0	13 ± 1	117
Yogurt (curd type)	13 ± 4	24 ± 1	85
Yogurt (liquid type)	12 ± 5	32 ± 14	167

[1] % increase = (TT folate values – CT folate values)/ CT folate values × 100.
[2] Folate values expressed by mean of three different samples (duplicate) and respective standard deviations. Adapted from Yon and Hyun (2003).

Studies of folate measurements in food usually showed folate values from raw vegetables, fruits, milk, and cereal-grain products (Aiso and Tamura 1998; Rader et al. 2000; Yon and Hyun 2003; Chew et al. 2012). Nevertheless, it is relevant to investigate the total amount of folate or the most important forms of this vitamin, as 5-methyltetrahydrofolate, present in other food products in order to determine real folate values for official nutritional tables.

Other analysis methods

As previously mentioned, microbiological assay is the method mostly used to deter- mine folate content in foods, especially because this technique is the only one rec- ognized as official (AOAC 2006). Another method also usually employed for folate quantification is the high performance liquid chromatography (HLPC). In both cases, the use of tri-enzyme extraction, based on the use of amylase, protease, and folate conjugase, is necessary to determine the real total folate and/or folate forms food contents.

Iyer and Tomar (2013) compared the folate values obtained from three methods employed for quantification of this vitamin. Microbiological assay, Enzyme Linked Immuno Sorbent Assay (ELISA), and HPLC were used. According to the authors of the study, HPLC was the most sensitive method for folic acid determination while microbiological assay was highly efficient, sensitive, and reproducible, able to estimate total folate, which has supported the potential use of microbiological assay for dietary folate estimation. ELISA showed lower response for some folate derivates except for folic acid and dihydro folic acid. HPLC and liquid chromatography coupled with mass spectrometry is another method currently applied (Phillips et al. 2011; Vishnumohan et al. 2011; Araya-Farias et al. 2014; Tyagi et al. 2015). These methods can distinguish several forms of folate present in food samples while microbiological assay determines only the total folate content.

Besides the methods discussed above, novel photosynthetic proteins-based devices - biosensors - have been developed for application in food analysis for folate measurement (Indyk 2011).

Conclusions

In this chapter, foods bio-enriched with natural folates produced by microorganisms were discussed as promising alternatives for the low intake of this vitamin and might be considered with more attention by the food industry. In general, the development of novel fermented foods with increased folate content due to a fermentative process would raise the commercial and nutritional value of these products and could replace food fortifications using chemically synthesized vitamin, which would therefore become unattractive targets. The production of folate by microorganisms, such as LAB, is strain-dependent and can be affected by environment conditions, such as pH. Therefore, the proper selection of folate-producing strains might be useful for the development of new functional foods. Although most of the studies have assessed the production of folate in fermented milk, other food matrices might also be bio-enriched with natural folate, such as bread, kefir, vegetables, and fruit juices, since all these foods are produced through a fermentation process. Additionally, the folate content determination in foods using more sensitive methods, such as the trienzymatic treatment, would be stimulated since they may provide more accurate values of this vitamin in different food matrices.

Acknowledgments

The authors would like to thank CONICET and ANPCyT (Argentina), and FAPESP (project #2013/50506-8), CNPq, and CAPES for their financial support and fellowships.

References

AACC. (2000). AACC Method 86–47 Total Folate in Cereal Products - Microbiological Assay Using Trienzyme Extraction. *Approved Methods of the American Association of Cereal Chemists* (10th ed.). St Paul MN, American Association of Cereal Chemists.

Aiso, K. and Tamura, T. (1998). Trienzyme treatment for food folate analysis: optimal pH and incubation time for α-amylase and protease treatments. Journal of Nutritional Science and Vitaminology 44: 361–370.

AOAC. (2006). *AOAC Official Method 992.05 Total Folate (Pteroylglutamic Acid) in Infant Formula.* AOAC International.

Araya-Farias, M., Gaudreau, A., Rozoy, E. and Bazinet, L. (2014). Rapid HPLC-MS method for simultaneous determination of tea catechins and folates. Journal of Agricultural and Food Chemistry 62(19): 4241–4250.

Aryana, K.J. (2003). Folic acid fortified fat-free plain set yogurt. International Journal of Dairy Technology 56: 219-222.

ANVISA (2002). Resolução RDC n° 344, de 13 de dezembro de 2002. Aprova o regulamento técnico para a fortificação das farinhas de trigo e farinhas de milho com ferro e ácido fólico. Available at: http://portal.anvisa.gov.br/wps/wcm/connect/f851a500474580668c83dc3fbc4c6735/RDC_344_2002.pdf?MOD=AJPERES. Accessed Mai 2015.

Camilo, E., J. Zimmerman, J.B. Mason, B. Golner, R. Russell, J. Selhub and Rosenberg, I.H. (1996). Folate synthesized by bacteria in the human upper small intestine is assimilated by the host. Gastroenterology 110: 991–998.

Capozzi, V., P. Russo, M.T. Dueñas, P. López and G. Spano. (2012). Lactic acid bacteria producing B-group vitamins: a great potential for functional cereals products. Applied Microbiology and Biotechnology 96: 1383–1394.

Chew, S.C., S.P. Loh and G.L. Khor. (2012). Determination of folate content in commonly consumed Malaysian foods. International Food Research Journal 19(1). 189–197.

Chiang, F.F., S.C. Huang, H.M. Wang, F.P. Chen, and Y.C. Huang. (2015). High serum folate might have a potential dual effect on risk of colorectal cancer. Clinical Nutrition 34:986-990.

Cravo, M.L. and M.E. Camilo. (2000). Hyperhomocysteinemia in chronic alcoholism: relations to folic acid and vitamins B(6) and B(12) status. Nutrition 16: 296-302.

Crittenden, R.G., N.R. Martinez and M.J. Playne. (2003). Synthesis and utilization of folate by yoghurt starter cultures and probiotic bacteria. International Journal of Food Microbiology 80: 217–222.

Crider, K.S., L.B. Bailey, and R.J. Berry. (2011). Folic acid fortification – its history, effect, concerns, and future directions. Nutrients 3: 370-384.

D'Aimmo, M.R., P. Mattarelli, B. Biavati, N.G. Carlsson and T. Andlid. (2012). The potential of bifidobacteria as a source of natural folate. Journal of Applied Microbiology 112: 975–984.

D'Aimmo, M.R., M. Modesto, P. Mattarelli, B. Biavati. and T. Andlid. (2014). Biosynthesis and cellular content of folate in bifidobacteria across host species with different diets. Anaerobe 30: 169–177.

Dantas, J.A., A.S. Diniz and I.C.G. Arruda. (2010). Consumo alimentar e concentração intra-eritrocitárias de folato em mulheres do Recife, Nordeste do Brasil. Archivos Latino Americano de Nutricion 60(3): 227–234.

Delchier, N., C. Ringling, J.F. Maingonnat, M. Rychlik and C.M.G.C. Renard. (2014). Mechanisms of folate losses during processing: Diffusion vs. heat degradation. Food Chemistry 157: 439–447.

Divya, J.B. and K.M. Nampoothiri. (2015). Encapsulated *Lactococcus lactis* with enhanced gastrointestinal survival for the development of folate enriched functional foods. Bioresource Technology *in press* 188: 226-30.

Eitenmiller, R.R. and W.O. Landen, (1999). Folate. *In*: Vitamin Analysis for the Health and Food Sciences. R.R. Eitenmiller and W.O. Landen (eds.). CRC Press, Boca Raton, pp. 411-466.

Fajardo, V., E. Alonso-Aperte and G. Varela-Moreiras. (2012). Lack of data on folate in convenience foods: Should ready-to-eat products considered relevant for folate intake?
The European challenge. Journal of Food Composition and analysis 28(2).155–163.

Fajardo, V., E. Alonso-Aperte, and G. Varela-Moreiras. (2015). Folate content in fresh-cut
vegetable packed products by 96-well microliter plate microbiological assay. Food
Chemistry 169: 283-288.

FAO/WHO. Food and Agricultural Organization of the United Nations and World Health
Organization. Human vitamin and mineral requirements: report of a joint FAO/WHO
expert consulation. Rome, Italy: FAO, 290 p. 2002.

FAO/WHO (2006). Annex D - A procedure for estimating feasible fortification levels
for a mass fortification programme. In: L. Allen, B. de Benoist, O. Dary, R.Hurrell
(eds.), Guidelines on Food Fortification with Micronutrients. 1st ed. World Health
Organization, Portland, pp. 294-312.

Fujimori, E., C.F. Baldino, A.P.S. Sato, A.L.V. Borges and M.N. Gomes. (2013). Prevalência
e distribuição espacial de defeitos do tubo neural no Estado de são Paulo, Brasil,
antes e após a fortificação de farinhas com ácido fólico. Cadernos de Saúde Pública
29(1): 145–154.

Gangadharan, D. and K.M. Nampoothiri. (2011). Folate production using *Lactococcus
lactis* spp. cremoris with implications for fortification of skim milk and fruit juices.
Food Science and Technology 44: 1859–1864.

Gibson, G.R., H.M. Probert, J.A.E. Van Loo, R.A. Rastall and M.B. Roberfroid. (2004).
Dietary modulation of the human colonic microbiota: updating the concept of prebiotics. Nutrition Research Reviews, 17(2): 259–275.

Gibson, G.R., K. P. Scott, R. A. Rastall, K. M. Tuohy, A. Hotchkiss, A. Dubert-Ferrandon,
et al. (2010). Dietary prebiotics: current status and new definition. Food Science and
Technology Bulletin: Functional Foods, 7(1): 1–19.

Gregory, J.F. (1995). The bioavailability of folate. *In*: L.P. Bailey (ed.), Folates in Health
and Disease. Marcel Dekker, New York, pp. 195-235.

Hendler, S. and D. Rorvik, (2001). Supplement monographs. *In*: S. Hendler and D. Rorvik (eds.),
PDR for Nutritional Supplements. 1st ed. Medical Economics, Montvale, pp. 157-165.

Herbison, C.E., S. Hickling, K.L. Allen, T.A. O'Sullivan, M. Robinson, A.P. Bremmer, R.C.
Huang, L.J. Beilin, T.A. Mori, and W.H. Oddy. (2012). Low intake of B-vitamins is
associated with poor adolescent mental health and behavior. Preventive Medicine
55: 634-638.

Hermann, W. and R. Obeid. (2011). The mandatory fortification of staple foods with folic
acid. Deutsches Ärzteblatt International 108: 249-254.

Holasová, M., V. Fiedlerová, P. Roubal and M. Pechacová. (2005). Possibility of increasing
natural folate content in fermented milk products by fermentation and fruit component
addition. Csech Journal of Food Science 23(5): 196–201.

Hjortmo, S., J. Patring, J. Jastrebova and T. Andlid. (2008). Biofortification of folates in
white wheat bread by selection of strains and process. International Journal of Food
Microbiology 127: 32–36.

Hugenschmidt, S., S.M. Schwenninger and C. Lacroix. (2011). Concurrent high production
of natural folate and vitamin B_{12} using a co-culture process with *Lactobacillus plantarum* SM39 and *Propionibacterium freundenreichii* DF13. Process Biochemistry
46: 1063–1070.

Indyk, H.E. (2011). An optical biosensor assay for the determination of folate in milk and
nutritional dairy products. International Dairy Journal 21(10): 783–789.

Iyer, R., S.K. Tomar, (2009). Folate: A functional food constituent. Journal of Food Science
74(9): R114-122.

Iyer, R., S.K. Tomar. (2011). Dietary effect of folate-rich fermented milk produced by Streptococcus thermophilus strains on hemoglobin level. Nutrition 27: 994-997.

Iyer, R. and S.K. Tomar. (2013). Determination of folate/folic acid level in milk by microbiological assay, immune assay and high performance liquid chromatography. Journal of Dairy Research 80(2): 233–239.

Iyer, R., S.K. Tomar, S. Kapila, J. Mani and R. Singh. (2010). Probiotic properties of folate producing Streptococcus thermophilus strains. Food Research International 43: 103–110.

Iyer, R., S.K. Tomar, R. Singh and R. Sharma. (2009). Estimation of folate in milk by microbiological assay using tri-enzyme extraction method. Milchwissenschaft 64(2): 125–127.

Iwatani, Y., J. Arcot and A. Shrestha. (2003). Determination of folate contents in some Australian vegetables. Journal of Food Composition and Analysis 16(1): 37–48.

Kariluoto, S., M. Aittamaa, M. Korhola, H. Salovaara, L. Vahteristo and V. Piironen. (2006). Effects of yeasts and bacteria on the levels of folates in rye sourdoughs. International Journal of Food Microbiology 106: 137–143.

Kariluoto, S., M. Edelmann, L. Nyström, T. Sontag-Strohm, H. Salovaara, R. Kivelä, M. Herranen, M. Korhola, and V. Piironen. (2014). In situ enrichment of folate by microorganisms in beta-glucan rich oat and barley matrices. International Journal of Food Microbiology 176: 38-48.

Kariluoto, S., M. Edelmann, M. Herranen, A.M. Lampi, A. Shmelev, H. Salovaara, M. Korhola and V. Piironen. (2010). Production of folate by bacteria isolated from oat bran. International Journal of Food Microbiology 143: 41–47.

Laiño, J.E., J.G. LeBlanc and G.S. De Giori. (2012). Production of natural folates by lactic acid bacteria starter cultures isolated from artisanal Argentinean yogurts. Canadian Journal of Microbiology 58: 581–588.

Laiño, J.E., M.J. Del Valle, G.S. De Giori and J.G.L. LeBlanc. (2013a). Development of a high concentration yogurt naturally bio-enriched using selected lactic acid bacteria. Food Science and Technology 54: 1–5.

Laiño, J.D., G.S. De Giori and J.G. LeBlanc. (2013b). Folate production by lactic acid bacteria. *In*: Watson RR and VR (eds.), Bioactive Food as Dietary Interventions for Liver and Gastrointestinal Disease. Academic Press, San Diego, pp. 251–270.

Laiño, J.E., M.J. del Valle, G.S. de Giori and J.G.J. LeBlanc. (2014). Applicability of a *Lactobacillus amylovorus* strain as co-culture for natural folate bio-enrichment of fermented milk. International Journal of Food Microbiology 191: 10–16.

LeBlanc, J.G., C. Milani, G.S. De Giori, F. Sesma, D. Van Sinderen and M. Ventura. (2013). Bacteria as vitamin suppliers to their host: a gut microbiota perspective. Current Opinion in Biotechnology 24: 160–168.

LeBlanc, J.G., G.S. de Giori, E.J. Smid, J. Hugenholtz and F. Sesma. (2007). Folate production by lactic acid bacteria and other food-grade microorganisms. Communicating Current Research and Educational Topics and Trends in Applied Microbiology.

Lin, M.Y. and C.M. Young. (2000). Folate levels in cultures of lactic acid bacteria. International Dairy Journal 10: 409–413.

Liu, Y., Tomiuk, S., Rozoy, E., Simard, S., Bazinet, L., Green, Kitts, D.D. (2012). Thermal oxidation studies on reduced folate, L-5-methyltetrahydrofolic acid (L-5-MTHF) and strategies for stabilization using food matrices. Journal of Food Science 77: C236-C243.

López-Nicolás, R., C. Frontela-Saseta, R. Gonzaléz-Abellán, A. Barado-Piqueras, D. Perez-Conesa, and G. Ros-Berruezo. (2014). Folate fortification of white and whole-grain bread by adding Swiss chard and spinach. Acceptability by consumers. LWT – Food Science and Technology 59: 263-269.

Masuda, M., M. Ide, H. Utsumi, T. Niiro, Y. Shimamura and M. Murata. (2012). Production potency of folate, vitamin B_{12} and thiamine by lactic acid bacteria isolated from Japanese pickles. Bioscience, Biotechnology and Biochemistry 76: 2061–2067.

Maharaj, P.P.P., S. Prasad, R. Devi and R. Gopalan. (2015). Folate content and retention in commonly consumed vegetables in the South Pacific. Food Chemistry 182. 327–332.

McNulty, H. and K. Pentieva, (2004). Folate bioavailability. The Proceedings of the Nutrition Society 63: 529-536.

McPartlin, J., A. Halligan, , J.M. Scott, M. Darling, and D.G. Weir. (1993). Accelerated folate breakdown in pregnancy. Lancet 341: 148-149.

Melse-Boonstra, A., C.E. West, M.B. Katan, F.J. Kok, and P. Verhoef, (2004). Bioavailability of heptaglutamyl relative to monoglutamyl folic acid in healthy adults. The American Journal of Clinical Nutrition 79: 424-429.

Nygren-babol, L. and Jägerstad. (2012). Folate-binding protein in milk: a review of biochemistry, physiology and analytical methods. Critical Reviews in food Science and Nutrition 52(5): 410–425.

Nor, N.M., R. Mohamad, H.L. Foo and R.A. Rahim. (2010). Improvement of folate biosynthesis by lactic acid bacteria using response surface methodology. Food Technology and Biotechnology 48: 243–250.

Ohrvik, V.E. and C.M. Witthoft, (2011). Human folate bioavailability. Nutrients 3: 475-490.

Padalino, M., D. Perez-Conesa, R. López-Nicolás and C. Prontela-Saseta. (2012). Effect os fructoologosaccharides and galactooligosaccharides on the folate production of some folate-producing bactéria in media cultures or milk. International Dairy Journal 27: 27–33.

Phillips, K.M., D.M. Ruggio and D.B. Haytowitz. (2011). Folate composition of 10 types of mushrooms determined by liquid chromatography-mass spectrometry. Food Chemistry 129(2): 630–636.

Pompei, A., L. Cordisco, A. Amaretti, S. Zanoni, D. Matteuzzi and M. Rossi, (2007). Folate production by bifidobacteria as potential probiotic property. Applied and Enviromental Microbiology 73(1): 179–185.

Pounis, G., A.F. Castelnuovo, M. Lorgeril, V. Krogh, A. Siani, J. Arnout, F.P. Cappuccio, M. van Dongen, B. Zappacosta, M.B. Donati, G. Gaetano, and L. Iacoviello, (2014). Folate intake and folate serum levels in men and women from two European populations: the IMMIDIET project. Nutrition 30: 822-830.

Quinlivan, E.P. and Gregory, J.F., III. (2007). Reassessing folic acid consumption patterns in the United States (1999-2004): potential effect on neural tube defects and overexposure to folate. The American Journal of Nutrition 86: 1773-1779.

Rossi, M., A. Ameratti and S. Raimoundi. (2011). Folate productionby probiotic bacteria. Nutrients 3: 118–114.

Rahman, T., M.M.H. Chowdhury, M.T. Islam and M. Akhtaruzzaman. (2015). Microbiological assay of folic acid content in some selected Bangladeshi food stuffs. International Journal of Biology 7(2): 35–43.

Rader, J.I., C.M. Weaver and G. Angyal. (1998). Use of a microbiological assay with trienzyme extraction for measurement of pre-fortification levels of folates in enriched cereal-grain products. Food Chemistry 62(4): 451–465.

Rader, J., C.M. Weaver and G. Angyal. (2000). Total folate in enriched cereal-grain products in the United States following fortification. Food chemistry 70(3):275–289.

Samaniego-Vaesken, M.L., E. Alonso-Aperte, and G. Varela-Moreiras. (2012). Vitamin food fortification today. Food & Nutrition Research 56:5459. DOI: 10.3402/fnr. v56i0.5459.

Santos, L.M.P. and M.Z. Pereira. (2007). Efeito da fortificação com ácido fólico na redução dos efeitos do tubo neural. Cadernos de Saúde Pública 23(1): 17–24.

Santos, F., A. Wegkamp, W.M. de Vos, E.J. Smid, and J. Hugenholtz. (2008). High-level folate production in fermented foods by the B_{12} producer *Lactobacillus reuteri* JCM1112. Applied and Enviromental Microbiology 74: 3291–3294.

Sauberlich, H.F., M.J. Kretsch, J.H. Skala, H.L. Johnson, and P.C. Taylor. (1987). Folate requirement and metabolism in nonpregnant women. The American Journal of Clinical Nutrition 46: 1016-1028.

Scott, J.M. (1999). Folate and vitamin B_{12}. The Proceedings of the Nutrition Society 58: 441-448.

Shuaibi, A.M., G.P. Sevenhuysen, and J.D. House. (2009). The importance of using folate intake expressed as dietary folate equivalents in predicting folate status. Journal of Food Composition and Analysis 22: 38-43.

Silva, N. and B. Davis. (2013). Iron, B_{12}, and folate. Medicine 41: 204-207.

Smith, A.D., Y.I. Kim, and H. Refsum. (2008). Is folic acid good for everyone? The American Journal of Clinical Nutrition 87: S17-S33.

Soongsongkiat, M., P. Puwastien, S. Jittinandana, A. Dee-Uam and P. Sungpuag. (2010). Testing of folate conjugase from chicken pancreas vs. commercial enzyme and studying the effect of cooking on folate retention in Thai foods. Journal of food Composition and Analysis 23: 681–688.

Stamm, R.A. and L.A. Houghton. (2013). Nutrient intake values for folate during pregnancy and lactation vary widely around the world. Nutrients 5: 3920-3947.

Strandler, H.S., J. Patring, M. Jägerstad and J. Jastrebova. (2015). Challenges in the determination of unsubstituted food folates: Impact of stabilities and conversions on analytical results. Journal of Agricultural and Food chemistry 63(9): 2367–2377.

Sybesma, W., M. Starrenburg, L. Tijsseling, M.H.N. Hoefnagel and J. Hugenholts. (2003). Effects of cultivation conditions on folate production by lactic acid bacteria. Applied and Enviromental Microbiology 69(8): 4542–4548.

Tyagi, K., P. Upadhyaya, S. Sarma, V. Tamboli, Y. Sreelakshmi and R. Sharma. (2015). High performance liquid chromatography coupled to mass spectrometry for profiling and quantitative analysis of folate monoglutamates in tomato. Food Chemistry 179: 76–84.

Tomar, S.K., N. Srivatsa, R. Iyer and R. Singh. (2009). Estimation of folate production by *Streptococcus thermophilus* using modified microbiological assay. Milchwissenschaft 64(3): 260–263.

Tamura, T., Y. Mizuno, K.E. Johnston and R.A. Jacob. (1997). Food folate assay with protease, α-amylase and folate conjugase treatments. Journal of Agricultural and Food chemistry 45(1):135–139.

Truswell, A. and S. Kounnavong. (1997). Quantitative responses of serum folate to increasing intakes of folic acid in healthy women. European Journal of Clinical Nutrition 51: 839-845.

Uehara, S.K. and G. Rosa. (2010). Associação da deficiência de ácido fólico com alterações patológicas e estratégias para sua prevenção: uma visão crítica. Revista de Nutrição 23(5): 881–894.

Vishnumohan, S., J. Arcot and R. Pickford. (2011). Naturally-occorring folates in foods: Method development ans analysis using liquid chromatography-tandem mass spectrometry (LC-MS/MS). Food Chemistry 125: 736–742.

Yon, M., and T.H. Hyun. (2003). Folate content of foods commonly consumed in Korea measured after trienzyme extraction. Nutrition Research 23(6): 735–746.

Walkey, C.J., D.D. Kitts, Y. Liu, and H.J.J. van Vuuren. (2015). Bioengineering yeast to enhance folate levels in wine. Process Biochemistry 50: 205-210.

Wang, X., X. Qin, H. Demirtas, J. Li, G. Mao, Y. Huo, N. Sun, L. Liu, and X. Xu. (2007). Efficacy of folic acid supplementation in stroke prevention: a meta-analysis. Lancet 369: 1876-1882.

Wyk, J.V., R.C. Witthuhn and T.J. Britz. (2011). Optimisation of vitamin B_{12} and folate production by *Propionibacterium freundenreichii* strains in kefir. International Dairy Journal 21: 69–74.

Yang, Q.H., H.K. Carter, J. Mulinare, R.J. Berry, J.M. Friedman, and J.D. Erickson. (2007). Race-ethnicity differences in folic acid intake in women of childbearing age in the United States after folic acid fortification: findings from the National Health and Nutrition Examination Survey, 2001-2002. The American Journal of Clinical Nutrition 85: 1409-1416.

Zhao, R., Diop-Bove, N., Visentin, M., Goldman, I.D. (2011). Mechanisms of membrane transport of folates into cells and across epithelia. Annual Review of Nutrition 31: 177-201.

14

Characteristics and Production of Microbial Cultures

Juliano De Dea Lindner*

Abstract

The successful manufacture of all fermented products relies on the presence, growth, and metabolism of specific microorganisms. Non fermented products generally lack the desired organoleptic qualities that are present, and that consumers expect, in the fermented products. This is because microorganisms are responsible for producing an array of metabolic end-products and textural modifications, and replicating those effects by other means is simply not possible. Starter cultures are defined as a preparation containing large numbers of variable microorganisms, which may be added to accelerate and/or improve a fermentation process. Microorganisms selected to be used as starter cultures are expected to have some characteristics such as adapting easily to raw food matrix and process, developing nutritional value and digestibility, improve safety by preservative effect and sometimes add functionality. The use of starter cultures in the food fermentation industry is widely known, such as cheese, yogurt, bread, beer and wine. In many cases, the technology has evolved from a traditional, spontaneous fermentation to a controlled industrialized process based on the use of well defined microbial strains as cultures to conduct the fermentation. Starter manufacture is one of the most important and also one of the most difficult processes in the food biotechnology. Production failures can result in heavy financial loss, as modern industries process large quantities of raw materials. Very careful attention must therefore be paid to the manufacturing technology and choice of process equipment.

Introduction

During past decades, specialists in different areas have experimented and developed methods to isolate, select, investigate, improve and produce different natural microorganism strains to apply in food fermentation with different purposes. The availability of microorganisms with highly specific activity (technological and/ or functional characteristic) is the basic element for the development of fermented

Food Science and Technology Department, Federal University of Santa Catarina, Rod. Admar Gonzaga, 1346. Itacorubi. 88034-001, Florianópolis, SC, Brazil.
* Corresponding Author: juliano.lindner@ufsc.br

products. Nowadays, selected strains of bacteria, yeasts and moulds are widely applied in the food industry for preparation, improvement and preservation of food.

Fermentation is a biological process conducted by microorganisms. The successful manufacture of fermented products relies on the presence, growth and metabolism of specific microorganisms. What the consumers expect in a fermented product generally lacks in an unfermented one. They desire organoleptic qualities that are present only in a fermented product, because microorganisms are responsible for producing metabolic end-products and textural modifications.

Food microbiology began with the study of the natural fermentation process that occurs when the raw food matrix are held for a time. These products had long been used as inocula to produce dairy products, but the resulting fermentations were of uneven quality. For example, in the 1880s, Conn in the United States, Starch in Denmark and Weigmann in Germany demonstrated the advantages of using lactic acid bacteria (LAB) to culture cream for butter manufacture. Also, in the United States, the brothers Fleischmann, in the 1860s, began a yeast factory (first culture industry devoted to baker's yeast) producing a compressed yeast cake for commercial as well as home-baking markets use. Bread produced using this yeast culture was far superior to breads made using brewing yeasts, which was the common form at the time.

There are essentially three ways to induce food fermentation. The oldest method simply relies on the autochthonous microorganisms present in the raw material. Raw milk, for example, usually harbors the very LAB necessary to convert the matrix into cheese. Grapes contain likewise the yeasts responsible for fermenting sugars into ethanol, transforming juice into wine. Suitable conditions for microorganism growth are necessary and, in terms, select the microflora. Even if these requirements are satisfied, however, there is no guarantee that the product will meet the quality expectations, be safe to consume, or even be successfully produced. Once a successful fermentation has been achieved, a portion of that product could be transferred to use as fresh raw material to initiate a new fermentation. This method works for almost any fermented food and is still commonly used for beer, sour dough bread and vinegar. This ancient method based on natural fermentation is called back-slopping and is still practiced for small-scale production facilities in artisanal/ethnical type products. Due to limitations in infrastructure and technologies, rural areas in most countries have not been able to keep informed of global developments toward industrialization. At the same time, fermented foods play a major role in the diet of numerous regions in Latin America, Africa and Asia.

Experience has also shown that back-slopping, or the inoculation of raw materials with a residue from a previous batch, accelerates the initial phase of fermentation and, consequently, the risk of fermentation failure is reduced. This practice results in the promotion of desirable changes during the fermentation process and selects well-acclimated and predominant microorganisms with many of the desired traits necessary for successful production. Thus, an organism isolated from the back-slope could be purified, studied and characterized to be introduced as a culture for fermentation. The implication of this method that pure cultures could be obtained and used to start fermentations resulted in a modern way to produce fermented foods. The technology has evolved from a traditional, spontaneous fermentation to a controlled industrialized process based on the use of well-defined microbial strains as cultures to conduct the fermentation.

European Food and Feed Cultures Association (EFFCA) defined microbial food cultures (MFC) as live microorganisms used in food production. MFC are preparations that consist of one or more microbial strains, including components that are used to give support for survival, storage and application in the food matrix. These microbial preparations can consist of LAB, propionibacteria, surface-ripening bacteria, yeasts and moulds. Equivalently by definition, starter cultures (SC) consist in a preparation containing large numbers of variable selected microorganisms, which may be added to accelerate and improve a fermentation process. The term "starter" in this context is appropriate because the SC initiates and carries through the necessary changes in the raw material to yield the cultured product. The entire cultured products manufacture is dependent upon the activity of the starter.

Microorganisms selected to be used as SC are expected to have some characteristics such as: adapting easily to raw food matrix and process; carrying through every change to attain the desired body, texture and flavour in the final product; developing nutritional value and digestibility; improving safety by preservative effect retarding or inhibiting pathogenic and spoilage flora and thereby increasing shelf life; sometimes adding functionality, for example, in probiotic products the added cultures impart health-promoting properties to the consumer.

In fermented dairy products, SC have a multifunctional role. Their fundamental ability to produce acid aids in separation of curd from whey during cheese manufacture, modifies texture by proteolytic activity during ripening, enhances preservation by competition with microbial consorcium, produces low molecular weight compounds that contribute to flavour and aroma, etc. Some LAB and yeast strains are capable of degrading antinutritional factors, such as phytic acid and phenolic compounds. Incorporation of these microorganisms as SC in fermented matrices, therefore, upgrades the nutritional value of food (De Dea Lindner et al. 2013). Selected strains may improve protein digestibility and micronutrient bioavailability and contribute more specifically to nutritional enrichment through the biosynthesis of vitamins and aminoacids (Holzapfel 2002).

Bourdichon et al. (2012) revised and updated the taxonomy and the inventory of microorganisms used in food fermentations covering a wide range of food matrices (dairy, meat, fish, vegetables, legumes, cereals, beverages and vinegar), and reviewed the US and European regulatory frameworks for use of microorganisms in food fermentations. Those regulatory frameworks emphasised on "the history of use", "traditional food", or "general recognition of safety".

Today, defined (single- and multiple-strains) and undefined (mixed-strain and back-slopping) cultures are extensively used around the world to produce fermented food (Table 14.1). Food processors in most cases do not propagate the SC at the industrial plant and prefer the simpler approach of carrying out a direct vat inoculation (DVI) with commercial products (frozen and freeze-dried culture concentrates). Although this appears simple to carry out, there are critical manipulations to ensure adequate storage and inoculation of the cultures. Furthermore, the use of commercial SC has reduced the diversity of fermented products because the available new and interesting cultures are limited.

TABLE 14.1 Defined (single- and multiple-strains) and undefined (mixed-strains and back-slopping) cultures used to produce fermented food. – not used; + applied. Source: Holzapfel (2002) and De Dea Lindner et al. (2013)

Food	Single culture	Multiple cultures	Mixed cultures	Back-slopping
Beer	+	+	–	–
Bread	+	+	+	+
Cacao	–	+	+	–
Cheese	+	+	+	–
Coffee	–	+	+	–
Miso	+	–	–	–
Olive	+	–	+	–
Pickled cucumber	+	–	+	–
Sauerkraut	+	–	–	+
Sausage	+	+	–	+
Soy sauce	+	–	–	–
Spirits	+	+	–	–
Tea	–	+	+	–
Vinegar	+	+	–	–
Wine	+	+	–	–

Because of the fundamental functions of cultures in fermented products, the selection, propagation and handling of SC are of paramount importance in successful product manufacture and merchandizing. This holds true in the industrial production of SC, which would entail an additional burden in using the most optimum harvesting and preservative techniques that would ensure optimum functionality of the culture during application. Selection and use of MFC in food industry are challenging tasks that require a solid understanding of manipulation and factors that affect microbial interaction. The first task is to find a suitable microorganism for use in the desired process, isolating from the environment or modifying using molecular techniques.

Commercial production of SC grew rapidly and was widespread at the beginning of the 20th century. The advantages of using SC to initiate fermentation were convincing as before their use, fermentations were long in hours and much of the product was too poor in terms of quality. Slow fermentation was also a public health threat, because raw materials were not thermal treated. Production failures can result in heavy financial loss, as modern industries process large quantities of raw materials. SC manufacture is one of the most important and also one of the most difficult processes in the food biotechnology. Very careful attention must therefore be paid to the manufacturing technology and choice of process equipment.

As consumers increasingly look for fermented products to improve their health, food and beverage companies can leverage the natural wholesomeness of cultured products to provide flavourful, innovative solutions that support their organoleptic requirements or meet their needs for convenience. Growing consumer interest in healthy, portable, convenient food and beverages has played an important role in the growth of cultured products which in turn has created a wide-open field of opportunity for product developers to formulate innovations to meet the needs of today's consumers. Advances in technology, processes and packaging, as well as investments in research and development, can benefit food and beverage companies as they continue to identify opportunities for product development.

Innovation is one of the main strategies of the food industry to generate growth and economic vitality. Latin America is a dynamic market, as well as mature and expressive in the fermented food production sector. However, the industries that use biotechnological processes have a barrier to expansion: there is a shortage in mainland production of inputs such as cultures for fermented products.

The intent of this chapter is to describe our current understanding of what constitutes MFC, microorganisms involved and functions, the criteria for selecting and developing SC, as well as evaluating performance, the use and production of SC, manufacturing parameters to consider in the use of cultures and future prospects.

Starter Culture Functions

Currently, food industries face the challenge of meeting the demand of consumers for safe food with functional effects (e.g. probiotics), with long shelf life, which are minimally processed and produced in the concept of a clean label. Thus, the use of compounds produced by these microorganisms can provide these features to the fermented food with an attractive technology option to replace the use of additives.

Some fermentations require the use of SC composed of different microorganisms with different growth requirements. In such operations, associative action by components in the starter mixtures may be desired, and in other cases, the conditions need to be manipulated to curtail the growth and activity of one component, but favour the other or promote a balanced growth and activity of both components. All these events have to be carefully controlled to obtain consistently superior products.

The native or wild cultures are those present in foods and exhibit interesting characteristics, since they survive adverse environment conditions and predominate in this ecosystem. In order to survive in the environment, they need to resist in competition with other microorganisms, generally by the production of antimicrobial substances. They are cultures with greater ability to colonize a matrix of origin. Several studies have also reported the production of enzymes, flavouring compounds, vitamins and many physiological biomodulation therapeutic effects in humans (e.g. production of probiotic bioactive peptides) in isolated native cultures of different foods.

The cultures that produce antimicrobial metabolites may be employed in order to improve the preservation of food. Such substances are recognized, for example, as hydrogen peroxide, organic acids and bacteriocins. The use of bacteriocins as natural substitutes for chemical preservatives in foods has been extensively studied, since they are a biopreservative effective in the control of pathogenic and spoilage microorganisms (Pratush et al. 2012).

The basic role of dairy SC composed of LAB is to drive the fermentation process. They are applied in the production of a variety of dairy products (cheese, fermented milks and cream butter). A SC can provide more control and predictability in milk fermentation. The LAB SC are used because they are able to produce lactic acid from lactose. The pH decrease during the fermentation affects a number of aspects of the process: quality, texture and composition. The other important function of LAB SC is the production of aroma compounds which determine the specific identity of the cultured dairy product.

Yogurt manufacture exemplifies a process where synergism between two different starter components is desired. In cultured buttermilk, the conditions are manipulated to prevent dominance by acid-producing bacteria, so that the flavour-producing components in the starter mixture can function, assuring a balanced growth and activity of both components. Typically, the culture is a combination of microorganisms such as, for example, in the case of yogurt, *Lactobacillus delbrueckii* subsp. *bulgaricus* and *Streptococcus thermophilus*. The most important characteristics for yogurt cultures are rapid acidification, production of characteristic balanced flavour and ability to produce the desired texture. Excessively rapid acidification can result in overacidification and a harsh flavour. Acidification of yogurt is controlled by refrigeration, but the culture may continue to acidify slowly at cold temperatures.

Microorganisms are an essential component of all natural cheese varieties and play important roles during both cheese manufacture and ripening. The microflora of cheese may be divided in two groups: starter LAB and secondary microorganisms. The first are involved in acid production during manufacture and contribute to the ripening process, while the latter do not contribute to acid production during manufacture, but generally play a significant role during ripening (De Dea Lindner et al. 2008). While the manufacture of cheese can frequently involve fermentation by naturally present non starter microflora, there are many cheeses in which SC are used. Nowadays, multiplex real time PCR (mRT-PCR) is useful to rapidly screen microbial composition of SC for cheeses and to compare samples with potentially different technological properties. In the study of Bottari et al. (2013), the developed mRT-PCR was found to be effective for analysing species present in the samples with an average sensitivity down to less than 600 copies of DNA and, therefore, sensitive enough to detect even minor LAB community members of thermophilic SC. This type of study can add to a major increase in understanding that the SC contribution to cheese ecosystem could be harnessed to control cheese ripening and flavour formation.

The yeasts associated with the Kefir grains generate the needed alcohol and carbon dioxide in Kefir that provide essential flavour notes. In dahi, yeasts acquired through chance contamination and carried over by back-slopping practice are attributable to the yeastiness. The SC used in Viili contain a mould, *Geotrichum candidum*, which forms a layer or mat on the surface of the product (aerobic growth). The mould metabolizes lactic acid, induces a "layered mildness" to the product and also imparts a "musty" aroma.

A controlled coffee fermentation process by the use of SC may guarantee a standardized quality and reduce the economic loss for the producer. Only few studies have been reported the use of SC for coffee fermentation, although the attempt to control coffee fermentation has existed for over 40 years. Early studies reported the use of residual waters from a previous fermentation as starter. Only in 1966, a study conducted by Agate and Bhat (1966) effectively introduced a SC for coffee fermentation. They demonstrated that the incorporation of a mixture of three pure cultures of the *Saccharomyces* species accelerates the process of modifying flavours. In this case, an enzyme preparation from the *Saccharomyces* species was observed to hasten the mucilage-layer degradation. Evangelista et al. (2014) evaluated the use of yeasts as SC in coffee semi-dry processing. Arabica coffee was inoculated with one of the following starter cultures: *Saccharomyces cerevisiae* UFLA

YCN727, *Saccharomyces cerevisiae* UFLA YCN724, *Candida parapsilosis* UFLA YCN448 and *Pichia guilliermondii* UFLA YCN731. A total of 47 different volatiles compounds have been identified. The coffee inoculated with yeast had a caramel flavour that was not detected in the control. They cited that the use of SC during coffee fermentation is an interesting alternative for obtaining a beverage quality with distinctive flavour.

Cocoa beans (*Theobroma cacao* L.) are the raw material for chocolate production. Fermentation of cocoa pulp by microorganisms is necessary to develop chocolate flavour precursors. Yeasts conduct an alcoholic fermentation within the bean pulp that is essential for the production of good quality beans, giving typical chocolate features. However, the roles of bacteria such as LAB and acetic acid bacteria (AAB) in contributing to the quality of cocoa bean and chocolate are not fully understood. Ho et al. (2015), using controlled laboratory fermentations, investigated the contribution of LAB to cocoa bean fermentation. The yeasts *Hanseniaspora guilliermondii*, *Pichia kudriavzevii*, *Kluyveromyces marxianus* and *Saccharomyces cerevisiae*, the LAB *Lactobacillus plantarum*, *Lactobacillus pentosus* and *Lactobacillus fermentum* and the AAB *Acetobacter pasteurianus* and *Gluconobacter frateurii* were the major species found in control fermentations. They demonstrated that beans fermented in the presence or absence of LAB were fully fermented, had similar shell weights, and gave acceptable chocolates with no differences in sensory rankings. The conclusion was that LAB may not be necessary for successful cocoa fermentation.

In wine production, the contribution of yeasts has also been studied. High alcohol concentrations reduce the complexity of wine sensory properties. Wine industry is actively seeking technologies that facilitate the production of wines with lower alcohol content. However, commercially available wine yeasts produce very similar ethanol yields. Contreras et al. (2015), in order to achieve this aim, used yeast strains which are less efficient at transforming grape sugars into ethanol. Non-conventional yeasts, in particular non-*Saccharomyces* species, have shown potential for producing wines with lower alcohol content. These yeasts are present in the early stages of fermentation and in general are not capable of completing alcoholic fermentation. Among 48 non-*Saccharomyces* strains evaluated, two (*Torulaspora delbrueckii* AWRI1152 and *Zygosaccharomyces bailii* AWRI1578) enabled the production of wine with reduced ethanol concentration under limited aerobic conditions.

Starter Culture Microorganisms

An inventory of microorganisms used in food fermentations covering a wide range of food matrices (dairy, meat, fish, vegetables, legumes, cereals, beverages and vinegar) was published by Bourdichon et al. in 2012. This work was the result of a review and update of the list published in 2002 by the joint project between the International Dairy Federation (IDF) and the European Food and Feed Cultures Association (EFFCA). In 2002, the IDF inventory became a *de facto* reference for food cultures in practical use. However, as the focus was mainly on commercially available dairy cultures, there was an unmet need for a list with a wider scope. Bourdichon et al. (2012) also reviewed the literature for each species in order to maintain in the new inventory only microbial species that make desirable contributions to the food fermentations.

Currently, this list of microbial species with a documented presence in fermented foods is the better grouping of information that can be visualized as supplementary data in the article of Bourdichon et al. (2012) using the Digital Object Identifier System (DOI) 10.1016/j.ijfoodmicro.2011.12.030.

Not all of these microorganisms cited in the Bourdichon et al. (2012) list, however, are produced as commercial SC. Some of the products do not lend themselves to industrial applications, and for others, the potential market is simply too small or too specialized. For example, sauerkraut fermentation is mediated by the natural microbiota. Manufacturers can produce these products without a commercial SC and may prefer to maintain and propagate their own cultures. Many dairies and breweries also have personal and laboratory facilities to support internal culture spread. The production of fungal-derived fermented foods also frequently relies on house strains.

MFC have directly or indirectly come under various regulatory frameworks in the last years. These frameworks place emphasis on "the history of use in traditional food" or "general recognition of safety". Generally Recognized as Safe (GRAS) from the US and Qualified Presumption of Safety (QPS) from the European regulatory environmental status are applicable to MFC adequately shown to be safe under the conditions of its intended conditions of usage. When microorganisms with a safe history in food are employed for a different use, a new GRAS determination is needed. The QPS system was proposed to harmonize approaches to the safety assessment of microorganisms. The list of QPS is updated annually by the European Food Safety Authority (EFSA) (Herody et al. 2010).

Taxonomy and systematics constitute the basis for the regulatory framework for MFCs. Bourdichon et al. (2012) have also reviewed and updated the taxonomy of the microorganisms used in food fermentations in order to bring the taxonomy in agreement with the current standing in nomenclature. The detailed list of microorganisms (bacteria, fungi, filamentous fungi and yeasts) that include 62 genera and 264 species with demonstration in food usage can be found in this publication.

There is no simple definition of the species as a taxonomical unit. Generally, a polyphasic approach is required to quantify the degree of phenotypic and/or genotypic similarity between strains belonging to the same species. A special issue of the Journal Systematic and Applied Microbiology introduced by Schleifer et al. (2015) offered in its manuscripts an overview of some of the future perspectives that challenges both taxonomy and the characterization of the cultures diversity. As pointed out by Rosselló-Móra and Amann (2015), the history of species classification has been linked to the technological developments to retrieve information on the organisms. Molecular methods that produce genetic and genomic data can be used to establish taxa boundaries. The retrieval of genomic information by using Next Generation Sequencing methods (NGS) has started a revolution in the study of species diversity. Almost complete genome sequences may take supremacy in circumscribing species and even higher taxa.

The wide variety of fermented food products consumed around the world requires a diverse array of microorganisms. Although lactic acid-producing bacteria and ethanol-producing yeasts are certainly the most frequently used microorganisms in fermented foods, there are many other bacteria, yeast and fungi that contribute to essential flavour, texture, appearance and other functional properties to the finished products. In most cases, more than one organism or group of organisms is involved in the fermentation.

Bacterial

The most important group of bacteria used as SC is LAB. In fact, among commercially available SC, only a few are non-LAB. They receive this name due to their ability to convert fermentable carbohydrates in lactic acid via either homofermentative or heterofermentative metabolism. They are characterized by Gram-positive microorganisms that share several physiological and biochemical traits. In general, they are non-sporulating, catalase-negative, acid-tolerant and grow over a wide temperature range, although most LAB are either mesophiles or moderate thermophiles (Zhang et al. 2013). Likewise, they vary with respect to salt tolerance, osmotolerance, aerobiosis, and other environmental conditions, accounting, in part, for the diversity of habitats with which they are associated (Giraffa 2012).

Taxonomically, LAB species are found in two distinct phyla, namely Firmicutes and Actinobacteria. Within the Firmicutes phylum, LAB belong to the *Lactobacillales* order and include, among others, the following genera: *Aerococcus, Alloiococcus, Carnobacterium, Enterococcus, Lactobacillus, Lactococcus, Leuconostoc, Oenococcus, Pediococcus, Streptococcus, Symbiobacterium, Tetragenococcus, Vagococcus*, and *Weissella*, which are all low-GC content organisms (31–49%) (Giraffa 2012).

Seven of the LAB genera (*Lactobacillus, Lactococcus, Leuconostoc, Oenococcus, Pediococcus, Streptococcus*, and *Tetragenococcus*) are widely used in many industrial applications, especially as SC in fermented foods such as fermented milks, cheeses, butter, as well as in fermented meat products such as salami and fermented vegetables. Due to their ability to hydrolyze proteins, sugars and lipids, they have great influence on the sensory profile of the final product, resulting in improved texture, flavour and aroma (Pescuma et al. 2008).

In sourdough bread, for example, SC strains of *Lactobacillus sanfranciscansis* ferment maltose and lower the dough pH via production of lactic acid, but the culture also produces acetic acid and other flavour and aroma compounds. In contrast, one of the LAB included in sausage SC, *Pediococcus acidilactici*, essentially performs the production of lactic acid and reduces the meat pH to a level inhibitory to undesirable competitors (Hutkins 2006). For decades, LAB have been used in food preservation. They play a role in the conservation recognized and microbiological safety of fermented food, thereby promoting microbial stability of the final product. The protective effects are promoted by producing organic acids, CO_2, ethanol, hydrogen peroxide and diacetyl, antimicrobial components, such as fatty acids and bacteriocins (Giraffa 2012).

Although LAB are the most important group of SC bacteria, other non LAB also are included in various SC. In the cheese industry, for example, *Propionibacterium freudenreichii* subsp. *shermanii* is used to manufacture Swiss type cheeses and *Brevibacterium linens* is used in the production of surface ripened cheeses. On the Actinobacteria phyla, the genera relevant to fermented foods include *Bifidobacterium, Kocuria, Staphylococcus* and *Micrococcus*. *Bifidobacterium* actually serve as functional (probiotic purposes) culture. Species of *Kocuria* and the *Staphylococcus/ Micrococcus* group are used in fermented sausages with only one purpose, enhancing the desired flavour and color via production of the enzyme nitrate reductase.

It is worth emphasizing that fermented foods may contain many other microorganisms, whose presence occurs as a result of inadvertent contamination. So, the production of quality fermented foods requires close attention to the characterization, differentiation and maintenance of bacterial cultures involved in the production process.

Yeast and Moulds

Relatively few species of fungi among thousands are involved in food fermentations. Yeasts and moulds belong to the kingdom Fungi. The fungi relevant to fermented foods are classified as *Zygomycota, Ascomycota*, or as *Deuteromycetes*. These groups contain several important genera of yeasts, including *Saccharomyces, Kluyveromyces*, and *Zygosaccharomyces*, as well as the mould genera *Aspergillus, Penicillium*, and *Rhizopus*. The most common moulds used in fermented foods are *Penicillium* and *Aspergillus*.

Several different forms of yeast SC are available, ranging from the moist yeast cakes, used by the baking industry (the largest user of yeast SC) to the active dry yeast packages sold at retail to consumers. Wine, beer and alcoholic beverage industry also use yeast SC. Yeasts used for the bread fermentation are classified as *Saccharomyces cerevisae* and are screened based on their ability to quickly produce large amounts of CO_2. Many of the ethanol producing yeasts are also classified as *Saccharomyces cerevisae*, although they differ markedly from those strains used for bread manufacture (Hutkins 2006).

Today, many of the wines produced in the world use indigenous yeasts selected from the winegrowing area. The trend for most wine manufacturers, however, has been to use SC yeasts. These local yeasts are presumed to be more competitive than commercial yeasts because they are better adapted to the ecological and technological features of their own winegrowing region (Lopes et al. 2006). As with other SC, strain selection is based primarily on the desired flavour and sensory attributes of the final product. However, other technological traits, including the ability to grow at high sugar concentrations, to flocculate and to produce adequate alcohol levels, are also relevant and are taken into consideration.

One of the most significant technological advances in winemaking has been the control of the microbiological process by grape must inoculation using selected yeasts (Lopes et al. 2006). The use of selected yeast SC for wine fermentation has led to the production of more consistent wines. Yeast and fermentation conditions are claimed to be the principal factors that influence the flavours in wine (Fundira et al. 2002). The selection of wine yeasts is usually carried out within the species *Saccharomyces cerevisiae* according to a set of physiological features indicative of their potential usefulness for industrial wine production (Rainieri and Pretorius 2000). Furthermore, for the induction of the malolactic fermentation bacterial SC consists of *Oenococcus oeni*. In the future yeast strains other than *Saccharomyces cerevisiae* or LAB SC other than *Oenococcus oeni* (Krieger-Weber 2009) may also be applied.

Several works have studied the influence on wine quality of yeast added SC to an alcoholic fermentation, because higher alcohol and ester contents in the wine depend on yeast and fermentation temperature. Yeasts form and modify the important components of fermented beverages: volatile organic acids, aldehydes, alcohols and esters (Mateo et al. 2001). The production levels of the by-products are variable and yeast-strain specific. The yeast strain used during fermentation can have a great influence on the ultimate quality of the end product, making the choice of yeast strain crucial if good quality wines and distillates are to be assured.

The brewing industry is another potential user of yeast SC, although breweries generally maintain their own cultures. As for wine, strain selection is based on characteristics to the needs of the particular technological process and includes factors such as flocculation, flavour development and ethanol production rates. One of the most important issues in modern brewing processes is how to produce flavour and foam-stable beer (Bravi et al. 2009). The presence of metabolically active yeasts is a factor that has a great influence on fatty acid composition of lipids found in both brewing byproduct and finished beer.

Despite this, several strain improvement methods are described to enrich the aromatic profile or fermentation efficiency of *Saccharomyces cerevisiae* strains (Steensels et al. 2014), although these techniques also have their limitations, and non-*Saccharomyces* (or non-conventional) yeasts are becoming increasingly popular in the fermentation industry (Steensels et al. 2015). Many of these yeasts (e.g. *Brettanomyces*) can have a beneficial role by increasing the fermentation efficiency, lowering the spoilage risk or changing the flavour profile of the end product (Steensels and Verstrepen 2014). The role of *Brettanomyces* in the food industry is confusing and ambiguous. For example, *Brettanomyces* yeasts are considered spoilage in wine. However, the potential of *Brettanomyces* as a SC in industrial fermentation processes is increasingly recognized. Their unique flavour profile and amylase activity make them very well suited for the production of novel alcoholic beverages such as beers (Daenen et al. 2009).

Concerning the moulds, the dairy industry is the major user of this type of SC. Two groups of *Penicillium* sp. are used in cheese production. The white mould (*Penicillium camemberti* Thom, formerly two species, *Penicillium caseicolum* and *Penicillium camemberti*), which grows on the surface of soft white mould-ripened cheese (e.g. Camembert) and the blue mould (*Penicillium roqueforti*, formerly *Penicillium roqueforti* var. *roqueforti*), which grows in the interior of blue-veined cheeses (e.g. Gorgonzola). *Penicillium camemberti* and *Penicillium roqueforti* are lipolytic and proteolytic. Both produce methyl ketones and free fatty acids, but the much higher levels produced by *Penicillium roqueforti* give blue cheeses their distinctive flavour and aroma (Fernández-Bodega et al. 2009). *Penicillium roqueforti* has been traditionally used as a secondary starter. In several villages of Europe, high quality varieties of local blue cheeses are made, using traditional manufacture procedures. The excellent quality of these cheeses is in part due to the ripening process in natural caves or cellars. Strains of *Penicillium roqueforti* suitable for production of these varieties of cheese may originate from the indigenous strains associated with the ripening cheese, as cited above (local artisan), or have been selected and commercialized.

The gradual shift in fermented food production from small local producers to large-scale factories and increasing awareness of the risks for consumer safety has paved the way for industrialized production of mould SC. The industrialized SC are carefully selected among hundreds of candidates going through multi-stage concepts, including diverse analytical and biochemical investigations (Perrone et al. 2015). Industries of fermented products nowadays prefer to use mould SC rather than to propagate their own cultures. Thus, fungal SC are available for several types of cheeses, soy-derived fermented foods (e.g. tempeh, made using *Rhizopus microspores* subsp. *Oligosporus* and soy sauce, made using *Aspergillus sojae* and *Aspergillus oryzae*) and European-style sausages and hams.

Fungi have an important role in the production of dry-cured meat products, especially during the seasoning period. In general, both industrial and artisanal salami are quickly colonized by a mycobiota during seasoning, often with a strong predominance of *Penicillium* species that are involved in the improvement of the characteristics and taste and in the prevention of the growth of pathogenic, toxigenic or spoilage fungi. Perrone et al. (2015) isolated and identified two predominantly present *Penicillium* species from seasoning and storage areas of a salami plant in Italy. The taxonomic position of these strains in *Penicillium* was investigated and resulted in the discovery of a new *Penicillium* species, described as *Penicillium salamii*. In their study, *Penicillium salamii* predominated during seasoning in levels similar to the commonly known SC *Penicillium nalgiovense*, indicating that this species could be used as a fungal SC for dry-cured meat.

Criteria for Selecting and Developing Starter Cultures

When SC technology first developed, the actual microorganisms contained within most MFC were generally not well established in terms of strain or even species identity. Rather, a culture was used simply because it worked, meaning that it produced a good product with consistent properties. Such undefined cultures, however, are now less frequently used. Instead, the organisms present in modern commercial SC preparations are usually very well defined strains. They are also carefully selected based on the precise phenotypic criteria relevant for the particular product. For example, any potential new SC for cheese must produce at least acid, lack off-flavour development in milk and be phage-resistant (Carminati et al. 2010).

Pure culture fermentations also have disadvantages. Selecting one single strain with all beneficial characteristics necessary for an efficient and high-quality fermentation might prove difficult, since the diversity of wild, contaminating yeasts and bacterial species create a distinct fermentation and flavour profile and a lot of the complexity or subtle aromatic notes might be eliminated in pure culture fermentations.

Selection of strains for use in food fermentation is a highly complex and challenging task that requires a solid understanding of microorganism growth and manipulation, as well as microbial interactions with other organisms. The use of microorganisms follows a logical sequence. First, it is necessary to determine matrices to isolate microorganisms that carry out the desired process in the most efficient manner that potentially exhibit technological properties. Next, phenotypic traits and technological suitability evaluation have to be taken into consideration. After that, the taxonomic affiliation is determined using DNA-DNA hybridization and the analysis of 16S rDNA gene sequence. The name of the identified strain needs to be consistent with the list of "history of safe use". This microorganism then is used, in a controlled environment such as a fermenter to achieve specific goals.

Basically, the criteria for the selection and development of SC for the food fermentation industry can be considered under three common categories: (1) properties that affect the performance and efficiency of the fermentation process; (2) properties that determine food quality and character; (3) properties associated with the commercial production of MFC. Buchenhüskes (1993) and Champagne and Møllgaard (2008) (Table 14.2) summarized the selection criteria for LAB to be used for food fermentations (desired properties for industrial use): lack of pathogenic or toxic activity (e.g. production of biogenic amines); ability to accelerate metabolic activities (acidification or alcohol production); production of desired changes in the matrix (structural and sensorial); improvement in safety and reduce hygienic and toxicological risks; ability to dominate microbial consortium; ease of propagation; ease of preservation; stability of desirable properties during culturing and storage. Specific properties desired in a MFC depend on the product being produced.

TABLE 14.2 Main selection criteria of lactic acid cultures (not exhaustive). Adapted from Buchenhüskes (1993) and Champagne and Møllgaard (2008)

Technological properties	Other properties
Dominate microbial consortium	Be easy for propagation and preservation
Assimilate other non-conventional carbon fonts	Improve safety and reduce hygienic and toxicological risks
Produce desired changes in the matrix	Be absent of pathogenic or toxic activity
Produce flavour and polysaccharides metabolites	Produce antimicrobial substances (bacteriocins)
Induce proteolysis	Resist to bacteriophages
Resist to process pressure (acidity, NaCl concentration, oxygen)	Have desirable properties stability during culturing and storage

Isolation of microorganisms and strain identification

One of the main interests of microbiologists is to study the diversity and dynamics of microorganisms during the manufacture of fermented products and their storage period, as well as to correlate the occurrence of certain microbial species or strains to present specific sensory characteristics of the final product. Traditionally, knowledge of the bacterial diversity in fermented products is a result of studies of cultures based on the growth of microorganisms on selective media and subsequent identification of the genus and species using phenotypic characterization (De Dea Lindner et al. 2008; Neviani et al. 2009; Randazzo et al. 2009).

Commercial SC generally originate either from food substrates or from the processes in which they are applied. Nowadays, there is an effort in searching for potential MFC from the pool of wild strains recoverable from diverse raw matrices, undefined cultures and/or artisanal fermented products. There is continuing interest in bioprospecting in all niches of the world and major companies have been organized to continue to explore microbial diversity and identify microorganisms with new capabilities.

The first stage in the screening for microorganisms of potential industrial application as SC is their isolation. Isolation involves obtaining pure cultures followed by its assessment to determine the desired process for the desired product.

Dastager (2013) described the important criteria to choose an organism and the isolation methods.

Investigations of microbial populations, after growth in media and appropriate environment, typically reveal the prevalent development of microorganisms detectable by colony formation visually or spectrophotometrically (Steele et al. 2006). However, Cogan et al. (2007) underscore the difficulty of micro-colony count and the wild strains identification of disability and indigenous bacteria by phenotypic methods. The traditional culture media used in food analysis, which cherish the recovery of prevalent microorganisms in food matrix, can be very generic and not selective enough to discriminate the species present in the minority (Denis et al. 2001; Neviani et al. 2009). Many LAB, including indigenous (NSLAB – non starter LAB) may be present in food in vital conditions, but not cultivable (Gatti et al. 2008). For this reason, in cases in which the isolation and identification of microorganisms are needed, a nutritionally adequate culture medium and, if possible, of similar composition to the isolating matrix would be ideal to retrieve this microbiota of fundamental importance (Neviani et al. 2009).

For example, the LAB morphological study includes both colony characteristics (color, size, shape, texture) and cellular (cell shape, no endospores) (De Dea Lindner 2008). The description of the physiological and biochemical characters implies information about its properties, such as development at different temperatures, pH, salt concentration and ability to assimilate O_2, development in the presence of various substances, including antimicrobial, viability in the presence of various enzymes, use of organic compounds and especially fermentative capacity. However, as well as the difficulties in the characterization of the colonies, the morphological and the description of the physiological and biochemical characters also present difficulties when traditional microbiological techniques are used. The identification of LAB based on biochemical tests or on the use of quick kits may give false results, or even not identify the species.

LAB isolated from different sources usually belong to *Lactobacillus, Lactococcus, Pediococcus, Streptococcus* and *Enterococcus* (De Dea Lindner 2008). According to Dikes and von Holy (1994), mixed cultures of LAB present in a system are complex and usually consist of an association of lactobacilli, which have identification difficulties. For example, closely related bacteria, such as *Lactobacillus bulgaricus, Lactobacillus delbrueckii* subsp. *delbrueckii, Lactobacillus delbrueckii* subsp. *lactis, Lactobacillus rhamnosus, Lactobacillus plantarum, Lactobacillus casei* and *Lactobacillus paracasei*, share many biochemical and physiological properties.

Unlike the phenotypic and physiological methods, identification by molecular techniques is much more consistent, fast, reliable and reproducible, and can discriminate species that are closely indistinguishable groups as identified by traditional tests. Thus, many *Lactobacillus* species, among others, were reclassified using advanced molecular techniques based on the Polymerase Chain Reaction (PCR) (Dellaglio et al. 2004). Molecular typing techniques, combined with sensitive tools to identify species, are viable for true identification of these microorganisms. However, the use of molecular biology techniques, as culture independent techniques, has facilitated access to information on the taxonomy and ecology of LAB (Gatti et al. 2008; Juste et al. 2008).

Strain improvement

MFC produce a wide range of metabolic end products that function as preservatives, texturizers, stabilizers and flavouring and coloring agents. The science and technology of manipulating and improving microbial strains in order to enhance their metabolic capabilities for biotechnological applications are referred to as strain improvement (Demain and Davis 1998). The advent of modern industrial microbiology made possible the application of pure culture and strain development. With advances in biochemistry and genetic manipulative techniques, scientists took a more targeted approach towards developing MFC. Today, the large-scale production of food additives and selected SC serves as testimony to the important role of strain improvement in shaping the fermentation industries.

Parekh et al. (2000) reviewed various strategies employed in strain improvement, practical aspects of screening procedures and the overall impact of strains improvement on the economics of fermentation process. Several methods have been used to improve metabolic properties of MFC through DNA manipulation. These include mutation, selection and genetic recombination (transduction, conjugation and transformation). Several properties could be enhanced using genetic engineering. For example, transfer of bacteriocins production to microbial SC could improve the safety of fermented foods; cloning of additional copies of specific proteolytic enzyme involved in cheese ripening could accelerate the process (Dastager 2013).

The ability to genetically manipulate microorganisms to improve process and properties has grown into a billion dollar enterprise that has revolutionized many fields of industry. Although the use of recombinant DNA technology for microbial products is commonplace, a similar situation does not apply to the use of genetically modified cultures. Cogan et al. (2007) cited that, in some cases, genetic strain improvements can be effected by means that do not require the introduction and expression of rDNA molecules. They illustrated two examples of genetic improvements that reflect it and have already been implemented by industry involving the enhancing of phage resistance and diacetyl production in lactococci.

Stadhouders and de Vos (2014) described three categories of SC strains manipulated by DNA technology. The first is obtained by applying the natural gene transfer system (conjugation). Since the plasmids that encode industrial properties are naturally transferable, no limitations on application are expected, since the natural genetic barriers between species are not crossed. SC genetically improved by this method bypass the regulatory issues associated with recombinant DNA technology and, nowadays, this category of starters is marketed worldwide. The second category consists of microorganisms that have been constructed employing artificial gene transfer system. When the use is made of homologous and naturally occurring plasmids, no limitation may be expected either. Such strains are genetically modified, but this feature has been accepted. However, when artificial gene transfer system (third category) is used to clone and express heterologous genes (non-homologous genetic material), these genetically modified microorganism are subject to approval of regulatory bodies. To date and in our knowledge, no genetically modified SC, via third cited category, has been approved by any regulatory agency in the world.

Evaluating Starter Culture Performance

Objectively, the work of a SC is to carry out the desired fermentation, promptly and consistently, and to generate the expected flavour and texture properties relevant to the specific food product. Furthermore, particular requirements for a given SC, however, depend on the application for that culture. As industries expect particular product specifications, the specific traits and properties expressed by SC have also become quite demanding. For example, LAB SC used in dairy fermentations are now selected based not just on lactic acid production rates, but also on flavour and texture producing properties, salt sensitivity, compatibility with other strains, phage resistance and viability during production and shelf life (Hutkins 2006). As another example, traits desirable for brewing SC of *Saccharomyces cerevisiae* (e.g. flocculation and ethanol tolerance) are quite different from those important for bakers' yeasts SC of *Saccharomyces cerevisiae*.

For the measurement of the activity and evaluation of the performance, several tests are routinely carried out to assess properties of SC. The term activity refers to the ability of SC to produce desirable changes in fermented products and is a consequence of many factors, such as physiological state of cultures. Usually, activity measurements are confined to the ability of SC, for example, to acidify the milk (LAB), whereas for yeast cultures rates of CO_2 evolution may be appropriate.

Acidification activity (Fonseca et al. 2000) is considered the classic way to determine SC activity for LAB measuring the pH in the matrix at different time intervals. The capacity of a culture to adapt to a new environment and the biomass production may all be useful determinants. Activities other than cited above, such as proteolysis (Gatti et al. 2008), lipolysis and specific enzymes production, can also be determined by other means. Another critical test used specifically for dairy SC involves the determination of bacteriophage resistance (Leroy and De Vuyst 2004).

Performance characteristics for the yeast SC depend on the specific product. The ability of brewing yeasts to ferment all available fermentable sugars (attenuation) is a critical property that must be measured. For bakers' yeasts, the measure of growing phases and the production of carbon dioxide in the dough (gassing rate) is necessary to evaluate activity. In wine manufacture, wine yeasts are evaluated based on the results for the growth at low temperatures, as well as for the resistance to sulfiting agents. For some applications, wine yeast should also have high ethanol tolerance and osmotolerance.

One absolute requirement for multiple SC strains is that each microorganism must be compatible with all the others within a particular culture. For example, LAB are well-known producers of bacteriocins, hydrogen peroxide and others antimicrobial components that can inhibit other microorganisms. The same manner, some yeasts produce inhibitory proteinaceous compounds called "killer" factors that are active against members of the same species or closely related species, therefore influencing how mixed SC strain are prepared. The activities of these toxins are analogous to the activities of bacteriocins in bacterial strains (Lowes et al. 2000). El Baz and Shetaia (2005) described four different assays (methylene blue, thiazolyl blue tetrazolium bromide, bromo cresol purple and plate count agar) for the evaluation of killer toxin activity of six isolates of yeast, *Kluyveromyces lactis*, *Candida silvatica*, *Candida tropicalis*, *Rhodotorula mucilaginosa*, *Rhodotorula lactosa* and *Lipomyces tetrasporus*, in the assessment of the viability of *Saccharomyces cerevicae* sensitive strain.

How Starter Cultures are used in Industry

The food industry has changed in a very impressive manner in the past years and the average size of a typical production facility has increased several-fold. The fermented food industry is dominated by producers with large production capacity, although small (traditional-style) facilities still exist, as it is evident by the many breweries, wineries, bakery and artisanal dairies. Together with these modifications, the manner which fermented foods are produced has also changed. Nowadays, by commercial and financial pressure, fermented food production is strongly subject to time and scheduling demands. In a production operation, a slow fermentation may cause financial lost, problems with workers shift also affect the entire production schedule. The fermented food industry, as cited earlier, is the only food processing industry in which product success depends highly on the growth and activity of microorganisms.

There are two general ways that the cultures are used in fermented food and beverage industries. The first type, often referred to in the dairy industry as bulk cultures, is used to inoculate a bulk tank. The bulk SC is essentially the equivalent of several intermediate cultures that traditionally have been required to build up the culture. The dairy industry is by far the main user of bulk cultures routinely used to inoculate production vats throughout a manufacturing day. The dairies can buy the commercial SC in various forms: liquid, for propagation of mother culture; frozen concentrated and freeze-dried concentrated, for propagation of bulk starter; frozen superconcentrated in readily soluble form, for direct inoculation of the product.

Bulk culture can be prepared at the plant or from commercially available frozen concentrated or freeze-dried concentrated cultures. Preparing inoculum at the plant involves serial culturing, starting with a mother culture maintained in small amount of medium. The mother culture is used to inoculate successively larger amounts of medium until sufficient inoculum volume is obtained to prepare the bulk culture. Bulk culture handling at the plant carries an increased risk of phage contamination. In order to minimize this problem, bulk culture media used in the propagation may contain phosphate buffers and other agents that protect the cells from acid damage and from infection by lytic bacteriophages. Buffering activity supplied in the form of phosphate salts provides a phage-inhibitory function via calcium chelation (Hutkins 2006).

In contrast to the bulk cultures, some fermented foods are inoculated with a second type of culture. This type is usually most expensive and can simply be inoculated directly into the food substrate (direct-to-vat). Using the nominated direct-vat-set (DVS) cultures avoids the possibility that the starter will become contaminated with phage during preparation within the plant. Also, the use can reduce mixed strain compositional variability, assuring an appropriate strain balance. This is the normal means by which active bakers' yeast, meat and several dairy SC are added directly to the dough, the meat mixture and milk, respectively. In the case of frozen concentrated

SC, the containers are first thawed in water prior to use. Freeze-dried pellets can be added directly, but may require mixing in the vat to facilitate solubility/hydration.

According to Hutkins (2006), the use of DVS cultures has advantages: eliminates the labour, equipment and by-products associated with the preparation and maintenance of bulk SC system. Although they were initially produced as frozen concentrates and packaged in cans, they are now available as pourable pellets or lyophilized powders, making it easy to dispense the exact amount necessary for inoculation. However, when DVS cultures are used in large industrial volume, the SC must be superconcentrated to deliver a sufficient inoculum into the raw material. The industrial production operations, discussed in the next section, may indirectly reduce SC viability, leading to slow-starting fermentations. Improvements in concentration technologies have minimized some of these problems.

Industrial Production of Starter Cultures

Commercial SC industrial production requires specialized knowledge of microbiology, molecular biology and biotechnology. Furthermore, SC development, maintenance and distribution demand special logistic and economic considerations. Since the fermented food industry depends on the use of selected strains with known metabolic and technological properties, the introduction of commercial SC has undoubtedly improved the quality of the products and the overall process of standardization discussed in the section above.

The limited number of available strains with specific technological performance and the constant risk of bacteriophage attacks justify the continuous need to search for new commercial strains for industrial product diversification. Nowadays, the organisms present in modern commercial SC preparations are very well defined. They are carefully selected based on the precise phenotypic criteria relevant for the particular fermented product.

For the preparation of fermented products, Latin American industries acquire imported SC that sometimes have inadequate performance in local food matrix, mainly due to differences in the quality of raw materials, for example, in raw milk. The SC global market in 2014 was around 2.5 million tons and Latin America represented a small share of this with 100,000 tons (4%) and an annual growth of 2%.

Commercial SC are currently available for direct addition to production vats contain $\geq 10^9$ viable microorganisms, preserved in a form that could be readily and quickly activated in the matrix to perform the technological functions necessary to transform the raw food into the desired fermented product. To achieve that, the selected microorganism needs to be grown in a suitable media to high biomass volume and to concentrate the cells.

The amenable manufacturing processes for large-scale SC fundamentally follow four steps: (1) preparation of the growth medium; (2) biomass production; (3) biomass concentration; and (4) biomass preservation. All steps will contribute to the quality of the product. SC need to be tolerant of the stresses of concentration, packaging, storage and reactivation processes (Soubeyrand 2005). These requirements need to be reached without loss of the desirable fermentation properties.

Many of the manufacturing processes are considered stressful for microorganisms. Adverse environmental conditions include extreme temperatures, osmotic stress, mechanical stress, acidity of the medium, as well as nutrient-deficit stress at the end of culture. These hostile conditions for culture require selection of resistant strains, which should retain adequate viability, metabolic activity and genetic stability.

SC quality control is required to prepare the user for the importance of sanitization and strict adherence to use protocols. Quality control tests for commercial SC include the assessment of viable cell numbers, presence of pathogens and extraneous matter, technological performance as acid-producing and other functional activities, package integrity, accuracy of label information on the package, and shelf life of the product according to specification (Vedamuthu 2006).

Sip and Grajek (2010) described the production of probiotic cultures on commercial scale in a strongly concentrated form for introducing to the food as DVS preparations. Cultures of probiotic microorganisms are commercialized in the form of frozen concentrates, either freeze- or spray-dried. Probiotic cultures may be composed of single strain or a mixture of strains. Due to differences in growth requirements, they may not be cultured in the food together with commercial LAB SC, but in practice, this occurs. For example, in mixed SC for fermented milk containing *Lactobacillus acidophilus*, *Bifidobacterium* and *S. thermophilus*. They need to be added to the food in sufficient CFU per gram to provide a probiotic effect. Production technology for probiotic cultures is identical to that cited for SC and comprises the same basic stages (Lacroix and Yildirim 2007).

Growth Media

The composition of the media used to grow SC also varies, according to the microorganisms being grown and the nutrient requirements necessary for optimum biomass production. The choice of the components of the medium is critical, because it can influence the economic competitiveness of a particular process. Unfortunately, there is no single low-cost medium that can be universally used to grow all cultures, but the components used consist of food grade lower-cost agricultural by-products (e.g. crude plant hydrolysates, whey from cheese manufacture, brewery grains) and are used as economically sources of carbon and nitrogen.

Parekh et al. (2000) cited a number of studies that have been published describing media design strategies for fermentation. Furthermore, they reviewed the methods for media improvement and process optimization. Classical approaches to media designs are frequently employed where individual components such as nitrogen sources are alternatively tested as single variables. These methods seek to define a nutrient profile that mimics the elemental composition of the cells to be grown. As newer techniques to optimize process performance are being introduced, statistical experimental design is used to screen and optimize fermentation as well as evolutionary algorithms (genetic algorithms and particle swarming) and artificial neural networks.

Champagne et al. (1999) used an automated spectrophotometric method to evaluate the growth-promoting ability of yeast extracts on cultures of *Lactobacillus acidophilus* and *Lactococcus lactis* subsp. *cremoris*. The automated spectrophotometry method can ascertain what is the most appropriate amino supplement, as well as enable the screening of up to 50 media in a single assay.

The generally used components in media formulations are found in Table 14.3. For example, dairy SC are frequently produced using whey-based media (Champagne and Møllgaard 2008). Molasses can be used as the basal medium for other lactic cultures. Media most often need to be adjusted to specific genera. To achieve high biomass density and cell viability, it is important that inhibitory formed components (metabolic end-products) be neutralized. For example, theoretically, the higher the acid formed buffering capacity of the medium, the higher the final biomass attained. The levels and balance of other minor components like minerals and growth factors can be critical in medium formulation. The medium also may be designed so that major and minor nutritional components will not become limiting after a given time during the fermentation.

TABLE 14.3 Components of growth media used in starter culture production (Not exhaustive)

Source	Matrix
Carbon, aminoacids and energy	Non fat milk
	Whey
	Hydrolysates of milk and whey proteins
	Soy isolates and protein hydrolysates
	Meat hydrolysates and extracts
	Egg proteins
	Grains extract
	Potato infusions
	Yeast extracts and autolysates
	Sugars such as lactose, glucose, HFCS*, sucrose
	Molasses
	Agricultural wastes
Nitrogen	Corn steep liquor
	Ammonium salts
	Nitrates
Minerals	Inorganic chemicals
Vitamins	Water soluble vitamin mixes
Unsaturated fatty acids and surfactants	Polysorbates (tweens)
Buffer	Phosphates and carbonates
Antifoam agents	Higher alcohols, esters, vegetable oils
Antiacid agents	Gaseous NH_3, NH_4OH, Na_2CO_3, KOH
Hydrogen peroxide reducer	Catalase

* HFCS - High-fructose corn syrup.

The type of carbohydrate in the growth medium can affect subsequent survival during concentration and preservation processes. Conditions that enhance the intracellular accumulation of protective compounds not only improve survival, but also influence stability of the cells during storage.

Fermentation, Concentration and Preservation Technologies

SC for fermented food industries are mass-produced in fermenters. The process begins with an isolate stock culture which is grown in a small volume prior to inoculation into the production fermentor. After that, the inoculum is added to the medium and is incubated until the predetermined endpoint is reached (for LAB, late log phase or early stationary phase). The maintenance of the microbial stock and the process for preparing the inoculum for the large scale fermentor are under the management of the quality control section of the industry. Generally, the microbial harvest point coincides with the depletion of carbohydrates in the medium. When provided optimum growth conditions, maximum densities of 9-10 $\log_{10}$ cells per ml are typically obtained. However, the optimum harvest time depends on the specific microorganism.

After an optimum medium is formulated, the environment for microbial propagation in the culture system must be defined. This involves control in the operations of agitation, temperature, pH and oxygenation. Medium composition, aeration, pH adjustment, sampling and process monitoring must be carried out under rigorously controlled conditions. Buffers can be used to control pH and the O_2 flux and concentration must be controlled. Fermentation processes carried out under pH control have the disadvantage of generating SC with lower specific acidifying activities (Savoie et al. 2007).

Scale-up of strains from laboratorial flasks or bottles to fermenters needs to consider any changes that the new microorganism brings to the process (e.g. sensitivity to shear; sterilization of media). It is essential to assure that physical factors are not limiting microbial growth, especially during scale-up (Parekh et al. 2000).

Microorganisms can be grown in culture tubes, shake flasks and stirred fermenters or other mass culture systems. Kaur et al. (2013) reviewed the various types, shapes and sizes of bioreactors employed in submerged fermentations, with special reference to their use in the food industry. Three modes of operation of fermenters are very common: batch, continuous and feed-batch. Batch and feed-batch operation result in different physiological conditions for cultivated microorganisms. Continuous cell production systems have not been adopted by the food industry yet. The primary function of a fermenter is to provide a suitable operation under aseptic conditions in which a specific microorganism can efficiently grow and multiply. For example, the pneumatically agitated fermenters are better for cultivating moulds because of reduced shear.

A typical industrial fermentation unit requires a large capital investment and skilled operators. Computers are commonly used to monitor parameters. Such information is needed for precise process control. After ascertaining the production of biomass, the fermentor is cooled to break the metabolic process and the cells are harvested either by continuous centrifugation or cross-flow membrane filtration (ultrafiltration). However, these processes to obtain concentrated biomass, if in the presence of medium and metabolites such as organic acids and bacterial exopolysac-charide, may lead to low viability levels. Both technologies are heavily affected by products generated in fermentation. Unfortunately, these operations may generate damages to the cells, which result in lower survival during the application of preservation technologies (Champagne and Gardner 2002).

Champagne (2006) reviewed the characteristics of the fermentation process based on growing microentrapped lactic cultures in alginate beads. He describes that growing cells in alginate beads instead of free cells in a culture medium can offer an alternative process to address the problem cited in the last paragraph. Thus, the choice between the filtration and centrifugation must be based on the specific microorganisms and set-up process in the biotechnological factory. The cell concentrate is obtained in the form of a viscous liquid. A high degree of concentration is desirable in order to make an active product and minimize storage. A concentrated product also minimizes the cost of preservation techniques.

Sterile preparations of cryoprotectant agents (e.g. glycerol, sorbitol, mannitol, lactose, sucrose, trehalose, ascorbate, glutamate) are added to the cell concentrate to maintain cell viability and protect against freeze and stress damage (Carvalho et al. 2004). The cryoadditives also function during the subsequent storage of cells. Some microorganisms simply suffer a more serious loss in cell number and viability. For example, *Lactobacillus* sp. may be particularly sensitive to freeze-drying (Monnet et al. 2003). This is important for commercial SC for yogurt and other dairy products that are produced in freeze-dried forms.

Concentrated cells, in the pellet technology, are dropped directly into liquid nitrogen. These frozen particles of pure cultures can be mixed to obtain multiple-strain cultures of defined composition. The cells may be washed to remove the spent medium and resuspended in a suspension solution containing cryoprotectant agents before freeze. The frozen cells in form of pellets can be collected and subsequently packaged in high cell densities (10-12 $\log_{10}$ cells per gram). Thus, this concentration contains about the equivalent amount of cells that would normally be present in a grown bulk culture sufficient to inoculate thousands of kg of raw food. This was the most common form to deliver LAB cultures and it is how bulk dairy and meat SC were usually packaged.

Concentrated cells can also be obtained by one of the different dehydration processes. Freeze-dried cultures are a good alternative to frozen concentrates. The current freeze-drying technology can provide highly active SC that, like some frozen concentrated cultures, can be DVS. Freeze-drying has been recognized as being a more gentle process for dehydrating and preserving LAB SC. Preparation of freeze-dried cultures is, in the first steps, similar to that of frozen concentrates. After freezing, the concentrate culture is sublimated for hours under vacuum. The dry cells are then packaged under aseptic conditions, preferably in anaerobic atmosphere. Freeze-dried cultures contain from 9-12 $\log_{10}$ cells per gram. Nowadays, freeze-dried products are commercialized as frozen DVS SC, especially for cultured milk products.

Survival to freezing is critical for LAB. The freezing rate is critical in providing high viability cells. It must be stated that freeze-drying is not the only method of drying SC. Thus, spray-drying has also been carried out with variable success, but the survival of cultures using this technique is usually much lower than after freeze-drying. This reflects the result of simultaneous exposure to both thermal and dehydration stresses (Champagne et al. 1991). The following process parameters will affect survival levels during a spray-drying: type of atomization, air pressure and outlet temperature and adaptation of the cells (Champagne and Møllgaard 2008).

Modern Starter Culture Technologies

In recent years, science brought immense advances in the methods used to identify microorganisms, understand their physiology, sequence their genomes, as well as evaluate their proprieties. For example, natural genetic engineering is a class of process proposed by the molecular biologist James Shapiro, who employs environmental forced evolution and adaptive mutations to create microorganisms with new biological capabilities for use in industrial biotechnology.

The microbial systematics requires a range of modern methodologies for the comprehensive characterization of microorganisms. Studies on microbial dynamics and metabolic activities of different microbial species have provided a solid base to improve the SC industrial production. The use of appropriate SC may establish one major step towards improved quality of the fermentations.

The study of fermented food microbiota remains a major subject to gain insight on the metabolic and physiological properties of the cultures. The selection of strains with desired characteristics from a pool of microorganisms by traditional methods (e.g. culture dependent) is a difficult task. The omics age of microbiology offers a new knowledge-based approach to the exploitation of microorganisms for food fermentation. In this regard, whole-genome analysis and analytical chemistry should expand our knowledge on metabolic mechanisms in different cultures.

The selection of new SC with a special distinction would be facilitated with the recent development of microarray technology and comparative genomics. Strain collections would be efficiently screened for the presence of certain desired traits. DNA microarrays and high-throughput screening approaches such as gas chromatography-mass spectrometry (GC-MS) analysis allow the targeted selection of, for example, flavour-forming cultures for yogurt fermentation (Carminati et al. 2010).

Aldridge and Rhee (2014) reviewed technological developments in the area of metabolomics and their impact on the understanding of the fundamental roles of metabolism in cellular physiology. Recent advances in analytical chemistry platforms have begun to reveal an unexpected diversity in the composition, structure and regulation of metabolic networks. Mass spectrometry and nuclear magnetic resonance spectroscopy have made it possible to measure thousands of metabolites simultaneously.

Proteomics provides indirect measure of genome sequence data. Recent developments in applications of proteomics for analyzing microorganisms have paralleled the growing microbial genome sequence database, as well as the evolution of mass spectrometry instrumentation and bioinformatics. Model studies demonstrated the application of proteomics for bacterial strain-level identification and for characterization of environmentally-relevant, metabolically-diverse bacteria (Karlsson et al. 2015).

Holzapfel (2002) cited that select strains may be improved through the application of genetic technologies. Recombinant DNA technology may be applied in the production of custom-made SC which would meet the necessary technical and metabolic requirements for a specific fermentation (e.g. accelerated acid production, overproduction of bacteriocins or proteolytic enzymes). A large number of genetically optimized cultures already exist (e.g. undesirable metabolites production by GRAS moulds that may be suppressed by gene disruption technique), regulatory issues however preclude their use.

These modern tools and techniques which have been cited provide improved knowledge for obtaining the most suitable strains for specific substrates and situations and have opened up opportunities for the development of customized SC. However, traditional research is still needed to obtain quantitative data which may yield information about the relationship between the food matrix and SC functionality. The quality of industrial fermented products may be improved through the use of SC implemented on a multifunctional basis.

Future Prospects and Final Considerations

The knowledge of microbiology is critical to the industrial production of SC. The discovery and use of beneficial microbial cultures in fermented products has contributed to an immense development of the food industry and, consequently, to human life and nutrition. Our challenge as food microbiologists is to understand, as much as possible, the potential impact of immediate and longer-term impacts of microbial based technologies in the fermented food innovation.

At present the fermented food industry in Latin America requires the nationalization of technology for the production of large-scale SC. In addition, the isolation of autochthonous cultures from the production environment or raw materials for industrial prospecting preserves the genetic heritage of microorganisms that dominate the ecosystem due to the environmental and climatic diversity. In Brazil, for example, the companies acquire SC imported from multinational companies, which do not always provide adequate fermentation performance, mainly due to differences in the quality and composition of raw materials. Currently, there are no companies that locally produce cultures to meet the domestic market. The total value of the annual SC market in Brazil is concentrated in a few companies. The domestic production of cultures could contribute to the generation of quality products, as well as to the reduction of costs involved in the acquisition of imported products.

Applied research is needed to better improve SC in the present production technology and to obtain information about the relationship between the food matrix and microorganism functionality, thus contributing to optimal strain selection. Even around the world, many producers of fermented food still rely on spontaneous fermentation based on the indigenous food and environmental microflora. The recent developments of genomics, for example, will allow a more complete analysis for a better selection of strains for specific uses.

Moreover, genome analysis using bioinformatics tools should expand the knowledge about metabolic pathways and mechanisms in different strains. Bioinformatics can be used to search for essential components involved in fermentation in genomes, such as proteinases, for example. The use of genomics (e.g. DNA microarrays) and high-throughput screening approaches (e.g. GC-MS analysis) allow a facilitated selection of target cultures.

To date and in our knowledge, no genetically modified SC organisms, via the artificial gene transfer system, have been approved by any regulatory agency in the world. Furthermore, any specific criteria for approval have not been established so far. There are a number of issues that must be resolved before non-homologous genetically engineered SC could be used in food. In Latin America, a primary limitation is that genetic modification or improvement of microbial SC requires sophisticated equipment and expensive materials that may not be available in developing countries.

References

Agate, A.D. and J.V. Bhat. (1966). Role of pectinolytic yeasts in the degradation of mucilage layer of *Coffea robusta* cherries. Applied Microbiology 14: 256–260.

Aldridge, B.B. and K.Y. Rhee. (2014). Microbial metabolomics: innovation, application, insight. Current Opinion in Microbiology 19: 90–96.

Bottari, B., C. Agrimonti, M. Gatti, E. Neviani, and N. Marmiroli. (2013). Development of a multiplex real time PCR to detect thermophilic lactic acid bacteria in natural whey starters. International Journal of Food Microbiology 160: 290–297.

Bourdichon, F., S. Casaregola, C. Farrokh, J.C. Frisvad, M.L. Gerds, W.P. Hammes, J. Harnett, G. Huys, S. Laulund, A. Ouwehand, I.B. Powell, J.B. Prajapati, Y. Seto, E. Ter Schure, A. Van Boven, V. Vankerckhoven, A. Zgoda, S. Tuijtelaars and E. Bech Hansen. (2012). Food fermentations: Microorganisms with technological beneficial use. International Journal of Food Microbiology 154: 87–97.

Bravi, E., G. Perretti, P. Buzzini, R. Della Sera and A. Fantozzi. (2009). Technological steps and yeast biomass as factors affecting the lipid content of beer during the brewing process. Journal of Agriculture and Food Chemistry 57: 6279–6284.

Buckenhüskes, H.J. (1993). Selection criteria for lactic acid bacteria to be used as starter cultures for various food commodities. FEMS Microbiology Reviews 12: 253–272.

Carminati, D., G. Giraffa, A. Quiberoni, A. Binetti, V. Suárez and J. Reinheimer. (2010). Advances and Trends in Starter Cultures for Dairy Fermentations. *In:* F. Mozzi, R. R. Raya and G.M. Vignolo (eds.), Biotechnology of Lactic Acid Bacteria: Novel Applications. Blackwell Publishing, Oxford, pp. 177–192.

Carvalho, A.S., J. Silva, P. Ho, P. Teixeira, F.X. Malcata and P. Gibbs. (2004). Relevant factors for the preparation of freeze-dried lactic acid bacteria. International Dairy Journal 14: 835–847.

Champagne, C.P. (2006). Starter cultures biotechnology: the production of concentrated lactic cultures in alginate beads and their applications in the nutraceutical and food industries. Chemical Industry and Chemical Engineering Quarterly 12: 11–17.

Champagne, C.P. and N.J. Gardner. (2002). Effect of process parameters on the production and drying of *Leuconostoc mesenteroides* cultures. Journal of Industrial Microbiology & Biotechnology 28: 291–296.

Champagne, C.P. and H. Møllgaard. (2008). Production of probiotic cultures and their addition in fermented foods. *In:* E.R. Farnworth (eds.), Handbook of Fermented Functional Foods. CRC Press, 2nd ed, Boca Raton, pp. 513–532.

Champagne, C.P., N. Gardner, E. Brochu, and Y. Beaulieu. (1991). The freeze-drying of lactic acid bacteria: A review. Canada Institute of Food Scince and Technology Journal 24: 118–128.

Champagne, C.P., H. Gaudreau, N. Chartier, J. Conway and E. Fonchy. (1999). Evaluation of yeast extracts as growth media supplements for lactococci and lactobacilli using automated spectrophotometry. Journal of Genetic and Applied Microbiology 45: 17–21.

Cogan, T.M., T.P. Beresford, J. Steele, J. Broadbent, N.P Shah and Z. Ustunol. (2007). Advances in starter cultures and cultured foods. Journal of Dairy Science 90: 4005–4021.

Contreras, A., C. Hidalgoa, S. Schmidt, P.A. Henschke, C. Curtina and C. Varela. (2015). The application of non-*Saccharomyces* yeast in fermentations with limited aeration as a strategy for the production of wine with reduced alcohol content. International Journal of Food Microbiology 205: 7–15.

Daenen, L., D. Saison, D.P. De Schutter, L. De Cooman, K.J. Verstrepen, F. Delvaux, G. Derdelinckx and H. Verachtert. (2009). Bioflavouring of beer through fermentation, refermentation and plant parts addition. *In*: V. R. Preedy (ed.), Beer in Health and Disease Prevention. Elsevier, Amsterdam, pp. 33–49.

Dastager, S.G. (2013). Isolation, improvement, and preservation of microbial cultures. *In*: C.R. Soccol, A. Pandey and C. Larroche (eds.), Fermentation Processes Engineering in the Food Industry. CRC Press, Boca Raton, pp. 21–45.

De Dea Lindner, J. (2008). Traditional and innovative approaches to evaluate microbial contribution in long ripened fermented foods: the case of Parmigiano Reggiano cheese. Ph.D. Thesis, Università degli Studi di Parma, Italy.

De Dea Lindner, J., A.L.B. Penna, I. Mottin Demiate, C.T. Yamaguishi, A.R. Machado Prado and J.L. Parada. (2013). Fermented foods and human health benefits of fermented foods. *In*: C.R. Soccol, A. Pandey and C. Larroche (eds.), Fermentation Processes Engineering in the Food Industry. CRC Press, Boca Raton, pp. 263–297.

De Dea Lindner, J., V. Bernini, A. De Lorentiis, A. Pecorari, E. Neviani and M. Gatti. (2008). Parmigiano Reggiano cheese: evolution of cultivable and total lactic microflora and peptidase activities during manufacture and ripening. Dairy Science and Technology 88: 511–523.

Dellaglio, F., S. Torriani and G.E. Felis. (2004). Reclassification of *Lactobacillus cellobiosus* Rogosa et al. 1953 as a later synonym of *Lactobacillus fermentum* Beijerinck 1901. International Journal of Systematic and Evolutionary Microbiology, 54: 809–812.

Demain, A.L. and J.E. Davis. (1998). Manual of industrial microbiology and biotechnology. American Society Microbiology Press, Washington, DC.

Denis, C., M. Gueguenb, B. Henry and D. Levert. (2001). New media for the numeration of cheese surface bacteria. Le Lait, 81: 365–379.

Dikes, G.A., A. von Holy. (1994). Strain typing in the genus *Lactobacillus*. Letters in Applied Microbiology 19: 63–66.

El Baz, A.F. and Y.M. Shetaia. (2005). Evaluation of different assays for the activity of yeast killer toxin. International Journal of Agriculture Biology 7: 1003–1006.

Evangelista, S.R., M.G. Miguel, C. Cordeiro, S. de, C.F. Silva, A.C. Pinheiro and R.F. Schwan. (2014). Inoculation of starter cultures in a semi-dry coffee (*Coffea arabica*) fermentation process. Food Microbiology 44: 87–95.

Fernández-Bodega, M.A., E. Mauriz, A. Gómez and J.F. Martín. (2009). Proteolytic activity, mycotoxins and andrastin A in *Penicillium roqueforti* strains isolated from Cabrales, Valdeón and Bejes–Tresviso local varieties of blue-veined cheeses. International Journal of Food Microbiology 136: 18–25.

Fonseca, F., C. Beal and G. Corrieu. (2000). Method of quantifying the loss of acidification activity of lactic acid starters during freezing and frozen storage. Journal of Dairy Research 67: 83–90.

Fundira, M., M. Blom, I.S. Pretorius and P. Van Rensburg. (2002). Selection of yeast starter culture strains for the production of marula fruit wines and distillates. Journal of Agriculture and Food Chemistry 50: 1535–1542.

Gatti, M., J. De Dea Lindner, F. Gardini, G. Mucchetti, D. Bevacqua, M.E. Fornasari and E. Neviani. (2008). A model to assess lactic acid bacteria aminopeptidase activities in Parmigiano Reggiano cheese during ripening. Journal of Dairy Science 91: 4129–4137.

Giraffa, G. (2012). Selection and design of lactic acid bacteria probiotic cultures. Engineering in Life Sciences 12: 391–398.

Herody, C., Y. Soyeux, E. Bech Hansen and K. Gillies. (2010). The legal status of microbial food cultures in the European Union: an overview. European Food and Feed Law Review 5: 258–269.

Holzapfel, W.H. (2002). Appropriate starter culture technologies for small-scale fermentation in developing countries. International Journal of Food Microbiology, 75: 197–212.

Ho, V.T.T., J. Zhao and G. Fleet. (2015). The effect of lactic acid bacteria on cocoa bean fermentation. International Journal of Food Microbiology 205: 54–67.

Hutkins, R.W. (2006). Microbiology and Technology of Fermented Foods. Blackwell Publishing, Ames, 488 p.

Juste, A., B.P.H.J. Thomma and B. Lievens. (2008). Recent advances in molecular techniques to study microbial communities in food-associated matrices and processes. Food Microbiology 25: 745–761.

Karlsson, R., L. Gonzales-Siles, F. Boulund, L. Svensson-Stadler, S. Skovbjerg, A. Karlsson, M. Davidson, S. Hulth, E. Kristiansson and E.R.B. Moore. (2015). Proteotyping: Proteomic characterisation, classification and identification of microorganisms – A prospectus. Systematic and Applied Microbiology 38: 246–257.

Kaur, P., A. Vohra, and T. Satyanarayana. (2013). Laboratory and industrial bioreactors for submerged fermentations. *In*: C.R. Soccol, A. Pandey and C. Larroche (eds.), Fermentation Processes Engineering in the Food Industry. CRC Press, Boca Raton, pp. 165–179.

Krieger-Weber, S. (2009). Application of yeast and bacteria as starter cultures. *In:* H. König, G. Unden and J. Fröhlich (eds.), Biology of Microorganisms on Grapes, in Must and in Wine. Springer, Berlin, pp. 489–522.

Lacroix, C. and S. Yildirim. (2007). Fermentation technologies for the production of probiotics with high viability and functionality. Current Opinion in Biotechnology 18: 1–8.

Leroy, F. and L. De Vuyst. (2004). Lactic acid bacteria as functional starter cultures for the food fermentation industry. Trends in Food Science & Technology 15: 67–78.

Lopes, C.A., M.E. Rodríguez. Sangorrín, M., Querol, A. and Caballero, A.C. (2006). Patagonian wines: the selection of an indigenous yeast starter. Journal of Industrial Microbiology and Biotechnology 34: 539–546.

Lowes, K.F., C.A. Shearman, J. Payne, D. Mackenzie, D.B. Archer, R.J. Merry and M.J. Gasson. (2000). Prevention of yeast spoilage in feed and food by the yeast mycocin HMK. Applied Environmental Microbiology 66: 1066–1076.

Mateo, J.J., M. Jiménez, A. Pastor and T. Huerta. (2001). Yeast starter cultures affecting wine fermentation and volatiles. Food Research International 34: 307–314.

Monnet, C., C. Béal and G. Corrieu. (2003). Improvement of the resistance of *Lactobacillus delbrueckii* ssp. *bulgaricus* to freezing by natural selection. Journal of Dairy Science 86: 3048–3053.

Neviani, E., J. De Dea Lindner, V. Bernini and M. Gatti. (2009). Recovery and differentiation of long ripened cheese microflora through a new cheese-based cultural medium. Food Microbiology 26: 240–245.

Parekh, S., V.A. Vinci and R.J. Strobel. (2000). Improvement of microbial strains and fermentation processes. Applied Microbiology and Biotechnology 54: 287–301.

Perrone, G., R.A. Samson, J.C. Frisvad, A. Susca, N. Gunde-Cimerman, F. Epifani and J. Houbraken. (2015). *Penicillium salamii*, a new species occurring during seasoning of dry-cured meat. International Journal of Food Microbiology 193: 91–98.

Pescuma, M., E.M. Héberta, F. Mozzia, and G. Font de Valdez. (2008). Whey fermentation by thermophilic lactic acid bacteria: Evolution of carbohydrates and protein content. Food Microbiology 25: 442–451.

Pratush, A, A. Gupta, A. Kumar. and G. Vyas. (2012). Application of purified bacteriocin produced by *Lactococcus lactis* AP2 as food biopreservative in acidic foods. Annals Food Science and Technology 13: 82–87.

Rainieri, S. and Pretorius, I.S. (2000). Selection and improvement of wine yeasts. Annals of Microbiology 50: 15–30.

Randazzo, C.L., C. Caggia and E. Neviani. (2009). Application of molecular approaches to study lactic acid bacteria in artisanal cheeses. Journal of Microbiological Methods 78: 1–9.

Rosselló-Móra, R. and R. Amann. (2015) Past and future species definitions for Bacteria and Archaea. Systematic and Applied Microbioliology 38: 209–216.

Savoie, S., C.P. Champagne, S. Chiasson and P. Audet. (2007). Media and process parameters affecting growth, strain ratios and specific acidifying activities of a mixed lactic starter containing aroma-producing and probiotic strains. Journal of Applied Microbiology 103: 163–174.

Schleifer, K., R. Amann and R. Rosselló-Móra. (2015). Taxonomy in the age of genomics. Systematic and Applied Microbiology 38: 207–208.

Sip, A. and W. Grajek. (2010). Probiotics and prebiotics. *In*: J. Smith and E. Charter (eds.), Functional Food Product Development. Blackwell Publishing, Oxford, pp. 146–177.

Soubeyrand, V. (2005) Formation of micella containing solubilized sterols during rehydration of active dry yeasts improves their fermenting capacity. Journal of Agriculture and Food Chemistry 53: 8025–8032.

Stadhouders, J and W.M. de Vos. (2014). Starter cultures of cheese production. *In*: J. Green (ed.), Biotechnological Innovations in Food Processing. Butterworth-Heinemann, Oxford, pp. 77–111

Steele, J.L., M.F. Budinich, H. Cai, S.C. Curtis and J.R. Broadbent. (2006). Diversity and metabolic activity of *Lactobacillus casei* in ripening Cheddar cheese. Australian Journal of Dairy Technology 61: 53–60.

Steensels, J. and K.J. Verstrepen. (2014). Taming wild yeast: potential of conventional and nonconventional yeasts in industrial fermentations. Annual Reviews Microbiology 68: 61–80.

Steensels, J., L. Daenen, P. Malcorps, G. Derdelinckx, H. Verachtert and K.J. Verstrepen. (2015). *Brettanomyces yeasts* — From spoilage organisms to valuable contributors to industrial fermentations. International Journal of Food Microbiology 206: 24–38.

Steensels, J., T. Snoek, E. Meersman, M.P. Nicolino, K. Voordeckers and K.J. Verstrepen. (2014). Improving industrial yeast strains: exploiting natural and artificial diversity. FEMS Microbiology Reviews 38: 947-995.

Vedamuthu, E.R. (2006). Starter cultures for yogurt and fermented milks. *In*: R.C. Chandan, C.H. White, A. Kilara and Y.H. Hui (eds.), Manufacturing Yogurt and Fermented Milks. Blackwell Publishing, 1st ed, Ames, pp. 89–116.

Zhang, H., L. Xie, W. Zhang, W. Zhou, J. Su and J. Liu. (2013). The association of biofilm formation with antibiotic resistance in lactic acid bacteria from fermented foods. Journal of Food Safety 33: 114–120.

15

Novel Biotechnological and Therapeutic Applications for Wild Type and Genetically Engineered Lactic Acid Bacteria

Rodrigo Dias de Oliveira Carvalho, Fillipe Luiz Rosa do Carmo,
Pamela Mancha-Agresti, Marcela Santiago Pacheco de Azevedo,
Cassiana Severiano de Sousa, Sara Heloisa da Silva Tessalia
Diniz Luerce Saraiva, Mariana Martins Drumond, Bruno
*Campos Silva and Vasco Ariston de Carvalho Azevedo**

Abstract

The human gastrointestinal (GI) tract is colonized by a complex and dense microbial community that can be divided into three major phyla—Bacteroidetes, Firmicutes and Actinobacteria, which, under normal conditions, live in a symbiotic relationship with the host. However, it has been shown that a dysfunctional interaction between the microbiota and the host can lead to several intestinal disorders, thus being considered a field of growing interest by the scientific community. In this context, some studies have been carried out to elucidate functions of true resident bacteria, while other research has attempted to assess transient bacteria. In addition, some studies have focused on the group of lactic acid bacteria (LAB) that are widely used as starter cultures in food fermentation of a large variety of fermented foods. It has being reported that allochthonous LAB bacteria may have positive effects on the host when administered in adequate amounts, thereby allowing them to be classified as probiotics microorganisms. Our research group recently investigated the mechanisms underlying the protective effects of dairy *Lactobacillus delbrueckii* Lb CNRZ327 *in vitro* and *in vivo* assays and have deposited the complete genome of *Lactococcus lactis* NCDO 2118, which will enable a greater understanding of its intrinsic characteristics. Beyond the classical employment of LAB, our group gathered research works using genetically engineered LABs, more specifically lactococci and lactobacilli as mucosal delivery vectors for therapeutic proteins and DNA

UFMG, Av. Antônio Carlos, 6627, Belo Horizonte, MG, Brasil. CP 486 CEP 31270-901.
* Corresponding Author: vascoariston@gmail.com

vaccines. In this context, several studies have been conducted to develop new strains and efficient expression systems to use LAB as "cell factories" for the production of anti-inflammatory proteins, where we provide the recombinant *Lactococcus lactis* strains efficiency in the prevention of the intestinal damage associated with inflammatory bowel disease in murine models. Moreover, the use of LAB as cell carriers for the production and presentation of antigens has contributed significantly to the development of new vaccines. A growing number of publications on biotechnological or therapeutic employment of LAB has emerged showing their effectiveness against disease but also its safety and immune efficiency, the fact that there are varied ongoing studies with tests at different stages of clinical phase, strengthens our belief that their use will soon benefit the population against most diseases whose treatment and cure is difficult or non-existent.

Introduction

The human gastrointestinal (GI) tract is colonized by up to 10^{14} bacteria, ten times higher than the number of cells in the human body (Artis 2008; Ley et al. 2006). Most of these microorganisms are bacteria and fungi appear to be rare (Gill et al. 2006). The composition and density of bacterial populations vary along the GI of healthy adults (Wang et al. 2005; Zoetendal et al. 2006). Low numbers (10^3) of bacteria, mainly belonging to the streptococci and lactobacilli group, are present in the upper GI tract, while in contrast, much higher numbers reside in the lower compartments, where bacterial populations reach 10^{11}–10^{12} (Whitman et al. 1998). Bacterial species from both upper and lower GI are classified into three phyla—Bacteroidetes, Firmicutes and Actinobacteria (Eckburg et al. 2005).

Under normal conditions, the intestinal microbiota lives in a symbiotic relationship with the host and this interaction has become a field of growing interest for the scientific community, which is beginning to understand the diversity and function of this microbiota that plays an important role in human health. Metagenomic sequencing studies are currently being developed with the intent to (i) identify the different species that lives in the GI tract and (ii) understand their specific function that are thought to be essential for the proper functioning of the gut ecosystem. This includes functions known to be important for the host, such as degradation of complex polysaccharides, metabolism of mineral, carbohydrates, and lipids, synthesis of short chain fatty acids (SCFA), amino acids and vitamins, activation of bioactive food components, maturation and modulation of the immune system, as well as protection against potentially pathogenic species (Arumugam et al. 2011; Qin et al. 2010). It has been shown that a dysfunctional interaction between the microbiota of the gut and the mucosal immune system of the host can lead to inflammatory intestinal diseases known as inflammatory bowel diseases (IBDs) in genetically disposed individuals (Sartor 2006).

Some microbial members can be classified as true residents, indigenous or autochthonous species, which have a long-term association with the intestinal habitat forming a stable community. On the other hand, there are microbial species that under normal conditions do not colonize the intestine but they do occur in the GI, at least temporarily, as they are present in the food intake and disappear a few days after (Berg 1996). These transiting or allochthonous bacterial species are usually present in fermented food products like yogurt and cheeses. Studies aimed to assess differences among true resident and transient bacteria are still incipient, and recent reports focus on the group of lactic acid bacteria (LAB) that are widely used as starter cultures in food fermentation of a large variety of fermented foods (Reuter 2001). Examples that illustrate this class of bacteria are *Lactobacillus plantarum*, *Lactobacillus casei*, *Lactobacillus paracasei*, *Lactobacillus buchneri*, *Lactobacillus brevis*, *Lactobacillus rhamnosus*, *Lactobacillus fermentum* and the thermophilic dairy lactobacilli *Lactobacillus delbrueckii* and *Lactobacillus helveticus* (Marteau and Shanahan 2003).

Probiotic Lactic Acid Bacteria

It has being reported that LAB allochthonous bacteria may have positive effect on their host. Actually, several health beneficial effects have been attributed to this group of bacteria, and the hypothesis of Metchnikoff that claimed that certain bacteria present in fermented food products might have positive effects on the consumers, improving their life expectancy proved to be correct (Metchnikoff 1907). Thus, the World Health Organization (WHO) in 2001 defined this group of microorganisms as probiotics–live microorganisms when administered in adequate amounts confer a health benefit on the host (FAO/WHO 2002). Most of the probiotics used and exploited today are lactobacilli, especially *Lactobacillus acidophilus*, *Lactobacillus gasseri* and *Lactobacillus johsonii*, most of which have been isolated from the human GI tract. Other representatives include *Lactobacillus casei*, *Lactobacillus plantarum*, *Lactobacillus reuteri*, *Lactobacillus fermentum*, *Lactobacillus salivarius*, *Lactobacillus paracasei and Lactobacillus rhamnosus* (Borchers et al. 2009; FAO/WHO 2002; Ventura et al. 2009; Walter 2008). A number of other bacteria like *Escherichia coli* strain Nissle, *Enterococcus faecium*, *Enterococcus faecalis* and some species from the *Bifidobacterium* genus are considered as probiotics as well (Borchers et al. 2009; Ventura et al. 2009).

Many research projects have shown that probiotics can induce changes in the gut microbial species composition and diversity, suggesting that an increase in bacterial diversity may have a therapeutic role to attenuate intestinal inflammation. Probiotics can also inhibit growth of pathogens by producing antimicrobial compounds reducing their population at mucosal surfaces through competitive exclusion (Ljungh and Wadström 2006). Moreover, they are able to increase mucosal barrier function turning gut mucosa resistant to pathogens that in other conditions would be capable to translocate the epithelium and cause disease. Another characteristics of probiotics are that they can modulate inflammatory signaling pathways (for example nuclear factor kappa B, NF-κB"; in macrophages, dendritic cells (DCs) and intestinal epithelial cells (IECs) decreasing the secretion of pro-inflammatory cytokines, such as IL-8, TNF-α and INF-y (Fitzpatrick et al. 2008; Haller et al. 2002). Several studies have indicated that probiotics can induce the proliferation of regulatory DCs and T lymphocytes (Treg) establishing an anti-inflammatory environment with the predominance of TGF-β and IL-10 cytokines (Foligne et al. 2007; Di Giacinto et al. 2005; Pronio et al. 2008). Another very interesting feature is that certain probiotics were shown to be able to regulate apoptosis in IECs (Yan and Polk 2002).

Bacterial factors implied in the probiotic effect from different bacterial strains of LAB still remain to be identified. These molecules might be factors secreted by the bacterium, as the induction of probiotic effects usually does not require direct cell contact. For instance, Lipoteichoic acid (LTA), a secreted factor from selected probiotics, has been shown to have a potent anti-inflammatory effect *in vitro* (Kim et al. 2008). In 2008, Mazmanian et al. (2008) demonstrated that the anti-inflammatory effect of *Bacteroides fragilis* was due to a single microbial molecule, polysaccharide A, PSA. Another example is the soluble peptides from the probiotic mixture VSL#3 that were shown to block NF-κB pathway decreasing the secretion of pro-inflammatory cytokines by the host (Petrof et al. 2004).

As strains of probiotics are capable of reverting an inflammatory to an anti-inflammatory environment, some are being tested as a therapeutic tool to fight against inflammatory intestinal diseases, such as IBD (Jurjus et al. 2004). With this purpose, several models of experimental colitis have been described in order to understand the pathogenesis and exploit probiotics as treatment for IBD (Jurjus et al. 2004; Prantera et al. 2002).

Lactobacillus delbrueckii

Our research group recently screened a collection of dairy *Lactobacillus delbrueckii* and tested its immune modulation effect *in vitro* through the quantification of (NF-κB) activation in a human intestinal epithelial cell line. All strains showed anti-inflammatory effects that varied from strong to light and this effect was due to bacterial surface exposed proteins. One strain (Lb CNZ327) that exhibited an extraordinarily anti-inflammatory function in the *in vitro* assays and was able to significantly reduce macroscopic and microscopic symptoms of dextran sulfate sodium (DSS) induced colitis in mice (Santos Rocha et al. 2012). In order to investigate the mechanisms underlying the protective effects of Lb CNRZ327 *in vivo*, mice were administrated with DSS and many immunological parameters were measured. It was observed that Lb CNRZ327 strain modulated the production of TGF-β, IL-6, and IL-12 in the colonic tissue and of TGF-β and IL-6 in the spleen causing the expansion of CD4+Foxp3+ regulatory T cells in the cecal lymph nodes, modulating not only mucosal but also systemic immune responses (Santos Rocha et al. 2014). Despite positive results from some pre-clinical or clinical trials using probiotics as treatment for intestinal inflammatory diseases, like IBD, our knowledge on the use of this group of bacteria is still preliminary. Due to the variability activity of different probiotic strains, well-designed studies and research projects are required to employ probiotics in medical practice.

Lactococcus lactis

Most beneficial effects of probiotics comprising the group of LAB have often been attributed to bacterial strains included into *Lactocbacillus* and *Bifidobacteria* genus. However, little is known about the effects of bacteria that are constantly present in our diet, such as *Lactococcus lactis*. This species is a facultative heterofermentative and mesophilic bacteria (optimum growth temperature around 30 °C) whose participation in the dairy industry is very relevant, especially for cheese production. There are two *Lactococcus lactis* subspecies reported to date: *lactis* and *cremoris*, both are found naturally in plants, especially grass. But they are also artificially found in the fermented foods as yoghurt, bread and in some types of wines, once they are used as starter cultures (Carr et al. 2002).

In order to understand the beneficial effect of these bacteria and their mechanism of action some strains of *Lactococcus lactis* were selected for assessment of their immunomodulatory potential *in vitro*. For this purpose three strains were chosen: (i) *Lactococcus lactis* subsp. *lactis* IL1403, the first LAB sequenced and extensively used for the production of various metabolic products such as vitamin B, diacetyl and alanine, as well for the production of recombinant proteins (Bolotin et al. 2001; Kleerebezem et al. 2002); (ii) the *Lactococcus lactis* subsp. *cremoris* MG1363 strain, is most widely used in genetic and physiological research throughout the world and is employed in several biotechnological applications, such as oral vaccines or delivery of bioactive peptides to mucosal GI (Hanniffy et al. 2007); and (iii) the *Lactococcus lactis* subsp. *lactis* NCDO 2118 (LLNCDO2118) strain, isolated from frozen pea, it has been routinely used in our laboratory for cloning and expressing proteins. Recently, this later strain was described as a gamma-aminobutyric acid (GABA) producer (Mazzoli et al. 2010). GABA, the most widely distributed neurotransmitter in the central nervous system of vertebrates, is the product of L-glutamate decarboxylation mediated by the enzyme glutamate decarboxylase (GAD, EC 4.1.1.15) and is known to have positive effects on human health. This neurotransmitter is able to lower blood pressure in mildly hypertensive patients (Inoue et al. 2003), induce tranquilizer and diuretic effect (Jakobs et al. 1993; Wong et al. 2003), prevent diabetes (Hagiwara et al. 2004) and reduce the levels of inflammatory response in rheumatoid arthritis murine model (Tian et al. 2011).

The initial evaluation of *Lactococcus lactis* properties was performed in *in vitro* inflammation model, using intestinal epithelial Caco-2 cell line which, in culture, exhibit enterocytes characteristics (Pinto 1983). When these cells are stimulated with the pro-inflammatory cytokine IL-1β, transcriptional factor NF-κB is activated and consequently, the production of inflammatory mediators, including IL-8, TNF-α, IL-6, Cox2, iNOS. Our results demonstrated that LLNCDO2118 does not induce pro-inflammatory events, and the culture supernatant decreased the secretion of IL-8 levels by 45%, a cytokine which is overproduced in mucosal cells of IBD patients. The ability to inhibit IL-8 secretion or its pathway suggests an immunomodulatory effect of LLNCDO2118, and shows its potential use for IBD treatment, since its inhibition can result in improvement of symptoms of these intestinal diseases (Neurath et al. 1996) .

Later, LLNCDO2118 was evaluated *in vivo* for their potential in the prevention of ulcerative colitis (UC) chemically induced by DSS in mice. As UC is a chronic inflammation characterized by remission and recurrence periods, a protocol that mimics this behavior was employed (Travis et al. 2011). Thus, the animals were subjected to an initial 7-day cycle DSS ingestion followed by 7 days of rest (without DSS ingestion), allowing for a regression of symptoms as well as in the period of remission in UC, in which the treatment (*ad libitum*) with LLNCDO2118 strain was carried out. A second cycle of DSS was used to simulate the disease recurrence. This new DSS cycle started on 14th day and finished on the 21st, which were the days chosen to evaluate the effectiveness of the strain. On the 14th, the group treated with LLNCDO2118 showed improvement in clinical signs of colitis, particularly diarrhea, suggesting that this strain has a local effect, contributing to epithelial cells protection. The analysis carried out on day 21 showed that animals treated with LLNCDO2118 had a reestablishment of the colon size, as well as microscopic scores improvement. This beneficial effect was not due to the production of Secretory Immunoglobulin A (sIgA) levels, an important factor to prevent bacterial translocation (Malin et al. 1996; O'Sullivan 2001), since they were unchanged. However, the cytokine profile of the LLNCDO2118 treated group was able to maintain intermediate levels of anti-inflammatory cytokine IL-10 in colon tissue, while animals that did not receive the strain showed reduced IL-10 levels. Furthermore, LLNCDO2118 administration was associated with an early increase in IL-6 production in the same tissue. IL-6 is a cytokine which can present both a pro-inflammatory and anti-inflammatory effect. In this case we suggest that IL-6 is related to increased mucosal repair by epithelial restitution (Chalaris et al. 2010; Dann et al. 2008; Grivennikov et al. 2009; Podolsky 1999; Scheller et al. 2011). As the DSS induced colitis is caused by the loss of immunological tolerance against the commensal microbiota antigens, being tolerance maintained primarily by Treg cells (and its relative ratio to activated T cells), the regulatory cells levels could then be associated with *Lactococcus lactis* anti-inflammatory mechanism. Thus, after colitis induction, the activated T cells (CD69$^+$) were quantified, which only animals fed with LLNCDO2118 showed higher levels in the spleen, suggesting that some *Lactococcus lactis* product could be able to activate T cells. The CD4$^+$CD25$^+$CD45RB^{low} and CD4$^+$CD25$^+$LAP$^+$ cells was also analyzed in the mesenteric lymph node and spleen of animals treated with LLNCDO2118, since these specialized T cell response counterbalance pro-inflammatory ones (Bouma and Strober 2003; Strober et al. 2007). Although the anti-inflammatory activity of LLNCDO2118 did not increase CD4$^+$CD25$^+$CD45RB^{low} Treg, there was, however, induction of T cells characterized by the surface expression of peptides associated with latency (LAP) both in the mesenteric lymph nodes and spleen of animals treated with this strain. Similar results were observed following treatment with VSL # 3 probiotic which was also administered during the remission period of colitis induced by TNBS (trinitrobenzenesulfonic acid), which has been shown to increase CD4$^+$LAP$^+$ cells, which is essential for the VSL # 3 probiotic effect (Di Giacinto et al. 2005).

As the differentiation of effector T cells (activated) and regulatory are modulated by DCs (Chen 2006), the profile of DCs expressing CD103$^+$ surface marker was investigated. These tolerogenic DCs are related to the differentiation of naïve CD4$^+$ T cells into Tregs (Coombes and Powrie 2008). In the inflamed group, which received only DSS, there was an increase in the population of CD11c$^+$CD103$^+$ cells compared to the control group (non-inflamed). LLNCDO2118 administration was able to further increase the amount of these cells, suggesting the expansion of Tregs, such as CD4$^+$LAP$^+$. So it was proposed that a second effect of LLNCDO2118 was observed on 21st day of experiment after the second DSS cycle, which no longer had the presence of the strain in the intestines of animals. In this second stage it was observed that the immunomodulatory capacity of LLNCDO2118 clearly depended on the expansion / recruitment of regulatory cells and their products, resulting in a milder form of UC. Similarly, Nishitani et al (2009). observed that the *Lactococcus lactis* subsp. *cremoris* FC strain when co-cultured with Caco-2 stimulated cells, was able to significantly reduce expression of IL-8 mRNA, and also inhibit nuclear translocation of NF-κB using RAW264.7 cells *in vitro* model. Now, LLNCDO2118 immunomodulation effect on T cells is considered the best characterized mechanisms of action. Recently our research group has deposited the complete genome of this probiotic strain, which will also enable a greater understanding of their intrinsic characteristics (Oliveira et al. 2014).

Heterologous Protein Production

Genetic engineering strategies in LAB have been employed to improve carbohydrate fermentation, metabolite production, enzymatic activities, or conferring them the capacity to produce beneficial compounds such as bacteriocins, and exopolysaccharides, vitamins, antioxidant enzymes and anti-inflammatory molecules (LeBlanc et al. 2013). In this context, several studies has been conducted to develop new strains and efficient expression systems to use LAB as "cell factories" for the production of proteins (de Moreno et al. 2011).

Lactococcus lactis is the best characterized LAB group member, being regarded as a model organism for the production of heterologous proteins for (i) being an easy to handle microorganism, and (ii) being safe for human use. It was the first LAB to have its genome fully sequenced (Bolotin et al. 1999) thus, it has currently a large number of genetic tools for cloning and expressing (Guimarães et al. 2009; de Vos 1999).

Gene expression regulation in Lactococcus lactis

Transcription

Bacteria gene transcription starts when the sigma subunit (σ) of RNA polymerase recognizes a specific region located on the DNA. This region, called the "promoter" is located on an upstream sequence of gene or operon characterized by the presence of 2 consensus motives, -35 (TTGACA) and -10 (TATAAT) base pairs from the transcription start site. After recognition of these hexanucleotides, the transcription process is carried out (Bolotin et al. 1999). In *Lactococcus lactis* a number of promoters have been described through comparative and functional analysis of already identified genes (Kuipers et al. 1993). They feature sequences -35 and -10 similar to those found in *Escherichia coli* and *Bacillus subtilis* and also a "TG" (thymine-guanine) motif located on the first upstream base pair of the -10 sequence. The primary sigma factor in *Lactococcus lactis* is encoded by rpoD gene (Araya-Kojima et al. 1995; Bolotin et al. 2001) and shows homology to the σ^{70} and σ^{A} factors genes of *Escherichia coli* and *Bacillus subtilis,* respectively. The transcription stops in the 3' portion of genes and operons, where a palindromic sequence of nucleotides rich in guanine, cytosine and thymine, called "transcriptional terminators", signals the end of the process. Most genes and operons in *Lactococcus lactis* have such sequences.

Translation

Once transcription has occurred, the translation process initiates. In *Lactococcus lactis,* the translation start signals are also similar to those described in *Escherichia coli* and *Bacillus subtilis.* The ribosome attachment site or "RBS" is located in the 5' portion of mRNA to be translated, and is complementary to the sequence 3' to the 16S rRNA (3'CUUUCCUCC 5') of *Lactococcus lactis* (Chiaruttini and Milet 1993). Although most of the initiation codons are AUG, other codons, such as GUG were also observed (van de Guchte et al. 1992).

Genetic tools for the production of heterologous proteins in *Lactococcus lactis*

Heterologous proteins expression systems in *Lactococcus lactis* were obtained by the progress of genetic knowledge, the development of molecular biology techniques and studies of regulatory elements of gene expression, such as constitutive or inductive promoters (Miyoshi et al. 2004). This combination has allowed a variety of proteins from different sources to be cloned and highly expressed in *Lactococcus lactis* through several plasmidial vectors (Bermúdez-Humarán et al. 2011; Langella and Le Loir 1999; Mercenier et al. 2000).

Heterologous proteins expression systems and cell targeting

An early gene expression system for use in *Lactococcus lactis* was based on the promoter Plac and the regulatory gene *lacR* from the bacterial lactose operon. This operon is activated when the Plac promoter is induced in the presence of lactose and the transcription repressor gene (*lacR*) is suppressed in the same condition, allowing the target gene to be expressed (van Rooijen et al. 1992).

Subsequently another system was developed, consisting of three vectors that matched the lac operon elements and 2 more elements from the bacteriophage T7 of *Escherichia coli*, allowing a higher level of induction of the protein of interest (Wells et al. 1993). In this system, the gene coding RNA polymerase from phage T7 (T7 RNA pol) was placed under the control of Plac promoter in a first vector while in a second vector, the target protein is under control of the T7 promoter. Thus, this system works in a way that when lactose was added to the culture medium, the Plac induces the expression of T7 RNA pol, which activates expression of the gene of interest controlled by the T7 promoter. However, in order to the cell to be capable of metabolizing soluble lactose in the medium, a third vector containing the lac operon was necessary. Although this system allowed fine control of gene expression and higher levels of production, it became infeasible because it required three antibiotic resistance markers making it unsuitable for food and pharmaceutical industry employment (Wells et al. 1993).

Gene expression regulation studies on the common phages from the *Lactococcus* sp. were the basis for developing more simple expression systems like the "operator-repressor system" based on rlt, a *Lactococcus lactis* phage (Nauta et al. 1996). In the same vector, the gene coding of the protein of interest is placed under the control of P_{ORF5} phage promoter, which is repressed by the phage protein Rro. When added to the medium, the mutagen mitomycin C causes the proteolytic breakdown of the repressor protein Rro, and the consequent release of the PORF5 promoter. Free of repression, the promoter induces the expression from of the gene under its control. This system was tested using the *lac Z* reporter gene rom *Escherichia coli* and subsequently using the *acmA* gene (autolysin) from *Lactococcus* sp. However, the use of mitomycin C as inductor prevents the use of this system for protein production in fermenters as well in food products.

In another system, the genetic elements from phage φ31 were used to develop an expression system that matched the P_{15A10} promoter and the replication origin, ori31 (O'Sullivan 2001). Here, as in other systems, the gene of interest cloned under the control of P_{15A10}, and ori31, are in the same vector. After the start of φ31 phage infection, ori31 becomes target of the phage replication machinery and the amount of vector copies within the cell is increased. Due to this increase and due to the strength of P_{15A10} promoter, the gene of interest expression level is also increased. After cell lysis caused by phage replication, the protein molecules in question are released into the environment. The major disadvantage of this system is the need to obtain cell infection induction; which leads to the destruction of the cell culture, thus impeding their industrial use, in fermenters.

In this context, many studies have been conducted in order to develop safer and more suitable vectors for food industry. One of the most powerful expression systems already developed for use in food industry are based on genes involved in biosynthesis and regulation of the antimicrobial nisin, a peptide naturally secreted by several strains of *Lactococcus lactis.* Because of its antimicrobial properties, it is widely used as a natural food preservative. The Nisin Controlled Gene Expression–NICE system was developed in *Lactococcus lactis* where the genes *nis*R and *nis*K were inserted into the chromosome in the MG1363 strain, and the P_{nisA} promoter in the expression vector followed by multiple cloning sites (MCS) for insertion of genes of interest (Kuipers et al. 1993; Mierau and Kleerebezem 2005). In this system, induction of expression of heterologous proteins can be achieved by adding nisin in the extracellular medium, in which the *nis*K gene functions as a membrane sensor that recognizes the extracellular presence of nisin, while the signal is transferred to NisR by a phosphorylation process, turning NisR capable of binding to P_{nisA} promoter and consequently activate the gene of interest transcription.

This system is also versatile, making the heterologous protein able to accomplish their desired biological activity by addressing them properly to its final cell destination: (i) cytoplasm (ii) membrane or (iii) the extracellular medium. In bacteria, the protein secretion is accomplished by the addressing of specific sequences which encode a hydrophobic negatively charged signal peptide (SP) located at their amino-terminal portion (N-terminal). This SP is recognized and cleaved by the secretion machinery allowing translocation of the protein across the cell membrane, and thereby released in the extracellular medium. Another signal sequence is Cell Wall Anchor (CWA) that encodes a peptide composed of 30 amino acids which is located in the carboxy-terminal portion (C-terminus) of the protein. The CWA has a conserved motif (LPXTG) which is recognized by anchoring machinery. Thereby, the protein containing this motif is covalently attached to peptidoglycan present in the cell membrane (Le Loir et al. 1994; Mierau and Kleerebezem 2005; Piard et al. 1997).

Xylose-Inducible Expression System

In 2004, Miyoshi *and collaborators.* developed a new gene expression system for *Lactococcus lactis.* The system, called Xylose-Inducible Expression System (XIES) based on the xylose permease gene promoter (PxylT), from *Lactococcus lactis* NCDO2118, Described by Jamet and Renult (2001). In the presence of some sugars, as glucose, fructose and/or mannose, PxylT was shown to be repressed; otherwise, PxylT is transcriptionally activated by xylose in *Lactococcus lactis* (Miyoshi et al. 2004). Thereby, this promoter could be successively turned on by adding xylose and off by washing the cells and growing them on glucose (Jamet and Renault 2001). Myoshi and collaborators (2004) developed a new lactococcal XIES that also incorporates the ability to target heterologous proteins to cytoplasm or extracellular medium. This system contains two plasmids that are derived from two broad-host-range expression vectors, pCYT:Nuc and pSEC:Nuc that would send the protein to the cytoplasm or to the extracellular medium, respectively (Bermúdez-Humarán et al. 2003). The system combines the PxylT (Jamet 2001), the ribosome-binding site (RBS) and the signal peptide (SP) of the lactococcal secreted protein, Usp45 (van Asseldonk et al. 1990) and the *Staphylococcus aureus* nuclease gene (*nuc*) as the reporter (Le Loir et al. 1994; Shortle 1983) and was successfully applied to high-level Nuc production and correct protein targeting and was tested in the *Lactococcus lactis* subsp. *lactis* strain NCDO2118. These systems have great advantages once they are considered less expensive and safer for laboratory use as compared to the other available expressions methods (Azevedo et al. 2012).

De Azevedo et al. (2012) constructed the recombinant *Lactococcus lactis* strains that were able to produce and properly send the *Mycobacterium leprae* 65-kDa HSP (Hsp65) to the cytoplasm or to the extracellular medium, using XIES. Heat shock proteins (HSPs) expression in host is induced by a wide variety of stresses (including high temperature, anoxia, and ethanol) (Lindquist and Craig 1988). Hsp65 are also known to play a major role in immune modulation, controlling autoimmune responses. Some authors showed that oral administration of a recombinant *Lactococcus lactis* strain that produces and releases LPS-free Hsp65 prevented the development of experimental autoimmune encephalomyelitis (EAE) in C57BL/6 mice, reduced the incidence of type I diabetes in non-obese diabetic mice and attenuated atherosclerosis in low-density lipoprotein receptor-deficient mice (Jing et al. 2011; Ma et al. 2014; Rezende et al. 2013). Some *Lactococcus lactis* produced recombinant Hsp65 that could be used for biotechnological and therapeutic applications. The rHsp65 protein was efficiently produced in both the cytoplasm and secreted forms to the extracellular medium, confirming the ability of the XIES to produce and correctly address recombinant proteins (Azevedo et al. 2012).

del Carmen et al. (2011) have used the XIES to express anti-inflammatory molecules as an alternative therapy against IBDs, like use of a fermented dairy product containing IL-10-producing *Lactococcus lactis* for the prevention and/or treatment of IBD using a rodent model of Crohn's disease (CD). IL-10 has a central role in down-regulating inflammatory cascades, which makes it a good candidate for use in the therapeutic intervention in inflammatory processes (Marinho et al. 2010). The XIES was effective in dairy matrix, as observed by significant increases of the cytokine in fermented milks with IL-10-producing strains (Cyt and Sec). Once, XIES is a LAB expression system that can be added into the food matrix that is tightly regulated by xylose, which is rarely found in conventional foods, it can act perfectly as an inductor. In this way IL-10 expression can be up- or down regulated, which is especially useful for expression studies of the cytokines in media or milk. The use of an inducible expression system is very interesting from a genetics point of view, because genes in this expression system are only expressed when required by adding the inducer (in this case xylose). A constitutive expression system would continuously produce the IL-10 (or other genes under its control) when it might not be required, especially in the cases of bacterial strains that could persist in the gastrointestinal tract (del Carmen et al. 2011). Induction with xylose increased the cytokine levels production by IL-10 *Lactococcus lactis* producers (>500 pg/ml for the Cyt strain and >1,000 pg/ml for the Sec strain) (Marinho et al. 2010). There is some controversial data in the literature about the more effective IL-10 producer *Lactococcus lactis* strain. It was demonstrated by Marinho et al. (2010) that *Lactococcus lactis*- producing IL-10 in the cytoplasm showed a higher immunomodulatory potential in a murine lung inflammation model, hypothesizing that the recombinant IL-10 produced in the cytoplasmic form stored IL-10 for a longer period of time and is slowly released in the tissue when the bacterial host lysis occurs. However, Del Carmen et al. (2011) found that the secreting IL-10 strain showed a higher anti-inflammatory effect compared to the cytoplasmatic producing IL-10 strain. This could be due to the fact that this cytokine and the *Lactococcus lactis* Sec strain are probably both protected by the food matrix (milk), resulting in a more efficient delivery of IL-10 in the gut. The IL-10 produced by these lactococci strains was able to induce an anti-inflammatory effect in our TNBS model and this effect was attributed to IL-10 producers while the wild-type (wt) strains did not exert any effect. Prevention of intestinal damages (macroscopic and microscopic) was also observed in mice that received the milks fermented by both of these strains (Cyt and Sec producers) thereby proving the anti-inflammatory effect of these products. The promising results obtained in these studies showed that the employment of fermented milks as a new form of administration of IL-10-producing *Lactococcus lactis* could represent an alternative treatment for IBDs (del Carmen et al. 2011; Marinho et al. 2010).

In another study, researchers evaluated the production and the delivery of 15-lipoxygenase-1 (15-LOX-1) by *Lactococcus lactis* NCDO2118 containing XIES in a TNBS- induced colitis model in mice (Saraiva et al. 2015). 15-LOX-1 enzyme is found in endothelial/epithelial cells that plays a key role in the oxidative metabolism of arachidonic acid responsible for lipoxins production, lipid mediators with potent anti-inflammatory actions (Lee et al. 2011; Serhan 2005). They concluded that 15-LOX-1 producing *Lactococcus lactis* was effective in the prevention of the intestinal damage associated to inflammatory bowel disease in a murine model, proving the effectiveness and efficiency of XIES (Saraiva et al. 2015).

Production of several staphylococcal proteins in *Lactococcus lactis* has been reported. However, these studies were not dedicated to the development of an antigen for oral vaccination, most of them were dedicated to expression-secretion systems development, such as staphylococcal nuclease to be used as a reporter protein, used latter as a reporter protein in XIES (Le Loir et al. 1994), for the characterization of staphylococcal virulence factors as ClfA and FnbA (Que et al. 2001), ClfB (Clarke et al. 2009), IsdA (Innocentin et al. 2009), or to increase adhesion properties of recombinant *Lactococcus lactis* strains (Harro et al. 2012). Asensi et al. (2013) were the first to evaluate the production of a staphylococcal antigen in a recombinant LAB strain to be used for oral vaccination where two recombinant *Lactococcus lactis* strains allowed the production of Staphylococcal enterotoxin type B (rSEB), a potent superantigenic exotoxin, either cytoplasmatic or secreted in the intestinal mucosa of mice, using XIES. Oral immunization with the recombinant strains induced a protective immune response against a lethal challenge with *Staphylococcus aureus* ATCC 14458, an SEB producer strain, in murine model.

DNA Vaccines

Vaccination is one of the main tools to combat and eradicate diverse pathogenic and/or infectious agents widespread around the world.

DNA vaccines are the third generation vaccine, which utilizes genetically engineered DNA to produce an immunologic response. The first study about this platform started in the early 1990s when Wolff et al. (1992) observed that injection of a "naked" plasmid DNA encoding foreign antigens in mice made their muscle cells capable of expressing these same antigens since this first publication, the use of DNA as a strategy for vaccination has progressed very quickly. The Norwegian Biotechnology Advisory Board defines this platform as "the intentional transfer of genetic material (DNA or RNA) to somatic cells for the purpose of influencing the immune system" (The Norwegian Biotechnology Advisory Board 2003).

This platform has the propriety to induce humoral and cellular immune response against different kind of microorganisms such as parasites, bacteria and disease-producing viruses (Ulmer et al. 1993; Wolff et al. 1992). Moreover, it was also utilized on several tumor models (Cheng et al. 2005). The components of DNA vaccine are: (i) the plasmid backbone which contains a bacterial origin of replication needed for the vector's maintenance and propagation inside the bacteria; the cytosine-phosphate-guanine unmethylated (CpG) motifs, called immunostimulatory sequences (ISS), these motifs could be a contribution to DNA immunogenicity. In the mammalian genome CpG have a low frequency and are mainly methylated, but bacterial DNA contains many unmethylated CpG motifs allowing this motif to be recognized by mammals as a pathogen associated molecular pattern (PAMP). In this way, CpG motifs are the ones in charge of increasing the magnitude of the immune response because they can interact with Toll-like receptors (TLR), such as TLR9, adding adjuvant activity (Tudor et al. 2005), a resistance marker, required to permit selective growth of the bacteria that carries the plasmid; (ii) the transcriptional unit, essential for eukaryotic expression, which harbors a promoter/enhancer region, introns with functional splicing donor and acceptor sites, as well as the open reading frame (ORF) encoding the antigenic protein of interest, and the polyadenylation sequence (poly A), signal required for efficient and correct transcription termination of the ORF and transfer of the stable mRNA from the nucleus to the cytoplasm (Azevedo et al. 1999; Kowalczyk and Ertl 1999).

After an intramuscular or intradermal injection, the naked plasmid DNA will transfect somatic cells, such as myocytes and keratinocytes, and/or resident Antigen Presenting Cells (APCs) like DCs and macrophages located in the lamina propria (Kutzler and Weiner 2008; Liu 2011). Due to the fact that antigens are expressed intracellularly, both humoral and cell-mediated immunity can be activated to generate a broad immune protection. After the transcription and translation of the transgene the host-synthesized antigens become the target of immune surveillance in the context of both major histocompatibility complexes (MHC) class I and class II molecules of APCs. The APC cells have the propriety to move to the draining lymph nodes where they present the antigenic peptide-MHC complexes to stimulate naïve T cells. In the other hand, B cells are activated, beginning the antibody production cascades. Although plasmid DNA vaccines vectors can induce antibody and CD4+ T cell helper responses, they are particularly suited to induce CD8+ T cell responses (Anderson and Schneider 2007). The CD8+ T cells, the cytotoxic T lymphocytes, which are important in controlling infections (Leifert and Whitton 2000) induced by DNA vaccine, can occur in two main pathways: (i) the direct DNA transfection of the APCs like as dendritic cells (DCs) and (ii) cross-presentation approach, when somatic cells such as myocytes are transfected with DNA and the expressed antigens are taken up by the APCs, or when the transfected apoptotic cells are phagocytosed by the APCs (Leifert and whitton 2000, Xu et al. 2014).

DNA vaccines have an extensive range of features that give them many advantages over other vaccination platforms like traditional vaccines developed against pathogens, including either killed or attenuated pathogenic agents. DNA vaccines are relatively cheap and easy to produce, which is an important feature when considering an emerging pandemic threat (Liu 2011). An essential concern about vaccine products is safety. DNA vaccine are considered safe because they lack the risk of reversion to a disease causing state or secondary infection. Also the risk of integration of the plasmid into the host cell causing insertional mutagenesis, which may lead to the inactivation of tumor suppressor genes or activation of oncogenes, is found to be significantly lower than the spontaneous mutation rate (Nichols et al. 1995; Wang et al. 2004). No adverse effects have been reported either tolerance to the antigen or autoimmunity (Liu and Ulmer 2005), animal studies showed that there is no increase in anti-nuclear or anti-DNA antibodies after DNA vaccination. There has been no evidence that autoimmunity is associated with DNA vaccines (Le et al. 2000; Tavel et al. 2007).

Vaccine manufacturing is a simple and low cost method as it requires only the use of cloning techniques in order to clone the protein of interest. They are stable at room temperature, easy to store and transport, presents thermal stability and have a long shelf life (Grunwald and Ulbert 2015; Pereira et al. 2014). A systemic inflammation, which might conduce the increase of cardiovascular risk, is a rising concern about vaccination in general (Gherardi and Authier 2012; Ramakrishnan et al. 2012), although DNA vaccines are still contemplated as a relatively new approach to vaccination, and its potential to induce systemic inflammation must not be overlooked (Xu et al. 2014). Many studies have shown that DNA vaccines are generally satisfactory with an acceptably good safety profile, and no systemic inflammation has been reported (Goepfert et al. 2011; Jaoko et al. 2008; Kalams et al. 2012; Ledgerwood et al. 2011) .

Therefore, DNA vaccines portray as a smart tool due to its property to induce all three points of adaptive immunity: antibodies, helper T cells (TH) and cytotoxic T lymphocytes (CTLs), as well as being capable of stimulating innate immune responses (Li et al. 2012) with safety. However, among the disadvantages, the poor immunogenicity of naked-DNA platform when is administrated in large animals (Kim et al. 2010) can be highlighted along with the necessity of using adjuvants besides the gene encoding of the protein. These platform are limited to protein immunogens and are not useful for non-protein based antigens such as bacterial polysaccharides (Kuby et al. 2007).

To circumvent this problem a delivery vehicle is needed to protect the DNA vaccine against endonucleases degradation. Thus, pathogenic bacteria attenuated strains appear as an interesting delivery method as they have innate tropism for specific tissues of host, which makes them attractive to use as a vehicle delivery to DNA vaccine. The use of bacteria as a delivery vector has numerous benefits: they can maintain the plasmid in a high copy number, they are easy to manufacture, they are less laborious and the cost is low as there is no need to amplify and purify the plasmid before handling (Becker et al. 2008; Schoen et al. 2004), and large-size plasmid can be housed inside the bacteria, permitting the insertion of multiple genes of interest (Hoebe et al. 2004; Seow and Wood 2009). Another important features is the possibility of these vectors being used for mucosal administration, without the use of a needle, thus having the ability to stimulate both mucosal and systemic immune responses (Srivastava and Liu 2003).

In the intestinal mucosa, the bacteria carrying a DNA vaccine are able to cross the intestinal barrier, mainly via specialized epithelial cells called Microfold cells (M cells). M cells overlying Peyer's patches (PPs) whose lymphoid follicles are isolated while draining gut mesenteric lymph nodes are considered more accessible to antigens and bacteria present in the luminal compartment. DCs located in the PPs, are another pathway that bacteria have to access the body. Immature DC are able to open the tight junctions between epithelial cells, extend their dendrites outside the epithelium and directly sample bacteria, thereby monitoring the contents of the intestinal lumen (Rescigno et al. 2001). The IECs lining mucosal surfaces, can be invaded by bacteria through bacterial proteins called invasins. This characteristic refers to the capacity of attenuated pathogenic vectors to deliver DNA vaccines as they are able to naturally produce invasins.

Once inside the cells, bacteria vector have the ability to escape from the phagolysosome vesicles by the secretion of a variety of phospholipases and pore-forming cytolysins and enter the cytoplasm of the host cells (Hoebe et al. 2004; Schoen et al. 2004). The microtubules net are used by the plasmid to reach the nucleus. In the nucleus, using the host cell's transcription machinery, the protein of interest carried by the plasmid can be encoded, translated, and secreted afterwards (Grillot-Courvalin et al. 1999; Schoen et al. 2004) by the cell or be presented on the surface of epithelial cell or DCs. The MHC class-II, from APCs, presents the exogenous proteins, turning naïve T cells activated into CD4+ T cells. Furthermore, the exogenous protein may also be processed into small peptides, which are then presented on the surface of MHC class-I molecules to CD8+ T cells, and stimulate them (Saha et al. 2011).

The pattern recognition receptors (Toll-like and Nod-like receptors) expressed by IECs, B-lymphocytes and DCs located in the sub epithelial lamina propria are the other components of immunity used by bacteria. The bacterial components known as microbe-associated molecular patterns (MAMPs) are recognized by pattern recognition receptors and trigger intracellular signaling pathways that lead to cytokine secretion and immune cell activation (Barbosa and Rescigno 2010; Steinhagen et al. 2011). The bacterial recognition by the immune system modulates the innate immune response, thereby supporting a vigorous and lasting adaptive response (Hoebe et al. 2004).

Enteropathogenic bacteria like *Salmonella* typhi, *Listeria monocytogenes, Shigella flexneri, Yersinia enterocolitica* and *Escherichia coli* are the species that are most widely used as bacterial delivery systems into mammalian cells (Schoen et al. 2004) because of their natural tropism for macrophages as well as DCs in the lymphoid tissue of the intestinal mucosal surface (Becker et al. 2008).

The method that use senteropathogenic species as a bacterial carrier is being considered an advantage because of their capacity to infect human colonic mucosa after oral administration. However, they need to be attenuated or inactivated as they present the risk to revert to the virulent phenotype, there by compromising its safety. Therefore, World and Health Organization (WHO) does not recommend their use in children and immunocompromised individuals. Thus, to counteract this severe problem, the use of non-pathogenic bacteria, such as LAB as vectors for genetic immunization has been investigated (Wells and Mercenier 2008).

Lactic acid bacteria vehicles for DNA vaccine delivery

Regarding *Lactococcus lactis* as a vehicle to deliver DNA vaccines, many interesting features can be highlighted: (i) it was proved in different laboratories all over the world that they can carry recombinant plasmids and express antigens and therapeutic molecules at different cellular localizations (Le Loir et al. 2001; Wells et al. 1993); (ii) it was successfully demonstrated that *Lactococcus lactis* can deliver DNA into eukaryotic cells and *in vivo* to mice IECs (Chatel et al. 2008a; Guimarães et al. 2005a; Innocentin et al. 2009); (iii) they can induce both systemic and mucosal immunity when administrated at mucosa surfaces (Chang et al. 2003; Robinson et al. 1997); (iv) they can resist the acid environment of the stomach, are able to survive into the gastrointestinal tract, ensuring recombinant protein or plasmid delivery (Pereira et al. 2014). Regarding its extraordinary safety profile (Salminen et al. 1998), because *Lactococcus lactis* is not very immunogenic, it can be orally administrated several times (Guimarães et al. 2006). All these characteristics makes it a good option for being used in immunization programs (Macauley-Patrick et al. 2005).

Research that used wild-type (wt) *Lactococcus lactis* as a vector for genetic immunizationhave demonstrated both *in vitro* (Guimarães et al. 2006) and *in vivo* (Chatel et al. 2008a) that the percentage of gene transferred observed was low, as well as a low and transitory Th1-type immune response after immunization trials (Chatel et al. 2008a). To solve this problem scientist developed recombinant *Lactococcus lactis* expressing different invasins to improve bacterial interaction with IECs (Azevedo et al. 2012; Guimarães et al. 2005; Innocentin et al. 2009).

Regarding non-invasive LABs, Guimarães et al. (2006) carried out *in vitro* studies using non-invasin strains of *Lactococcus lactis* as DNA delivering vehicles. The pLIG:BLG plasmid, containing an eukaryotic expression cassette with the cDNA of the bovine -lactoglobulin (BLG) under the control of the human cytomegalovirus eukaryotic promoter (Pcmv) was used to transform *Lactococcus lactis* MG1363. Caco-2 human cells were co-incubated with purified pLIG:BLG, MG1363 (pLIG:BLG), MG1363 and a mix of MG1363(pLIG) and pLIG:BLG. Only the cells that were co-incubated with MG1363 (pLIG:BLG) exhibited the presence of BLG cDNA and the subsequent expression of BLG. This result indicated that there was the delivery of the BLG cDNA to the mammalian epithelial cells. The authors suggested that after the co-culture, some bacteria are internalized and lysed by the host phagolysosome and, consequently, the BLG cDNA was released in the cytosol.

After these *in vitro* results Chatel et al. (2008) described for the first time *in vivo* the transfer of functional genetic material from non-invasive food-grade transiting from bacteria to host. The delivery of a eukaryotic expression plasmid coding of the BLG to the epithelial cells of the intestinal membrane of mice using *Lactococcus lactis* is possible. This demonstrates the capacity of using these bacteria in the delivery of DNA vaccines. In this study, mice were submitted to intragastrically gavage. The BLG cDNA was detected in the epithelial membrane of the small intestine in

40% of the mice. Moreover, the BLG was produced by 53% of them. In addition, the BLG production was responsible for inducing a protective immune response when the mice were sensitized with cow's milk proteins. In this case, the induction of a Th1 immune response counteracting a Th2 response was observed. The delivery of a functional plasmid by *Lactococcus lactis* to the mice intestinal wall provides us with the understanding of the host-bacterium interaction and the modulation of host immune response due to the delivered DNA.

Regarding the use of invasive LABs, Guimarães et al. (2005) constructed a *Lactococcus lactis* strain capable of invading epithelial cells by cloning and expressing the internalin A gene (*inlA*) of *Listeria monocytogenes* under the control of a native promoter. Western Blot and immunofluorescence experiments showed that the cell wall anchored form of InlA was efficiently exhibited by the recombinant lactococci, that favored the internalization of *Lactococcus lactis inlA+* in Caco-2 cells. Invasivity test showed that *Lactococcus lactisinlA+* was 100 times more invasive than for wt *Lactococcus lactis*. Moreover, *Lactococcus lactis inlA+* could deliver the eukaryotic expression plasmid coding the Green fluorescent protein (GFP) gene to Caco-2 cells, as it was possible to detect the GFP in 1% of the invaded cells. Finally, *in vivo* studies using *Lactococcus lactis inlA+* for oral inoculation of guinea pigs revealed that *Lactococcus lactis inlA+* was able to penetrate intestinal cells.

With these invasive lactococci, DNA delivery by this bacterium can be measured. In order to achieve this, a new vector has been developed resulting from the co-integration of two replicons: one from *Eschericha coli* and the other from *Lactococcus lactis,* named vaccination using lactic acid bacteria (pValac). The pValac is formed by the fusion of (i) cytomegalovirus promoter (CMV), that allows the expression of the antigen of interest in eukaryotic cells, (ii) polyadenylation sequences from the Bovine Growth Hormone (BGH), essential to stabilize the RNA transcript, (iii) origins of replication that allow its propagation in both *Escherichia coli* and *Lactococcus lactis* hosts, and (iv) a chloramphenicol resistance gene for selection of strains harboring the plasmid. The functionality of pValac was observed after transfecting plasmids harboring the *gfp* ORF into mammalian cells, PK15. The PK15 cells were able to express GFP.

Although the interesting results obtained by the utilization of *Lactococcus lactis inlA+*, *in vivo* experimental studies are limited to guinea pigs or mutated mice, InlA cannot bind the murine E-cadherin. Thus, Innocentin et al. (2009) performed comparative studies using both *Lactococcus lactis* expressing the Fibronectin-Binding Protein A of *Staphylococcus aureus* (LL-FnBPA+) as a InlA. In this study, it was verified that LL-FnBPA+ or the truncated form coding only C and D domains of FnBPA (LL-CD+) were internalized by the Caco-2 intestinal epithelial cells as efficiently as *Lactococcus lactis inlA+*. Also in this study, it was evidenced for the first time that lactococci can be internalized in high levels and they as heterogeneously distributed in the cell monolayer. Finally, studies were performed using *Lactococcus lactis* InlA, *Lactococcus lactis* FnBPA and *Lactococcus lactis* CD carrying GFP and all of them were able to trigger GFP expression in Caco-2 cells.

Pontes et al. (2012) in *in vitro* and *in vivo* studies used invasive *Lactococcus lactis* expressing FnBPA of *Staphylococcus aureus* (LL-FnBPA+) and demonstrated that the production of FnBPA increased the plasmid transfer to Caco-2 cells (Pontes et al. 2012). When the invasiveness of Caco-2 cells by LL-FnBPA+ carrying the pValacBLG plasmid (LL-FnBPA+BLG) or not (LL-FnBPA+) was compared with the LL-wide type (LL-wt) and LL-BLG, it was observed that LL-FnBPA+BLG and LL-FnBPA+ were 10 times more invasive than LL-wt and LL-BLG. After the Caco-2 cells were co-incubated with LL-FnBPA+BLG and LL-BLG. It was found that the cells incubated with LL-FnBPA+BLG produced 30 times more BLG than the cells co-incubated with the non-invasive strain. Moreover, using BLG and GFP under the control of a eukaryotic promoter, the potential of LL-FnBPA+ as a DNA vaccine delivery vehicle was characterized *in vivo*. After the oral administration of LL-FnBPA+BLG and LL-BLG to mice, it was detected the plasmid transfer to enterocytes had no difference between both strains. The same result was observed when LL-FnBPA+GFP were used. Regarding the expression of BLG by mice, the oral administration of LL-FnBPA+BLG led to an increase in the number of mice able to produce BLG, but there was no difference in the levels of the BLG produced. In other words, *Lactococcus lactis* increased the plasmid transfer but not the quantity of plasmid transferred.

As mentioned before, InlA cannot bind the murine E-cadherin. Moreover, FnBPA requires an adequate amount of fibronectin to be used by the integrins. Therefore, to bypass these problems and to better understand the steps of DNA transfer to mammalian cells, De Azevedo et al. (2012), engineered *Lactococcus lactis* to express a mutated form of InlA (mInlA+) which allowed the affinity to murine E-cadherin and, consequently, *in vivo* experiments using conventional mice. The results of the tests with Caco-2 cells demonstrated that LL-mInlA+ were 1000 times more invasive than LL. To analyze the role of this strain of *Lactococcus lactis* as a DNA delivery vector, a plasmid carrying the BLG cDNA (pValacBLG) was used and the transfer to intestinal epithelial cells (IECs) was measured. *In vitro* results showed that LL-mInlA+BLG were 10 times more invasive than LL-BLG. *In vivo*, after oral administration of LL-mInlA+BLG and LL-BLG, the number of mice producing BLG in isolated enterocytes was slightly higher in mice administered with LL-mInlA+BLG than with LL-BLG.

Our research group was very interested to know whether uptake of *Lactococcus lactis* DNA vaccines by DCs could also lead to antigen expression, as observed in IECs, as they are unique in their ability to induce antigen-specific T cell responses. We demonstrated that both non-invasive and invasive lactococci could transfect bone-marrow DCs (BMDCs), inducing the secretion of the pro-inflammatory cytokine IL-12. This plasmid transfer to BMDCs was also measured through a polarized monolayer of IECs, mimicking the situation found in the GI tract. Co-incubation of strains in this co-culture model showed that DCs incubated with LL-mInlA+ containing pValac:BLG could express significant levels of BLG, suggesting that DCs could sample bacteria containing the DNA vaccine across the epithelial barrier and express the antigen (de Azevedo et al. 2015).

With reference to the IBDs, del Carmem et al. (2014) used LL-FnBPA+ carrying pValac:il-10. The Interleukin-10 (il-10) is an important anti-inflammatory cytokine involved in the intestinal immune system (del Carmen et al. 2014). Transfection and invasiveness assays using cell cultures showed the functionality of the plasmid and the invasive strain. Fluorescence microscopy using mice confirmed the *in vitro* results. After that, a trinitrobenzene sulfonic acid (TNBS) model for induction of intestinal inflammation in mouse was performed. Mice that received LL-FnBPA+ carrying pValac:il-10 plasmid exhibited lower damage scores by macroscopic and microscopic analysis of the large intestine, lower microbial translocation to liver and the anti-inflammatory/pro-inflammatory cytokine ratios were increased more than the mice that received *Lactococcus lactis* FnBPA+ without the pValac:il-10 plasmid. These results suggest that this DNA delivery strategy was efficient in preventing inflammation in this colitis murine model.

Continuing in the IBDs research line, Zurita-Turk et al. (2014) used LL-FnBPA+pValac:Il-10, LL-FnBPA+, LL-pValac:Il-10 and LL-wt in a different colitis model, the dextran sodium sulphate (DSS) model for induction of intestinal inflammation (Zurita-Turk et al. 2014). The results showed that both LL-FnBPA+pValac:Il-10 and LL-pValac:Il-10 were able to diminish the intestinal inflammation. Therefore, both strains delivered the eukaryotic expression vector to host cells directly at the sites of inflammation and lead *in situ* IL-10 production and its anti-inflammatory properties.

Christophe et al. (2015), working with another LAB, *Lactobacillus plantarum*, and aiming to increase the DNA delivery by these bacteria, constructed a strain targeting DEC-205, a receptor located at the surface of dendritic cells (Christophe et al. 2015). The objective was to increase the bacterial uptake and, consequently, improve the delivery of the cDNA to immune cells. For that, anti-DEC-205 antibody (aDec) was displayed at the surface of *Lactobacillus plantarum* using a covalent anchoring of aDec to the cell membrane, a covalent anchoring to the cell wall and a non-covalent anchoring to the cell wall. The results show that aDec was successfully expressed in the three strains, but surface location of the antibody could only be demonstrated for the strains with a covalent than a non-covalent anchoring to the cell wall. To verify the plasmid transfer, a plasmid for GFP expression under the control of a eukaryotic promoter was used to transform the three strains. GFP expression in DC cells was increased when using the strains producing cell-wall anchored aDec. However, *in vivo* tests using the mouse model exhibited a higher expression of GFP when the strain with a covalent anchoring to the cell membrane was used. It seems to be that the more embedded location of aDec in this strain is beneficial when cells are exposed to the gastro-intestinal tract conditions.

For further reading on this topic, the following works can be of great value: Almeida et al. (2014) evaluated the invasiveness of recombinant strains of *Lactococcus lactis* expressing FnBPA under the control of its constitutive promoter or driven by the strong NICE system (Almeida et al. 2014; Pontes et al. 2014) compared immune responses elicited by DNA immunization using LL-FnBPA+BLG and LL-BLG and they verified that the immune response could be modified by production of invasins on the cell surface (Pontes et al. 2014). They showed that intranasal or oral DNA administration using invasive LL-FnBPA+BLG elicited a TH2 primary immune response whereas the LL-BLG elicited a classical TH1 immune response; Pontes et al. (2012) revised not only the expression of heterologous protein but also the delivery systems developed for *Lactococcus lactis*, and its use as an oral vaccine carrier (Pontes et al. 2012); Bermúdez-Humarán et al. (2011) gathered research works using LABs, more specifically lactococci and lactobacilli, as mucosal delivery vectors for therapeutic proteins and DNA vaccines (Bermúdez-Humarán et al. 2011).

Recently, our team has developed another vector called pExu, to be used in *Lactococcus lactis*. This vector will be also used in genetic immunization like the pValac.

DNA vaccines in clinical trial phase and already licensed for use

Since the early 1990s, when studies of DNA vaccines were started, and till date, more than 18,000 scientific papers have been published on this subject. Among these, almost 500 were published in the first half of 2015's (PUBMED 2015). However, while research has advanced, currently, only four DNA vaccines are licensed and commercially available in the world, all of them for veterinary use.

The first two prophylactic vaccines based on recombinant DNA technology have been approved for use and licensed in 2005. The first one against horses West Nile virus (West Nile- Innovation®) (Davidson et al. 2005; Davis et al. 2001) and the second one against salmonids infectious hematopoietic necrosis virus (IHNV) (Apex-IHN®) (Anderson et al. 1996; Garver et al. 2005).

In 2008, a gene therapy based on the same technology has been licensed for pigs' treatment in Australia. Administration of a single dose of LifeTide® (plasmid containing the GHRH gene–growth hormone releasing hormone), in reproductive age females was able to reduce perinatal morbidity and mortality, thereby increasing productivity (Khan et al. 2010a,b).

The latter permit a DNA vaccine occurred in 2010. Once pt™ was developed to be used as immunotherapy for melanoma in dogs and its effectiveness is related to antibodies production that prevent the development and aggravation of the disease (Bergman et al. 2003; Liao et al. 2006).

While the number of studies with DNA-based vaccines are high, and currently there are already licensed treatment for veterinary use and any type of vaccine or treatment based on this technology are available for use in humans.

According to clinical trials (www.clinicaltrials.gov), currently there are 33 studies involving DNA vaccines in clinical trial phase worldwide. The vast majority of studies, more than 57%, are related to cancers. The testing HIV vaccines are also significant, accounting for over 27% of all ongoing studies. The other 21% are related to other diseases such as Ebola virus, HPV, hepatitis, etc.

As the licensing process for the commercialization of DNA vaccines in humans is long and meticulous, taking into account not only effectiveness against disease but also their safety and immune efficiency, the fact that there are already licensed DNA vaccines for veterinary use and various ongoing studies with tests at different stages of clinical phase, makes us believe that vaccinology based on recombinant DNA technology is a tool that will soon benefit the population against most diseases whose treatment and cure is difficult or non-existent.

Acknowledgments

The authors would like to thank CNPq and CAPES for their financial support and fellowships.

References

Almeida, J.F., D. Mariat, V. Azevedo, A. Miyoshi, A. de Moreno de LeBlanc, S. Del Carmen, R. Martin, P. Langella, J.G. LeBlanc and J.M. Chatel. (2014). Correlation between fibronectin binding protein A expression level at the surface of recombinant *Lactococcus lactis* and plasmid transfer in vitro and in vivo. BMC Microbiol. 14: 248.

Anderson, R.J., and J. Schneider. (2007). Plasmid DNA and viral vector-based vaccines for the treatment of cancer. Vaccine 25, Supplement 2: B24–B34.

Anderson, E.D., D.V. Mourich, S.C. Fahrenkrug, S. LaPatra, J. Shepherd and J.A. Leong. (1996). Genetic immunization of rainbow trout (Oncorhynchus mykiss) against infectious hematopoietic necrosis virus. Mol. Mar. Biol. Biotechnol. 5: 114–122.

Araya-Kojima, T., N. Ishibashi, S. Shimamura, K. Tanaka and H. Takahashi. (1995). Identification and molecular analysis of Lactococcus lactis rpoD operon. Biosci. Biotechnol. Biochem. 59: 73–77.

Artis, D. (2008). Epithelial-cell recognition of commensal bacteria and maintenance of immune homeostasis in the gut. Nat. Rev. Immunol. 8: 411–420.

Arumugam, M., J. Raes, E. Pelletier, D. Le Paslier, T. Yamada, D.R. Mende, G.R. Fernandes, J. Tap, T. Bruls, J.M. Batto, et al. (2011). Enterotypes of the human gut microbiome. Nature 473: 174–180.

Asensi et al (2013). Oral immunization with Lactococcus lactis secreting attenuated recombinant staphylococcal enterotoxin B induces a protective immune response in a murine model. Microbial Cell Factories 12:32.

Azevedo, M. de, J. Karczewski, F. Lefévre, V. Azevedo, A. Miyoshi, J.M. Wells, P. Langella and J.M. Chatel. (2012). In vitro and in vivo characterization of DNA delivery using recombinant Lactococcus lactis expressing a mutated form of L. monocytogenes Internalin A. BMC Microbiol. 12: 299.

Azevedo, V., G. Levitus, A. Miyoshi, A.L. Cândido, A.M. Goes and S.C. Oliveira. (1999). Main features of DNA-based immunization vectors. Braz. J. Med. Biol. Res. 32: 147–153.

Barbosa, T., and M. Rescigno. (2010). Host-bacteria interactions in the intestine: homeostasis to chronic inflammation. Wiley Interdiscip. Rev. Syst. Biol. Med. 2: 80–97.

Becker, P.D., M. Noerder and C.A. Guzmán. (2008). Genetic immunization: bacteria as DNA vaccine delivery vehicles. Hum. Vaccin. 4: 189–202.

Berg, R.D. (1996). The indigenous gastrointestinal microflora. Trends Microbiol. 4: 430–435.

Bergman, P.J., J. McKnight, A. Novosad, S. Charney, J. Farrelly, D. Craft, M. Wulderk, Y. Jeffers, M. Sadelain, A.E. Hohenhaus, et al. (2003). Long-term survival of dogs with advanced malignant melanoma after DNA vaccination with xenogeneic human tyrosinase: a phase I trial. Clin. Cancer Res. 9: 1284–1290.

Bermúdez-Humarán, L.G., P. Langella, J. Commissaire, S. Gilbert, Y. Le Loir, R. L'Haridon and G. Corthier. (2003). Controlled intra- or extracellular production of staphylococcal nuclease and ovine omega interferon in Lactococcus lactis. FEMS Microbiol. Lett. 224: 307–313.

Bermúdez-Humarán, L.G., P. Kharrat, J.M. Chatel and P. Langella. (2011). Lactococci and lactobacilli as mucosal delivery vectors for therapeutic proteins and DNA vaccines. Microb. Cell Factories 10 Suppl 1: S4.

Bolotin, A., S. Mauger, K. Malarme, S.D. Ehrlich and A. Sorokin. (1999). Low-redundancy sequencing of the entire Lactococcus lactis IL1403 genome. Antonie Van Leeuwenhoek 76: 27–76.

Bolotin, A., P. Wincker, S. Mauger, O. Jaillon, K. Malarme, J. Weissenbach, S.D. Ehrlich and A. Sorokin. (2001). The complete genome sequence of the lactic acid bacterium Lactococcus lactis ssp. lactis IL1403. Genome Res. 11: 731–753.

Borchers, A.T., C. Selmi, F.J. Meyers, C.L. Keen and M.E. Gershwin. (2009). Probiotics and immunity. J. Gastroenterol. 44: 26–46.

Bouma, G., and W. Strober. (2003). The immunological and genetic basis of inflammatory bowel disease. Nat. Rev. Immunol. 3: 521–533.

Carr, F.J., D. Chill and N. Maida. (2002). The lactic acid bacteria: a literature survey. Crit. Rev. Microbiol. 28: 281–370.

Chalaris, A., N. Adam, C. Sina, P. Rosenstiel, J. Lehmann-Koch, P. Schirmacher, D. Hartmann, J. Cichy, O. Gavrilova, S. Schreiber, et al. (2010). Critical role of the disintegrin metalloprotease ADAM17 for intestinal inflammation and regeneration in mice. J. Exp. Med. 207: 1617–1624.

Chang, T.L.Y., C.H. Chang, D.A. Simpson, Q. Xu, P.K. Martin, L.A. Lagenaur, G.K. Schoolnik, D.D. Ho, S.L. Hillier, M. Holodniy, et al. (2003). Inhibition of HIV infectivity by a natural human isolate of *Lactobacillus jensenii* engineered to express functional two-domain CD4. Proc. Natl. Acad. Sci. U. S. A. 100: 11672–11677.

Chatel, J.M., L. Pothelune, S. Ah-Leung, G. Corthier, J.M. Wal and P. Langella. (2008). In vivo transfer of plasmid from food-grade transiting lactococci to murine epithelial cells. Gene Ther. 15: 1184–1190.

Chen, W. (2006). Dendritic cells and (CD4+)CD25+ T regulatory cells: crosstalk between two professionals in immunity versus tolerance. Front. Biosci. J. Virtual Libr. 11: 1360–1370.

Cheng, W.F., C.F. Hung, C.A. Chen, C.N. Lee, Y.N. Su, C.Y. Chai, D.A.K. Boyd, C.Y. Hsieh and T.C. Wu. (2005). Characterization of DNA vaccines encoding the domains of calreticulin for their ability to elicit tumor-specific immunity and antiangiogenesis. Vaccine 23: 3864–3874.

Chiaruttini, C., and M. Milet. (1993). Gene organization, primary structure and RNA processing analysis of a ribosomal RNA operon in *Lactococcus lactis*. J. Mol. Biol. 230: 57–76.

Christophe, M., K. Kuczkowska, P. Langella, V.G.H. Eijsink, G. Mathiesen and J.M. Chatel. (2015). Surface display of an anti-DEC-205 single chain Fv fragment in *Lactobacillus plantarum* increases internalization and plasmid transfer to dendritic cells in vitro and in vivo. Microb. Cell Factories 14: 95.

Clarke, S.R., G. Andre, E.J. Walsh, Y.F. Dufrêne, T.J. Foster and S.J. Foster. (2009). Iron-regulated surface determinant protein A mediates adhesion of *Staphylococcus aureus* to human corneocyte envelope proteins. Infect. Immun. 77: 2408–2416.

Coombes, J.L., and F. Powrie. (2008). Dendritic cells in intestinal immune regulation. Nat. Rev. Immunol. 8: 435–446.

Dann, S.M., M.E. Spehlmann, D.C. Hammond, M. Iimura, K. Hase, L.J. Choi, E. Hanson and L. Eckmann. (2008). IL-6-dependent mucosal protection prevents establishment of a microbial niche for attaching/effacing lesion-forming enteric bacterial pathogens. J. Immunol. Baltim. Md 1950 180: 6816–6826.

Davidson, A.H., J.L. Traub-Dargatz, R.M. Rodeheaver, E.N. Ostlund, D.D. Pedersen, R.G. Moorhead, J.B. Stricklin, R.D. Dewell, S.D. Roach, R.E. Long, et al. (2005). Immunologic responses to West Nile virus in vaccinated and clinically affected horses. J. Am. Vet. Med. Assoc. 226: 240–245.

Davis, B.S., G.J. Chang, B. Cropp, J.T. Roehrig, D.A. Martin, C.J. Mitchell, R. Bowen and M.L. Bunning. (2001). West Nile virus recombinant DNA vaccine protects mouse and horse from virus challenge and expresses in vitro a noninfectious recombinant antigen that can be used in enzyme-linked immunosorbent assays. J. Virol. 75: 4040–4047.

de Azevedo, M.S.P., Rocha, C.S., Electo, N., Pontes, D.S., Molfetta, J.B., Gonçalves, E.D.C., Azevedo, V., Silva, C.L., and Miyoshi, A. (2012). Cytoplasmic and extracellular expression of pharmaceutical-grade mycobacterial 65-kDa heat shock protein in *Lactococcus lactis*. Genet. Mol. Res. 11, 1146–1157.

de Azevedo, M., M. Meijerink, N. Taverne, V.B. Pereira, J.G. LeBlanc, V. Azevedo, A. Miyoshi, P. Langella, J.M. Wells and J.M. Chatel. (2015). Recombinant invasive Lactococcus lactis can transfer DNA vaccines either directly to dendritic cells or across an epithelial cell monolayer. Vaccine 33: 4807–4812.

de Vos, W.M. (1999). Gene expression systems for lactic acid bacteria. Curr. Opin. Microbiol. 2: 289–295.

del Carmen, S., A. de Moreno de LeBlanc, G. Perdigon, V. Bastos Pereira, A. Miyoshi, V. Azevedo and J.G. LeBlanc. (2011). Evaluation of the anti-inflammatory effect of milk fermented by a strain of IL-10-producing *Lactococcus lactis* using a murine model of Crohn's disease. J. Mol. Microbiol. Biotechnol. 21: 138–146.

del Carmen, S., R. Martín Rosique, T. Saraiva, M. Zurita-Turk, A. Miyoshi, V. Azevedo. A. de Moreno de LeBlanc, P. Langella, L.G. Bermúdez-Humarán and J.G. LeBlanc. (2014). Protective effects of lactococci strains delivering either IL-10 protein or cDNA in a TNBS-induced chronic colitis model. J. Clin. Gastroenterol. 48 Suppl 1: S12–S17.

Eckburg, P.B., E.M. Bik, C.N. Bernstein, E. Purdom, L. Dethlefsen, M. Sargent, S.R. Gill, K.E. Nelson and D.A. Relman. (2005). Diversity of the human intestinal microbial flora. Science 308: 1635–1638.

Fitzpatrick, L.R., J. Small, R.A. Hoerr, E.F. Bostwick, L. Maines and W.A. Koltun. (2008). *In vitro* and *in vivo* effects of the probiotic *Escherichia coli* strain M-17: immuno-modulation and attenuation of murine colitis. Br. J. Nutr. 100: 530–541.

Foligne, B., G. Zoumpopoulou, J. Dewulf, A. Ben Younes, F. Chareyre, J.C. Sirard, B. Pot and C. Grangette. (2007). A key role of dendritic cells in probiotic functionality. PloS One 2: e313.

Garver, K.A., S.E. LaPatra and G. Kurath. (2005). Efficacy of an infectious hematopoietic necrosis (IHN) virus DNA vaccine in Chinook Oncorhynchus tshawytscha and sock-eye O. nerka salmon. Dis. Aquat. Organ. 64: 13–22.

Gherardi, R.K. and F.J. Authier. (2012). Macrophagic myofasciitis: characterization and pathophysiology. Lupus 21: 184–189.

Di Giacinto, C., M. Marinaro, M. Sanchez, W. Strober and M. Boirivant. (2005). Probiotics ameliorate recurrent Th1-mediated murine colitis by inducing IL-10 and IL-10-dependent TGF-beta-bearing regulatory cells. J. Immunol. Baltim. Md 1950 174: 3237–3246.

Gill, S.R., M. Pop, R.T. DeBoy, P.B. Eckburg, P.J. Turnbaugh, B.S. Samuel, J.I. Gordon, D.A. Relman, C.M. Fraser-Liggett and K.E. Nelson. (2006). Metagenomic analysis of the human distal gut microbiome. Science 312: 1355–1359.

Goepfert, P.A., M.L. Elizaga, A. Sato, L. Qin, M. Cardinali, C.M. Hay, J. Hural, S.C. DeRosa, O.D. DeFawe, G.D. Tomaras, et al. (2011). Phase 1 safety and immunogenicity testing of DNA and recombinant modified vaccinia Ankara vaccines expressing HIV-1 virus-like particles. J. Infect. Dis. 203: 610–619.

Grillot-Courvalin, C., S. Goussard and P. Courvalin. (1999). Bacteria as gene delivery vectors for mammalian cells. Curr. Opin. Biotechnol. 10: 477–481.

Grivennikov, S., E. Karin, J. Terzic, D. Mucida, G.Y. Yu, S. Vallabhapurapu, J. Scheller, S. Rose-John, H. Cheroutre, L. Eckmann, et al. (2009). IL-6 and Stat3 are required for survival of intestinal epithelial cells and development of colitis-associated cancer. Cancer Cell 15: 103–113.

Grunwald, T., and S. Ulbert. (2015). Improvement of DNA vaccination by adjuvants and sophisticated delivery devices: vaccine-platforms for the battle against infectious diseases. Clin. Exp. Vaccine Res. 4: 1–10.

van de Guchte, M., F.J. van der Wal, J. Kok and G. Venema. (1992). Lysozyme expression in *Lactococcus lactis*. Appl. Microbiol. Biotechnol. 37: 216–224.

Guimarães, V., S. Innocentin, J.-M. Chatel, F. Lefèvre, P. Langella, V. Azevedo and A. Miyoshi. (2009). A new plasmid vector for DNA delivery using lactococci. Genet. Vaccines Ther. 7: 4.

Guimarães, V.D., J.E. Gabriel, F. Lefèvre, D. Cabanes, A. Gruss, P. Cossart, V. Azevedo and P. Langella. (2005). Internalin-expressing *Lactococcus lactis* is able to invade small intestine of guinea pigs and deliver DNA into mammalian epithelial cells. Microbes Infect. 7: 836–844.

Guimarães, V.D., S. Innocentin, F. Lefèvre, V. Azevedo, J.M. Wal, P. Langella and J.M.Chatel. (2006). Use of native lactococci as vehicles for delivery of DNA into mammalian epithelial cells. Appl. Environ. Microbiol. 72: 7091–7097.

Hagiwara, H., T. Seki and T. Ariga. (2004). The effect of pre-germinated brown rice intake on blood glucose and PAI-1 levels in streptozotocin-induced diabetic rats. Biosci. Biotechnol. Biochem. 68: 444–447.

Haller, D., M.P. Russo, R.B. Sartor and C. Jobin. (2002). IKKβ and phosphatidylinositol 3-Kinase/Akt participate in on-pathogenic Gram-negative enteric bacteria-induced RelA phosphorylation and NF-κB activation in both primary and intestinal epithelial cell lines. J. Biol. Chem. 277: 38168–38178.

Hanniffy, S.B., A.T. Carter, E. Hitchin and J.M. Wells. (2007). Mucosal delivery of a pneumococcal vaccine using *Lactococcus lactis* affords protection against respiratory infection. J. Infect. Dis. 195: 185–193.

Harro, C.D., R.F. Betts, J.S. Hartzel, M.T. Onorato, J. Lipka, S.S. Smugar and N.A. Kartsonis. (2012). The immunogenicity and safety of different formulations of a novel *Staphylococcus aureus* vaccine (V710): results of two Phase I studies. Vaccine 30: 1729–1736.

Hoebe, K., E. Janssen and B. Beutler. (2004). The interface between innate and adaptive immunity. Nat. Immunol. 5: 971–974.

Information, N.C. for B., U.S.N.L. Pike, of M. 8600 R., B. MD and Usa, 20894 Home - PubMed - NCBI.

Innocentin, S., V. Guimarães, A. Miyoshi, V. Azevedo, P. Langella, J.M. Chatel and F. Lefèvre. (2009). *Lactococcus lactis* expressing either *Staphylococcus aureus* fibronectin-binding protein A or *Listeria monocytogenes* internalin A can efficiently internalize and deliver DNA in human epithelial cells. Appl. Environ. Microbiol. 75: 4870–4878.

Inoue, K., T. Shirai, H. Ochiai, M. Kasao, K. Hayakawa, M. Kimura and H. Sansawa. (2003). Blood-pressure-lowering effect of a novel fermented milk containing gamma-aminobutyric acid (GABA) in mild hypertensives. Eur. J. Clin. Nutr. 57: 490–495.

Jakobs, C., J. Jaeken and K.M. Gibson. (1993). Inherited disorders of GABA metabolism. J. Inherit. Metab. Dis. 16: 704–715.

Jamet, E. (2001) Etude de l'expression et de la régulation des gènes impliqués dans le métabolisme carboné chez *Lactococcus lactis*. Ph.D. Thesis. Institut National Agronomique Paris-Grignon, Paris.

Jaoko, W., F.N. Nakwagala, O. Anzala, G.O. Manyonyi, J. Birungi, A. Nanvubya, F. Bashir, K. Bhatt, H. Ogutu, S. Wakasiaka, et al. (2008). Safety and immunogenicity of recombinant low-dosage HIV-1 A vaccine candidates vectored by plasmid pTHr DNA or modified vaccinia virus Ankara (MVA) in humans in East Africa. Vaccine 26: 2788–2795.

Jing, H., Yong, L., L. Haiyan, M. Yanjun, X. Yun, Z. Yu, L. Taiming, C. Rongyue, J. Liang. W. Jie, et al. (2011). Oral administration of *Lactococcus lactis* delivered heat shock protein 65 attenuates atherosclerosis in low-density lipoprotein receptor-deficient mice. Vaccine 29: 4102–4109.

Jurjus, A.R., N.N. Khoury and J.M. Reimund. (2004). Animal models of inflammatory bowel disease. J. Pharmacol. Toxicol. Methods 50: 81–92.

Kalams, S.A., S. Parker, X. Jin, M. Elizaga, B. Metch, M. Wang, J. Hural, M. Lubeck, J. Eldridge, M. Cardinali, et al. (2012). Safety and immunogenicity of an HIV-1 gag DNA vaccine with or without IL-12 and/or IL-15 plasmid cytokine adjuvant in healthy, HIV-1 uninfected adults. PloS One 7: e29231.

Khan, A.S., A.M. Bodles-Brakhop, M.L. Fiorotto and R. Draghia-Akli. (2010a). Effects of maternal plasmid GHRH treatment on offspring growth. Vaccine 28: 1905–1910.

Khan, A.S., R. Draghia-Akli, R.J. Shypailo, K.I. Ellis, H. Mersmann and M.L. Fiorotto. (2010b). A comparison of the growth responses following intramuscular GHRH plasmid administration versus daily growth hormone injections in young pigs. Mol. Ther. J. Am. Soc. Gene Ther. 18: 327–333.

Kim, D., C.F. Hung, T.C. Wu and Y.M. Park. (2010). DNA vaccine with α-galactosylceramide at prime phase enhances anti-tumor immunity after boosting with antigen-expressing dendritic cells. Vaccine 28: 7297–7305.

Kim, H.G., N.R. Kim, M.G. Gim, J.M. Lee, S.Y. Lee, M.Y. Ko, J.Y. Kim, S.H. Han and D.K. Chung. (2008). Lipoteichoic acid isolated from *Lactobacillus plantarum* inhibits lipopolysaccharide-induced TNF-alpha production in THP-1 cells and endotoxin shock in mice. J. Immunol. 180: 2553–2561.

Kleerebezem, M., I.C. Boels, M.N. Groot, I. Mierau, W. Sybesma and J. Hugenholtz. (2002). Metabolic engineering of *Lactococcus lactis*: the impact of genomics and metabolic modelling. J. Biotechnol. 98: 199–213.

Kowalczyk, D.W., and H.C.J. Ertl. (1999). Immune responses to DNA vaccines. Cell. Mol. Life Sci. 55: 751–770.

Kuby, J., T.J. Kindt, R.A. Goldsby and B.A. Osborne. (2007). Immunology Freeman, New York, New York.

Kuipers, O.P., M.M. Beerthuyzen, R.J. Siezen and W.M. De Vos. (1993). Characterization of the nisin gene cluster nisABTCIPR of *Lactococcus lactis*. Requirement of expression of the *nisA* and *nisI* genes for development of immunity. Eur. J. Biochem. FEBS 216: 281–291.

Kutzler, M.A. and D.B. Weiner. (2008). DNA vaccines: ready for prime time? Nat. Rev. Genet. 9: 776–788.

Langella, P. and Y. Le Loir. (1999). Heterologous protein secretion in *Lactococcus lactis*: a novel antigen delivery system. Braz. J. Med. Biol. Res. 32: 191–198.

Le, T.P., K.M. Coonan, R.C. Hedstrom, Y. Charoenvit, M. Sedegah, J.E. Epstein, S. Kumar, R. Wang, D.L. Doolan, J.D. Maguire, et al. (2000). Safety, tolerability and humoral immune responses after intramuscular administration of a malaria DNA vaccine to healthy adult volunteers. Vaccine 18: 1893–1901.

LeBlanc, J.G., C. Aubry, N.G. Cortes-Perez, A. de Moreno de LeBlanc, N. Vergnolle, P. Langella, V. Azevedo, J.M. Chatel, A. Miyoshi and L.G. Bermúdez-Humarán. (2013). Mucosal targeting of therapeutic molecules using genetically modified lactic acid bacteria: an update. FEMS Microbiol. Lett. 344: 1–9.

Ledgerwood, J.E., C.J. Wei, Z. Hu, I.J. Gordon, M.E. Enama, C.S. Hendel, P.M. McTamney, M.B. Pearce, H.M. Yassine, J.C. Boyington, et al. (2011). DNA priming and influenza vaccine immunogenicity: two phase 1 open label randomised clinical trials. Lancet Infect. Dis. 11: 916–924.

Lee, S. Il, X. Zuo and I. Shureiqi. (2011). 15-Lipoxygenase-1 as a tumor suppressor gene in colon cancer: is the verdict in? Cancer Metastasis Rev. 30: 481–491.

Leifert, J.A., and J.L. Whitton. (2000). Immune responses to DNA vaccines: Induction of CD8 T Cells. DNA Vaccines 82–104.

Ley, R.E., D.A. Peterson and J.I. Gordon. (2006). Ecological and evolutionary forces shaping microbial diversity in the human intestine. Cell 124: 837–848.

Li, L., Saade, F. and N. Petrovsky. (2012). The future of human DNA vaccines. J. Biotechnol. 162: 171–182.

Liao, J.C.F., P. Gregor, J.D. Wolchok, F. Orlandi, D. Craft, C. Leung, A.N. Houghton and P.J. Bergman. (2006). Vaccination with human tyrosinase DNA induces antibody responses in dogs with advanced melanoma. Cancer Immun. 6: 8.

Lindquist, S. and E.A. Craig. (1988). The heat-shock proteins. Annu. Rev. Genet. 22: 631–677.

Liu, M.A. (2011). DNA vaccines: an historical perspective and view to the future: The past and future of DNA vaccines. Immunol. Rev. 239: 62–84.

Liu, M.A. and J.B. Ulmer. (2005). Human clinical trials of plasmid DNA vaccines. Adv. Genet. 55: 25–40.

Ljungh, A. and T. Wadström. (2006). Lactic acid bacteria as probiotics. Curr. Issues Intest. Microbiol. 7: 73–89.

Le Loir, Y., Gruss, A., S.D. Ehrlich and P. Langella. (1994). Direct screening of recombinants in Gram-positive bacteria using the secreted staphylococcal nuclease as a reporter. J. Bacteriol. 176: 5135–5139.

Le Loir, Y., S. Nouaille, J., Commissaire, L. Brétigny, A. Gruss and P. Langella. (2001). Signal peptide and propeptide optimization for heterologous protein secretion in *Lactococcus lactis*. Appl. Environ. Microbiol. 67: 4119–4127.

Ma, Y., J. Liu, J. Hou, Y. Dong, Y. Lu, L. Jin, R. Cao, T. Li and J. Wu. (2014). Oral administration of recombinant *Lactococcus lactis* expressing HSP65 and tandemly repeated P277 reduces the incidence of type I diabetes in non-obese diabetic mice. PloS One 9: e105701.

Macauley-Patrick, S., M.L. Fazenda, B. McNeil and L.M. Harvey. (2005). Heterologous protein production using the *Pichia pastoris* expression system. Yeast Chichester Engl. 22: 249–270.

Malin, M., H. Suomalainen, M. Saxelin and E. Isolauri. (1996). Promotion of IgA immune response in patients with Crohn's disease by oral bacteriotherapy with *Lactobacillus* GG. Ann. Nutr. Metab. 40: 137–145.

Marinho, F. A., L.G. Pacífico, A. Miyoshi, V. Azevedo, Y. Le Loir, V.D. Guimarães, P. Langella, G.D. Cassali, C.T. Fonseca and S.C. Oliveira. (2010). An intranasal administration of *Lactococcus lactis* strains expressing recombinant interleukin-10 modulates acute allergic airway inflammation in a murine model. Clin. Exp. Allergy 40: 1541–1551.

Marteau, P., and F. Shanahan. (2003). Basic aspects and pharmacology of probiotics: an overview of pharmacokinetics, mechanisms of action and side-effects. Best Pract. Res. Clin. Gastroenterol. 17: 725–740.

Mazmanian, S.K., J.L. Round and D.L. Kasper. (2008). A microbial symbiosis factor prevents intestinal inflammatory disease. Nature 453: 620–625.

Mazzoli, R., E. Pessione, M. Dufour, V. Laroute, M.G. Giuffrida, C. Giunta, M. Cocaign-Bousquet and P. Loubière. (2010). Glutamate-induced metabolic changes in *Lactococcus lactis* NCDO 2118 during GABA production: combined transcriptomic and proteomic analysis. Amino Acids 39: 727–737.

Mercenier, A., H. Müller-Alouf and C. Grangette. (2000). Lactic acid bacteria as live vaccines. Curr. Issues Mol. Biol. 2: 17–25.

Metchnikoff, E. (1907). The prolongation of life; optimistic studies. P.C. Mitchell (ed.), G.P. Putnam's Sons, New York & London.

Mierau, I. and M. Kleerebezem (2005). 10 years of the nisin-controlled gene expression system (NICE) in *Lactococcus lactis*. Appl. Microbiol. Biotechnol. 68: 705–717.

Miyoshi, A., E. Jamet, J. Commissaire, P. Renault, P. Langella and V. Azevedo. (2004). A xylose-inducible expression system for *Lactococcus lactis*. FEMS Microbiol. Lett. 239: 205–212.

de Moreno de LeBlanc, A., S. del Carmen, M. Zurita-Turk, C. Santos Rocha, M. van de Guchte, V. Azevedo, A. Miyoshi and J.G. LeBlanc. (2011). Importance of IL-10 modulation by probiotic microorganisms in gastrointestinal inflammatory diseases. ISRN Gastroenterol. 2011.

Nauta, A., D. van Sinderen, H. Karsens, E. Smit, G. Venema and J. Kok. (1996). Inducible gene expression mediated by a repressor-operator system isolated from *Lactococcus lactis* bacteriophage r1t. Mol. Microbiol. 19: 1331–1341.

Neurath, M.F., S. Pettersson, K.H. Meyer zum Büschenfelde and W. Strober. (1996). Local administration of antisense phosphorothioate oligonucleotides to the p65 subunit of NF-kappa B abrogates established experimental colitis in mice. Nat. Med. 2: 998–1004.

Nichols, W.W., B.J. Ledwith, S.V. Manam and P.J. Troilo. (1995). Potential DNA vaccine integration into host cell genome. Ann. N. Y. Acad. Sci. 772: 30–39.

Nishitani, Y., T. Tanoue, K. Yamada, T. Ishida, M. Yoshida, T. Azuma and M. Mizuno. (2009). *Lactococcus lactis* subsp. cremoris FC alleviates symptoms of colitis induced by dextran sulfate sodium in mice. Int. Immunopharmacol. 9: 1444–1451.

Oliveira, L.C., T.D.L. Saraiva, S.C. Soares, R.T.J. Ramos, P.H.C.G. Sá, A.R. Carneiro, F. Miranda, M. Freire, W. Renan, A.F.O. Júnior, et al. (2014). Genome sequence of *Lactococcus lactis* subsp. lactis NCDO 2118, a GABA-Producing strain. Genome Announc. 2.

O'Sullivan, D.J. (2001). Screening of intestinal microflora for effective probiotic bacteria. J. Agric. Food Chem. 49: 1751–1760.

Pereira, V.B., M. Zurita-Turk, T.D.L. Saraiva, C.P. De Castro, B.M. Souza, P. Mancha Agresti, F.A. Lima, V.N. Pfeiffer, M.S.P. Azevedo, C.S. Rocha, et al. (2014). DNA vaccines approach: from concepts to applications. World J. Vaccines 04: 50–71.

Petrof, E.O., K. Kojima, M.J. Ropeleski, M.W. Musch, Y. Tao, C. De Simone and E.B. Chang. (2004). Probiotics inhibit nuclear factor-kappaB and induce heat shock proteins in colonic epithelial cells through proteasome inhibition. Gastroenterology 127: 1474–1487.

Piard, J.C., I. Hautefort, V.A. Fischetti, S.D. Ehrlich, M. Fons and A. Gruss. (1997). Cell wall anchoring of the *Streptococcus pyogenes* M6 protein in various lactic acid bacteria. J. Bacteriol. 179: 3068–3072.

Pinto, M. (1983). Enterocyte-like differentiation and polarization of the human colon carcinoma cell line Caco-2 in culture. Biol Cell 47: 323–330.

Podolsky, D.K. (1999). Mucosal immunity and inflammation. V. Innate mechanisms of mucosal defense and repair: the best offense is a good defense. Am. J. Physiol. 277: G495–G499.

Pontes, D., S. Innocentin, S. Del Carmen, J.F. Almeida, J.G. Leblanc, A. de Moreno de Leblanc, S. Blugeon, C. Cherbuy, F. Lefèvre, V. Azevedo, et al. (2012). Production of fibronectin binding protein A at the surface of *Lactococcus lactis* increases plasmid transfer *in vitro* and *in vivo*. PloS One *7*, e44892.

Pontes, D., M. Azevedo, S. Innocentin, S. Blugeon, F. Lefévre, V. Azevedo, A. Miyoshi, P. Courtin, M.P. Chapot-Chartier, P. Langella, et al. (2014). Immune response elicited by DNA vaccination using *Lactococcus lactis* is modified by the production of surface exposed pathogenic protein. PloS One *9*, e84509.

Prantera, C., M.L. Scribano, G. Falasco, A. Andreoli and C. Luzi. (2002). Ineffectiveness of probiotics in preventing recurrence after curative resection for Crohn's disease: a randomised controlled trial with *Lactobacillus* GG. Gut 51: 405–409.

Pronio, A., C. Montesani, C. Butteroni, S. Vecchione, G. Mumolo, A. Vestri, D. Vitolo and M. Boirivant. (2008). Probiotic administration in patients with ileal pouch-anal anastomosis for ulcerative colitis is associated with expansion of mucosal regulatory cells. Inflamm. Bowel Dis. 14: 662–668.

Qin, J., R. Li, J. Raes, M. Arumugam, K.S. Burgdorf, C. Manichanh, T. Nielsen, N. Pons, F. Levenez, T. Yamada, et al. (2010). A human gut microbial gene catalogue established by metagenomic sequencing. Nature 464: 59–65.

Que, Y.A., P. François, J.A. Haefliger, J.M. Entenza, P. Vaudaux, and P. Moreillon. (2001). Reassessing the role of *Staphylococcus aureus* clumping factor and fibronectin-binding protein by expression in *Lactococcus lactis*. Infect. Immun. 69: 6296–6302.

Ramakrishnan, A., K.N. Althoff, J.A. Lopez, C.L. Coles and J.H. Bream. (2012). Differential serum cytokine responses to inactivated and live attenuated seasonal influenza vaccines. Cytokine 60: 661–666.

Rescigno, M., M. Urbano, B. Valzasina, M. Francolini, G. Rotta, R. Bonasio, F. Granucci, J.P. Kraehenbuhl and P. Ricciardi-Castagnoli. (2001). Dendritic cells express tight junction proteins and penetrate gut epithelial monolayers to sample bacteria. Nat. Immunol. 2: 361–367.

Reuter, G. (2001). The *Lactobacillus* and *Bifidobacterium* microflora of the human intestine: composition and succession. Curr. Issues Intest. Microbiol. 2: 43–53.

Rezende, R.M., R.P. Oliveira, S.R. Medeiros, A.C. Gomes-Santos, A.C. Alves, F.G. Loli, M.A.F. Guimarães, S.S. Amaral, A.P. da Cunha, H.L. Weiner, et al. (2013). Hsp65-producing *Lactococcus lactis* prevents experimental autoimmune encephalomyelitis in mice by inducing CD4+LAP+ regulatory T cells. J. Autoimmun. 40: 45–57.

Robinson, K., L.M. Chamberlain, K.M. Schofield, J.M. Wells and R.W. Le Page. (1997). Oral vaccination of mice against tetanus with recombinant *Lactococcus lactis*. Nat. Biotechnol. 15: 653–657.

Saha, R., S. Killian and R.S. Donofrio. (2011). DNA vaccines: a mini review. Recent Pat. DNA Gene Seq. 5: 92–96.

Salminen, S., A. von Wright, L. Morelli, P. Marteau, D. Brassart, W.M. de Vos, R. Fondén, M. Saxelin, K. Collins, G. Mogensen, et al. (1998). Demonstration of safety of probiotics -- a review. Int. J. Food Microbiol. 44: 93–106.

Santos Rocha, C., O. Lakhdari, H.M. Blottière, S. Blugeon, H. Sokol, L.G. Bermúdez-Humarán, V. Azevedo, A. Miyoshi, J. Doré, P. Langella, et al. (2012). Anti-inflammatory properties of dairy lactobacilli. Inflamm. Bowel Dis. 18: 657–666.

Santos Rocha, C., A.C. Gomes-Santos, T. Garcias Moreira, M. de Azevedo, T. Diniz Luerce, M. Mariadassou, A.P. Longaray Delamare, P. Langella, E. Maguin, V. Azevedo, et al. (2014). Local and systemic immune mechanisms underlying the anti-colitis effects of the dairy bacterium *Lactobacillus delbrueckii*. PLoS One 9: e85923.

Saraiva, T.D.L., K. Morais, V.B. Pereira, M. de Azevedo, C.S. Rocha, C.C. Prosperi, A.C. Gomes-Santos, L. Bermudez-Humaran, A.M.C. Faria, H.M. Blottiere, et al. (2015). Milk fermented with a 15-lipoxygenase-1-producing *Lactococcus lactis* alleviates symptoms of colitis in a murine model. Curr. Pharm. Biotechnol. 16: 424–429.

Sartor, R.B. (2006). Mechanisms of Disease: pathogenesis of Crohn's disease and ulcerative colitis. Nat. Clin. Pract. gastroenterol. Hepatol. 3, 390–407.

Scheller, J., A. Chalaris, D. Schmidt-Arras and S. Rose-John. (2011). The pro- and anti-inflammatory properties of the cytokine interleukin-6. Biochim. Biophys. Acta 1813: 878–888.

Schoen, C., J. Stritzker, W. Goebel and S. Pilgrim. (2004). Bacteria as DNA vaccine carriers for genetic immunization. Int. J. Med. Microbiol. 294: 319–335.

Seow, Y. and M.J. Wood. (2009). Biological gene delivery vehicles: beyond viral vectors. Mol. Ther. 17: 767–777.

Serhan, C.N. (2005). Lipoxins and aspirin-triggered 15-epi-lipoxins are the first lipid mediators of endogenous anti-inflammation and resolution. Prostaglandins Leukot. Essent. Fatty Acids 73: 141–162.

Shortle, D. (1983). A genetic system for analysis of staphylococcal nuclease. Gene 22: 181–189.

Srivastava, I.K. and M.A. Liu. (2003). Gene vaccines. Ann. Intern. Med. 138: 550–559.

Steinhagen, F., T. Kinjo, C. Bode and D.M. Klinman. (2011). TLR-based immune adjuvants. Vaccine 29: 3341–3355.

Strober, W., I. Fuss and P. Mannon. (2007). The fundamental basis of inflammatory bowel disease. J. Clin. Invest. 117: 514–521.

Tavel, J.A., J.E. Martin, G.G. Kelly, M.E. Enama, J.M. Shen, P.L. Gomez, C.A. Andrews, R.A. Koup, R.T. Bailer, J.A. Stein, et al. (2007). Safety and immunogenicity of a Gag-Pol candidate HIV-1 DNA vaccine administered by a needle-free device in HIV-1-seronegative subjects. J. Acquir. Immune Defic. Syndr. 1999 44: 601–605.

The Norwegian Biotechnology Advisory Board (2003). Regulation of DNA vaccines and gene therapy on animals.

Tian, J., J. Yong, H. Dang and D.L. Kaufman. (2011). Oral GABA treatment downregulates inflammatory responses in a mouse model of rheumatoid arthritis. Autoimmunity 44: 465–470.

Travis, S.P.L., P.D.R. Higgins, T. Orchard, C.J. Van Der Woude, R. Panaccione, A. Bitton, C. O'Morain, J. Panés, A. Sturm, W. Reinisch, et al. (2011). Review article: defining remission in ulcerative colitis. Aliment. Pharmacol. Ther. 34: 113–124.

Tudor, D., C. Dubuquoy, V. Gaboriau, F. Lefèvre, B. Charley and S. Riffault. (2005). TLR9 pathway is involved in adjuvant effects of plasmid DNA-based vaccines. Vaccine 23: 1258–1264.

Ulmer, J.B., J.J. Donnelly, S.E. Parker, G.H. Rhodes, P.L. Felgner, V.J. Dwarki, S.H. Gromkowski, R.R. Deck, C.M. DeWitt and A. Friedman. (1993). Heterologous protection against influenza by injection of DNA encoding a viral protein. Science 259: 1745–1749.

Van Rooijen, R.J., M.J. Gasson and W.M. de Vos. (1992). Characterization of the *Lactococcus lactis* lactose operon promoter: contribution of flanking sequences and LacR repressor to promoter activity. J. Bacteriol. 174: 2273–2280.

Van Asseldonk, M., G. Rutten, M. Oteman, R.J. Siezen, W.M. de Vos and G. Simons. (1990). Cloning of *usp45*, a gene encoding a secreted protein from *Lactococcus lactis* subsp. *lactis* MG1363. Gene 95: 155–160.

Ventura, M., S. O'Flaherty, M.J. Claesson, F. Turroni, T.R. Klaenhammer, D. van Sinderen and P.W. O'Toole. (2009). Genome-scale analyses of health-promoting bacteria: probiogenomics. Nat. Rev. Microbiol. 7: 61–71.

WHO. (2002). Guidelines for the Evaluation of Probiotics in Food: Joint FAO/WHO Working Group meeting, London Ontario, Canada, 30 April-1 May

Walter, J. (2008). Ecological role of lactobacilli in the gastrointestinal tract: Implications for Fundamental and Biomedical Research. Appl. Environ. Microbiol. 74: 4985–4996.

Wang, M., S. Ahrné, B. Jeppsson and G. Molin. (2005). Comparison of bacterial diversity along the human intestinal tract by direct cloning and sequencing of 16S rRNA genes. FEMS Microbiol. Ecol. 54: 219–231.

Wang, Z., P.J. Troilo, X. Wang, T.G. Griffiths, S.J. Pacchione, A.B. Barnum, L.B. Harper, C.J. Pauley, Z. Niu, L. Denisova, et al. (2004). Detection of integration of plasmid DNA into host genomic DNA following intramuscular injection and electroporation. Gene Ther. 11: 711–721.

Wells, J.M. and A. Mercenier. (2008). Mucosal delivery of therapeutic and prophylactic molecules using lactic acid bacteria. Nat. Rev. Microbiol. 6: 349–362.

Wells, J.M., P.W. Wilson, P.M. Norton, M.J. Gasson and R.W. Le Page. (1993). *Lactococcus lactis*: high-level expression of tetanus toxin fragment C and protection against lethal challenge. Mol. Microbiol. 8: 1155–1162.

Whitman, W.B., D.C. Coleman and W.J. Wiebe. (1998). Prokaryotes: The unseen majority. Proc. Natl. Acad. Sci. 95: 6578–6583.

Wolff, J.A., M.E. Dowty, S. Jiao, G. Repetto, R.K. Berg, J.J. Ludtke, P. Williams and D.B. Slautterback. (1992). Expression of naked plasmids by cultured myotubes and entry of plasmids into T tubules and caveolae of mammalian skeletal muscle. J. Cell Sci. 103: 1249–1259.

Wong, C.G.T., T. Bottiglieri and O.C. Snead. (2003). GABA, gamma-hydroxybutyric acid, and neurological disease. Ann. Neurol. 54 Suppl 6: S3–S12.

Xu, Y., P.W. Yuen and J. Lam. (2014). Intranasal DNA vaccine for protection against respiratory infectious diseases: The delivery perspectives. Pharmaceutics 6: 378–415.

Yan, F. and D.B. Polk. (2002). Probiotic bacterium prevents cytokine-induced apoptosis in intestinal epithelial cells. J. Biol. Chem. 277: 50959–50965.

Zoetendal, E.G., E.E. Vaughan and W.M. de Vos. (2006). A microbial world within us. Mol. Microbiol. 59: 1639–1650.

Zurita-Turk, M., S. del Carmen, A.C. Santos, V.B. Pereira, D.C. Cara, S.Y. Leclercq, A. dM de LeBlanc, V. Azevedo, J.M. Chatel, J.G. LeBlanc, et al. (2014). *Lactococcus lactis* carrying the pValac DNA expression vector coding for IL-10 reduces inflammation in a murine model of experimental colitis. BMC Biotechnol. 14: 73.

(2002). Probiotics in food: health and nutritional properties and guidelines for evaluation (Rome: Food and Agriculture Organization of the United Nations : World Health Organization).

Index